Environmental Science

Series editors: R. Allan · U. Förstner · W. Salomons

Springer

Berlin
Heidelberg
New York
Barcelona
Hong Kong
London
Milan
Paris
Singapore
Tokyo

Marc Lucotte · Roger Schetagne · Normand Thérien
Claude Langlois · Alain Tremblay (Eds.)

Mercury in the Biogeochemical Cycle

Natural Environments and Hydroelectric Reservoirs of Northern Québec (Canada)

With 69 Figures and 46 Tables

 Springer

Editors

Dr. Mark Lucotte
Université du Québec à Montréal
C.P. 8888, Succursalle Centre-ville
Montréal, Québec, Canada
H3C 3P8

Roger Schetagne
Hydro-Québec
855 Ste-Catherines east, 18th floor
Montréal, Québec, Canada
H2L 4P5

Dr. Normand Thérien
Université de Sherbrooke
Dept. of Chemical Engineering
Faculty of Engineering
Sherbrooke, Québec, Canada
J1K 2R1

Claude Langlois
Environment Canada
105 McGill street, 4th floor
Montréal, Québec, Canada
H2Y 2E7

Dr. Alain Tremblay
Hydro-Québec
855 Ste-Catherines east, 18th floor
Montréal, Québec, Canada
H2L 4P5

ISBN 3-540-65755-X Springer-Verlag Berlin Heidelberg New York

CIP data applied for

Die Deutsche Bibliothek - CIP-Einheitsaufnahme
Mercury in the biogeochemical cycle : natural environments and hydroelectric reservoirs of northern Québec (Canada) ; with 46 tables / Marc Lucotte ... (ed.). - Berlin ; Heidelberg ; New York ; Barcelona ; Hong Kong ; London ; Milan ; Paris ; Singapore ; Tokyo : Springer, 1999
(Environmental science)
ISBN 3-540-65755-X

© Springer-Verlag Berlin Heidelberg 1999
Printed in Germany

Typesetting: Camera-ready by editors
Cover layout: Struve & Partner, Heidelberg
SPIN: 10706080   32 / 3020 - 5 4 3 2 1 0 - Printed on acid-free paper

Foreword

Nowadays, major environmental issues are the object of large public debates despite the fact that scientific knowledge is often insufficient to draw unequivocal conclusions. Such is the case in the ongoing debate regarding the specific contributions of anthropogenic greenhouse gas emissions and of natural climate changes to global warming. At least 10 to 20 years of additional observations will be required, before we will be able to conclude, with certainty, on this subject. In the mean time, and as directed by their immediate interests, people will continue to promote contradictory opinions. The media are, in part, responsible for perpetuating such debates in that they convey indiscriminately the opinion of highly credible scientists as that of dogmatic researchers, the latter, unfortunately too often expressing working hypotheses as established facts. Naturally, in a similarly misinformed manner, pressure groups tend to support the researcher whose opinions most closely represent either their particular ideological battles or their economic interests and, hence, in their own way, add further to the confusion and obscurity of the debate.

Only a few years ago, mercury (Hg)contamination in hydroelectric reservoirs was the object of such media and social biases. At the time, analytical data used to support the discourse were themselves uncertain and numerous hypotheses, often times fanciful, were proposed and hastily "delivered" to the public. Thankfully, in this case, the scientific progress of the past decade has made it possible to present now a dispassionate view of the question and, hence, to address the issue without its former irrational aspects. The review that follows falls within the scope of such a perspective. It presents an exhaustive review of what is known about the biogeochemical cycling of Hg in hydroelectric reservoirs as well as in natural aquatic systems of the boreal forest region of northern Québec. This work constitutes the North American counterpart to a parallel publication by Scandinavian researchers. The convergence of both studies permits the drawing of important conclusions.

Indeed, the food chain of numerous aquatic ecosystems is contaminated by Hg. The contamination can even reach, in certain organisms, such as predatory fish and aquatic mammals, levels that exceed those prescribed by governmental agencies. This contamination is unfortunately already widespread in natural lakes, but it is particularly noticeable in hydroelectric reservoirs during the first few decades following their impoundment. With respect to natural lakes, the role of increasing anthropogenic Hg emissions cannot be ignored. The present study clearly shows that the atmospheric fluxes of this metal have increased by a factor of three over the past century. In the second case, i.e., that of hydroelectric reservoirs, the extensive flooding of soils, whose various horizons accumulate both natural and anthropogenic heavy metal pollutants, also plays a critical role. In either case, the stimulation of bacterial activity by the availability of organic matter appears to be a determining factor. In fact, the availability of organic matter is the key to the bacterial processes leading to the methylation of Hg, that is, to its transformation from its inorganic state (essentially nontoxic) to an organo-metallic form that readily bioaccumulates in the food chain.

The various chapters of this volume represent benchmark contributions not only for a better understanding of the biogeochemical cycling of Hg but also, more generally, for a better management of the associated social issues. From a practical point of view, one can conclude with some relief that the enhanced Hg contamination of hydroelectric reservoirs subsides in a period of, at most, only a few decades, and that during this transitory period Hg levels in a number of the fishery resources remain below tolerable limits. Unfortunately, the long term contamination of aquatic systems, be they natural or developed, as a consequence of increasing atmospheric contaminant fluxes, remains disconcerting and will demand action on an international scale to be rectified.

Nevertheless, the following work will constitute an important reference in the scientific literature for many years. It is exhaustive, scholarly, solidly supported, and marks the conclusion of a university-industry collaborative effort initiated nearly ten years ago, under the aegis of the Hydro-Québec-Université du Québec à Montréal *Chaire de recherche en environnement* , with the support of the National Science and Engineering Research Council of Canada and the collaboration of the Université de Sherbrooke and the Canadian Wildlife Service. The findings presented in this volume are the outcome of an exemplary collaboration between university researchers and Hydro-Québec scientists. Many findings central to this work were successfully concluded and sanctioned in the form of Master's and Ph.D. theses, all of which constitute important annexes in support of the present document. The reader who wishes to get deeper insights about given aspects of the biogeochemistry of Hg, may refer to these complementary works. However, as it is, the present volume is detailed enough to provide answers to most questions one could have about the Hg contamination of the boreal forest domain of Québec.

Claude Hillaire-Marcel
Titulaire de la Chaire de recherche en environnement
Hydro-Québec-CRSNG-UQAM

Contents

Mercury and Methylmercury in Natural Ecosystems of Northern Québec

Mercury Dynamics at the Flooded Soil-Water Interface in the Reservoirs

Evolution of Mercury Concentrations in Aquatic Organisms from Hydroelectric Reservoirs

Mercury Toxicity for Wildlife Resources

Editorial Committee

Marc Lucotte, Ph.D.
Université du Québec à Montréal
Chaire de recherche en environnement
Hydro-Québec/CRSNG/UQAM
C.P. 8888, Succusale Centre-ville
Montréal, Québec, Canada
H3C 3P8

Roger Schetagne, B.Sc.
Hydro-Québec
Hydraulique et Environnement
855, Ste-Catherine Est
18e étage
Montréal, Québec, Canada
H2L 4P5

Alain Tremblay, Ph.D.
Université du Québec à Montréal
Hydro-Québec
Hydraulique et Environnement
855, Ste-Catherine Est
18e étage
Montréal, Québec, Canada
H2L 4P5

Claude Langlois, M.Sc.
Environnement Canada
Direction de la protection de
l'environnement
105, rue McGill
4e étage
Montréal, Québec, Canada
H2Y 2E7

Normand Thérien, Ph.D. ing.
Université de Sherbrooke
Département de génie chimique
Faculté de génie
Sherbrooke, Québec, Canada
J1K 2R1

External Reviewers

The editorial committee would like to thank the external reviewers for their highly relevant comments and useful suggestions to improve the monograph.

Matti Verta, Ph.D.
Finnish Environment Institute
P.O. Box 140
Fin-00251, Helsinki
Kesakuta 6, 00260 Helsinki
Finland

Drew Bodaly, Ph.D.
Freshwater Institute
501 University Cres.
Winnipeg, Manitoba, Canada
R3T 2N6

Brian Rood, Ph.D.
Director
Environmental Sciences
Mercer University
Macon, GA 31207
USA

List of Contributors

Maxime Bégin, M.Sc.
Université du Québec à Montréal
Chaire de recherche en environnement
Hydro-Québec/CRSNG/UQAM
C.P. 8888, Succursale Centre-ville
Montréal, Québec, Canada
H3C 3P8

Bernard Caron, M.Sc.
Université du Québec à Montréal
Chaire de recherche en environnement
Hydro-Québec/CRSNG/UQAM
C.P. 8888, Succursale Centre-ville
Montréal, Québec, Canada
H3C 3P8

Jean-Luc Desgranges, Ph.D.
Service canadien de la faune
Environnement Canada
C.P. 10100
Sainte-Foy, Québec, Canada
G1V 4H5

Martin Kainz, M.Sc. (Ph. D. student)
Université du Québec à Montréal
Chaire de recherche en environnement
Hydro-Québec/CRSNG/UQAM
C.P. 8888, Succursale Centre-ville
Montréal, Québec, Canada
H3C 3P8

René Langis, M.Sc.
CH2M Hill
1111 Broadway, bureau 1200
P.O. Box 12681
Oakland, California
94604-2681
USA

Marcel Laperle, B.Sc.
8, rue Marcel Potvin
Mont Shefford
R.R. 3
Granby, Québec, Canada
J2G 9J9

Danielle Messier, M.Sc.
Hydro-Québec
Hydraulique et Environnement
855, Ste-Catherine Est
18ième étage
Montréal, Québec, Canada
H2L 4P5

Shelagh Montgomery, M.Sc.
(Ph.D. student)
Université du Québec à Montréal
Chaire de recherche en environnement
Hydro-Québec/CRSNG/UQAM
C.P. 8888, Succursale Centre-ville
Montréal, Québec, Canada
H3C 3P8

François Morneau, M.Sc.
63, rue Champagne
Saint-Basile-le-Grand,
Québec, Canada
J3N 1C2

Ken Morrison, Ph.D.
Hatch et associés
5, Place Ville-Marie
Bureau 200
Montréal, Québec, Canada
H3B 2G2

Pierre Pichet, Ph.D.
Université du Québec à Montréal
Département de chimie
C.P. 8888, Succursale Centre-ville
Montréal, Québec, Canada
H3C 3P8

Isabelle Rheault, B.Sc.
Université du Québec à Montréal
Chaire de recherche en environnement
Hydro-Québec/CRSNG/UQAM
C.P. 8888, Succursale Centre-ville
Montréal, Québec, Canada
H3C 3P8

Jean Rodrigue, M.Sc.
Service canadien de la faune
Environnement Canada
C.P. 10100
Sainte-Foy, Québec, Canada
G1V 4H5

Julie Sbeghen, M.Sc.
Hydro-Québec
Direction Environnement
75, boul. René-Lévesque Ouest
18ᵉ étage
Montréal, Québec, Canada
H2Z 1A4

Bernard Tardif, M.Sc.
Service canadien de la faune
Environnement Canada
C.P. 10100
Sainte-Foy, Québec, Canada
G1V 4H5

Richard Verdon, M.Sc.
Hydro-Québec
Hydraulique et Environnement
855, Ste-Catherine Est
18ᵉ étage
Montréal, Québec, Canada
H2L 4P5

Résumé et synthèse

Dans cet ouvrage, nous présentons une synthèse des études portant sur la problématique du mercure (Hg) dans les milieux aquatiques naturels et les réservoirs hydroélectriques du Nord du Québec. Cette synthèse est basée sur plus de vingt ans de suivi environnemental au complexe La Grande réalisé par Hydro-Québec, ainsi que sur les résultats d'études menées pendant plus de dix ans par des équipes de recherches de l'Université du Québec à Montréal, de l'Université de Sherbrooke, de la Faculté de Médecine vétérinaire de l'Université de Montréal, du Service canadien de la faune et d'Hydro-Québec.

Le mercure dans les écosystèmes naturels du Nord du Québec

Sources du mercure

Bien qu'il n'existe aucune source industrielle ou municipale directe de Hg, ce métal lourd est omniprésent dans le Nord du Québec. On le retrouve dans des systèmes aquatiques et terrestres naturels situés à des centaines, voire même à un millier de kilomètres des centres industriels les plus proches. Le Hg aéroporté d'origine naturelle s'est progressivement accumulé dans les couches organiques des sols depuis le début de leur formation après la dernière glaciation, il y a de cela quelque 5 000 à 8 000 ans. Outre ce Hg naturel, le Hg anthropique aéroporté a également commencé à s'accumuler dans les sols du Nord du Québec au cours du siècle dernier. Dans l'horizon humique des sols vierges, le Hg est essentiellement présent sous sa forme inorganique, la forme organique méthylique ne représentant généralement que moins de 1 %. Le Hg semble peu mobile dans les sols, étant fermement lié à la matière organique humique. Cependant, une partie du Hg nouvellement déposé est entraînée vers les systèmes lacustres par l'eau de ruissellement, avec des matières organiques, sans s'incorporer aux horizons humiques des sols.

Dans les écosystèmes lacustres, les sédiments constituent le principal réservoir de Hg. La quantité de Hg entraînée dans un lac paraît directement proportionnelle à la quantité de carbone lessivée des sols du bassin versant environnant. Tout comme dans les horizons humiques des sols, le Hg est relativement stable dans les sédiments lacustres, étant fermement fixé à la matière organique présente, qui est principalement d'origine terrestre. Seule une légère remobilisation diagénétique post-sédimentaire est observée. Cette faible mobilité du Hg dans les sédiments permet de reconstituer l'historique de dépôts de ce métal lourd dans les régions éloignées. Depuis le début de l'ère industrielle à la fin du siècle dernier, les taux de dépôt de Hg atmosphérique dans le Nord du Québec ont augmenté par des facteurs variant de deux à trois, comme le révèlent les mesures effectuées dans des sédiments lacustres du 45e au 54e parallèle.

Une série d'indices indépendants, montrant des valeurs relativement constantes entre les 45e et 54e parallèles de latitude nord au Québec, suggèrent que la région entière est soumise à des dépôts de Hg aéroporté d'une intensité relativement uniforme et d'une source commune, probablement en provenance du secteur industriel des Grands Lacs. Parmi ces indices, notons le facteur d'enrichissement anthropique normalisé en fonction du contenu en carbone des sédiments lacustres, les concentrations de Hg dans différentes plantes inférieures et supérieures, la teneur en Hg normalisée en fonction du contenu en carbone des horizons humiques des sols, et les rapports de Hg/Al et Hg/Si dans les lichens. Au nord du 54e parallèle, le Hg atmosphérique semble provenir d'une source additionnelle, tel que le suggèrent les valeurs différentes obtenues pour les rapports de Hg/Al et Hg/Si dans les lichens. Cette région éloignée est peut-être sous l'influence du Hg volatilisé depuis la Baie d'Hudson avoisinante ou en provenance de l'Eurasie. De la même façon, les concentrations de Hg mesurées dans plusieurs espèces de poissons et dans une variété d'oiseaux ne révèlent aucun gradient systématique du nord au sud entre le 45e et 56e parallèle.

Les sédiments et la colonne d'eau

Dans le Nord du Québec, les concentrations de Hg dans les sédiments lacustres varient généralement de 50 à 300 ng g^{-1} (poids sec) et correspondent à des valeurs de 10 000 à 100 000 fois plus élevées que celles mesurées dans la colonne d'eau. Dans l'eau non filtrée, les concentrations sont généralement bien inférieures à 10 ng L^{-1}. Même si les concentrations dans les sédiments lacustres sont semblables à celles mesurées dans les sols, les organismes aquatiques bioaccumulent beaucoup plus de Hg que leurs congénères de position trophique équivalente des écosystèmes terrestres. Cette situation est attribuable à la biodisponibilité nettement plus élevée de ce métal lourd dans les milieux aquatiques principalement parce que les formes inorganiques de Hg sont transformées en méthylmercure (MeHg) par l'activité microbienne se produisant à différents niveaux du système aquatique. Dans les sédiments lacustres, le MeHg représente généralement moins de 2 % de la teneur en Hg total. Bien que seule une petite fraction de la charge totale de Hg dans un lac soit transformée en MeHg, sa bioamplification le long de la chaîne alimentaire est telle, qu'un facteur d'augmentation de l'ordre de 150 s'observe entre les concentrations en Hg mesurées dans le plancton de faible niveau trophique (25 ng g^{-1}, poids sec) et celles enregistrées dans les gros poissons piscivores (0,6 mg kg^{-1}, poids humide). Contrairement au Hg inorganique, le MeHg est facilement assimilé par les organismes vivants en raison de sa grande affinité pour les protéines et de son faible taux d'élimination, principalement chez les animaux à sang froid que sont les poissons.

Les invertébrés

Les concentrations de Hg total mesurées chez les invertébrés de plus de 20 lacs naturels du Nord du Québec couvrent une vaste gamme de valeurs, de 25 à 575 ng g^{-1} (poids sec) dans le plancton et de 31 à 790 ng g^{-1} (poids sec) dans les larves d'insectes. Dans le Nord du Québec, ces écarts ne s'expliqueraient pas par des variations de l'intensité des dépôts de Hg aéroporté, puisqu'elle semble être relativement uniforme sur la région étudiée. La proportion du Hg sous la forme méthylique augmente le long de la chaîne alimentaire des invertébrés et des facteurs de bioamplification d'environ 3 sont observés d'un niveau trophique à l'autre.

Des analyses statistiques indiquent que les variations dans les concentrations de Hg total et de MeHg chez les invertébrés peuvent être expliquées par le comportement alimentaire (niveau trophique) et, dans une plus faible mesure, par des paramètres de qualité de l'eau, tels que la couleur et le carbone organique dissous, et par la température de l'eau.

Les poissons

Les concentrations de Hg total mesurées chez les poissons de plus 180 stations d'échantillonnages, situées dans des lacs et rivières naturels du Nord du Québec, sont relativement élevées comparativement à celles enregistrées dans d'autres régions de l'Amérique du Nord. Les concentrations mesurées chez les espèces non piscivores, telles que le meunier rouge (*Catostomus catostomus*) et le grand corégone (*Coregonus clupeaformis*), sont inférieures à la norme canadienne de mise en marché des produits de la pêche de 0,5 mg kg^{-1}. Par contre, les teneurs enregistrées chez les espèces piscivores excèdent souvent cette norme. La variabilité dans les teneurs obtenues d'un lac à l'autre d'une même région est importante pour toutes les espèces de poissons. Les concentrations moyennes estimées pour une longueur standardisée varient souvent par des facteurs de 3 à 4 pour des lacs voisins. Les concentrations moyennes dans les poissons non piscivores de 400 mm varient de 0,05 à 0,30 mg kg^{-1} (poids humide). En ce qui concerne les espèces piscivores, les concentrations moyennes chez le doré (*Stizostedion vitreum*) de 400 mm, ainsi que chez le grand brochet (*Esox lucius*) de 700 mm, s'échelonnent entre 0,30 et 1,41 mg kg^{-1} (poids humide) d'un lac à l'autre. Les concentrations les plus élevées obtenues chez toutes les espèces étudiées proviennent de plans d'eau ayant une teneur organique élevée, mesurée par la couleur, la teneur en carbone organique total et dissous, ainsi que par la concentration de tanins. La biodisponibilité du Hg à la base de la chaîne alimentaire serait plus élevée dans ces milieux.

Un facteur moyen de bioamplification d'environ 5 a été obtenu entre les concentrations moyennes chez le grand corégone (de 400 mm), lequel se nourrit d'organismes benthiques, et celles enregistrées chez le brochet (700 mm) ou le doré (400 mm), qui sont des poissons piscivores. De plus, nos données montrent que le facteur de bioamplification obtenu pour ces espèces de poissons est assez variable d'un lac à l'autre. Cette variabilité pourrait s'expliquer par des différences dans les structures des communautés de poissons, se traduisant par des proies différentes pour les poissons prédateurs, et par conséquent, par des teneurs en Hg différentes.

Des écarts considérables de concentrations de Hg ont également été observés pour des poissons de même espèce et de même longueur provenant d'un même lac. Par exemple, les touladis (*Salvelinus namaycush*) de 575 ± 8 mm capturés la même année dans le lac Hazeur montrent des concentrations de Hg variant entre 0,67 et 1,28 mg kg^{-1}. La physiologie, la croissance ou les habitudes alimentaires individuelles pourraient expliquer une telle variabilité. En appui à cette dernière hypothèse, des études de contenus stomacaux révèlent que le grand brochet provenant du secteur ouest de la région du complexe La Grande se nourrit d'une grande variété de poissons de niveaux trophiques différents, présentant des teneurs en Hg très différentes. Ce fait montre bien l'importance du régime alimentaire dans le processus de bioaccumulation du Hg.

Les oiseaux aquatiques

L'importance du régime alimentaire, et en conséquence du niveau trophique, est également démontrée par les concentrations mesurées chez les oiseaux aquatiques du Nord du Québec. En effet, elles révèlent clairement une progression des teneurs en Hg total dans les muscles, le foie et les plumes en passant des espèces herbivores aux espèces piscivores. Les concentrations en Hg dans les muscles sont typiquement égales ou inférieures à 0,05 mg kg^{-1} (poids humide) chez les espèces herbivores, alors qu'elles varient de 0,16 à 0,21 mg kg^{-1} chez les espèces se nourrissant d'organismes benthiques aquatiques, et s'échelonnent entre 0,8 et 1,6 mg kg^{-1} chez les espèces partiellement ou strictement piscivores. Ces données montrent également l'influence du système aquatique sur les teneurs en Hg de la faune, par le processus de méthylation et de biodisponibilité accrue du Hg dans la chaîne alimentaire aquatique. Aussi, les concentrations dans les poissons et les oiseaux aquatiques du Nord du Québec sont équivalentes chez les espèces ayant des habitudes alimentaires semblables.

Les mammifères

Les concentrations de Hg enregistrées dans les mammifères terrestres du Nord du Québec démontrent également l'importance du milieu aquatique sur l'accumulation en Hg, puisque les espèces piscivores présentent les concentrations les plus élevées. Les poissons jouent le rôle de véhicule de transfert entre les deux écosystèmes. Les concentrations dans les muscles des espèces strictement herbivores, telles que le lièvre d'Amérique (*Lepus americanus*) ou le caribou (*Rangifer caribou*), sont généralement inférieures ou égales à 0,05 mg kg^{-1}. Les concentrations correspondantes chez les carnivores, tels que l'hermine (*Mustela erminia*), la martre (*Martes americana*) et le renard roux (*Vulpes fulva*), varient de 0,15 et 0,30 mg kg^{-1}, alors que les concentrations atteignent environ 2,5 mg kg^{-1} chez les espèces partiellement piscivores, comme le vison (*Mustela vison*).

L'importance du régime alimentaire est également démontrée chez les mammifères marins capturés au large de la baie d'Hudson. Les concentrations de Hg dans les muscles des espèces benthivores ou partiellement piscivores, comme le phoque annelé (*Phoca hispida*) et le phoque barbu (*Erignathus barbatus*), varient de 0,1 à 0,7 mg kg^{-1} (poids humide), alors qu'elles s'échelonnent entre 0,9 et 6,2 mg kg^{-1} chez les bélugas (*Delphinapterus leucas*), qui sont principalement piscivores. Les concentrations moyennes dans le foie se situent entre 2 et 5 mg kg^{-1} chez les phoques et atteignent 20 mg kg^{-1} chez les bélugas.

Toutes ces données démontrent que le risque pour la santé relié à la consommation de ressources riches en Hg au Nord du Québec se limite essentiellement aux espèces piscivores, qu'il s'agisse de poissons, d'oiseaux aquatiques ou de faune terrestre ou marine.

Nous comprenons maintenant assez bien les processus par lesquels le Hg inorganique aéroporté rejoint le milieu aquatique, y est transformé en MeHg et ensuite bioamplifié tout au long de la chaîne alimentaire, des invertébrés aux poissons et jusqu'à la faune piscivore. Cependant, les effets de l'augmentation récente, par un facteur d'environ 2 à 3, du Hg d'origine atmosphérique sur les niveaux de Hg chez les poissons et la faune piscivore demeurent inconnus. Bien qu'aucune tendance temporelle des teneurs en Hg dans les poissons n'ait été observée chez les espèces étudiées dans 5 lacs naturels pour des périodes allant jusqu'à douze ans, ce laps de temps est trop court pour confirmer ou réfuter toute relation directe, compte tenu de la grande variabilité des teneurs en Hg des poissons.

En raison de la baisse des teneurs en Hg observées chez les poissons et d'autres organismes aquatiques provenant de sites contaminés par des activités industrielles, une fois les apports de Hg réduits (comme dans le cas des rivières English-Wabigoon et Saguenay), on peut supposer qu'il existe une relation entre les dépôts de Hg aéroporté et les concentrations chez les poissons et la faune. Cette relation signifierait que les concentrations de Hg chez les poissons et la faune du Nord du Québec n'ont pas toujours été aussi élevées. Elle indiquerait également que les niveaux diminueraient si les émissions anthropiques de Hg dans l'atmosphère étaient réduites. La fermeture récente d'industries de l'ancienne Allemagne de l'Est a occasionné une baisse importante de niveaux de dépôt atmosphérique de Hg sur les lacs de la Suède, comme le prouve la chute des concentrations de Hg dans la plupart des sédiments récents des lacs de la région. Dans ce cas particulier, la baisse des concentrations de Hg prévue chez les poissons démontrerait clairement que la charge de Hg dans les organismes aquatiques est influencée par les taux de dépôt atmosphérique locaux de ce métal lourd.

La problématique du mercure au complexe hydroélectrique La Grande

Le suivi des teneurs en Hg des poissons du complexe La Grande a débuté vers la fin des années 1970, soit avant la mise en eau du premier réservoir. À l'instar des autres réservoirs au Canada et à l'étranger, les teneurs en Hg des poissons des réservoirs du complexe La Grande ont augmenté de façon marquée et rapide au début des années 1980, immédiatement après la mise en eau. Des programmes de recherche ont été mis sur pied à la fin des années 1980 afin d'élucider les mécanismes responsables de ce phénomène.

La méthylation et le transfert passif depuis les sols et la végétation inondés vers la colonne d'eau

Les études *in vitro* et *in vivo*, présentées dans cette monographie, ont démontré l'influence de la végétation et de la matière organique terrestre inondées sur le processus de méthylation et de libération du Hg vers la colonne d'eau et sur son transfert subséquent le long de la chaîne alimentaire aquatique.

Des expériences *in vitro* ont révélé une libération significative de Hg total et de MeHg depuis une variété de sols et de végétaux inondés vers les eaux sus-jacentes. Elles ont également permis d'évaluer l'ampleur, l'importance relative et la durée de cette libération selon les types de sol et de végétation inondés, ainsi que selon différentes conditions d'oxygène dissous, de pH et de température des eaux sus-jacentes. Les taux de diffusion obtenus à 20°C indiquent que pratiquement tout le Hg libérable serait effectivement libéré en moins d'un an et principalement au cours des premiers mois d'inondation. Nonobstant les conditions environnementales appliquées, la quantité totale de Hg libéré après environ 6 mois d'inondation varie entre 4 et 40 ng g^{-1} pour les différents types de végétation et entre 4 et 45 ng m^{-2} pour l'humus des sols. Les quantités correspondantes de MeHg varient de 1 à 5 ng g^{-1} pour la végétation et de 0,3 à 1 ng m^{-2} pour l'humus des sols. Une proportion significative du contenu initial en Hg total de la partie verte de la végétation inondée est libérée dans la colonne d'eau en 6 mois, soit de 20 % jusqu'à 50 à 100 % selon le type de végétation. En ce qui concerne les sols inondés, seule une petite fraction du contenu initial en Hg est libérée dans la colonne d'eau après un an d'inondation. En fait, il est impossible d'en déterminer le pourcentage réel libéré en raison de la grande variabilité de la charge initiale en Hg des échantillons. Il est à noter, que les sols renferment une biomasse nettement plus élevée de matières organiques que la végétation.

Les mesures *in situ* enregistrées dans les sols inondés des réservoirs du complexe La Grande confirment que seule une petite partie du contenu initial en Hg des sols inondés est libérée vers la colonne d'eau. En effet, après un peu plus d'une décennie, la majorité des sols inondés ne montrent aucune perte significative de leur charge en Hg, à l'exception des sols érodés dans la zone de marnage en bordure des réservoirs. Aussi, l'examen *in situ* des sols et de la végétation inondée ne révèle que très peu de changements structurels, à l'exception de la biodégradation partielle de la matière verte inondée. Même si l'activité bactérienne est suffisamment intense pour provoquer une forte consommation en oxygène dans les sols inondés et les eaux profondes des réservoirs pendant quelques années, les diminutions du contenu en carbone des sols inondés ne peuvent être mesurées, celles-ci étant plus petites que la variabilité inhérente des teneurs en carbone dans les sols avoisinants.

La méthylation progressive au fil des années du Hg présent dans les sols inondés, associée à l'activité bactérienne, constitue le changement majeur observé dans les sols à la suite de la mise en eau des réservoirs. Alors que moins de 1 % du Hg total dans les sols naturels est sous la forme organique, ce pourcentage atteint environ 10 à 30 % après 13 ans d'inondation pour les tourbières et les sols podzoliques inondés respectivement. Alors que les taux de méthylation du Hg présent dans les tourbières et les sols podzoliques sont semblables au cours de la première année d'inondation, ils ne se maintiennent par la suite que dans les sols podzoliques. On explique cette observation par le fait qu'à la suite d'une inondation, la matière organique des sols podzoliques semble être plus vulnérable à la dégradation bactérienne que la matière organique des tourbières, qui est déjà saturée d'eau avant l'inondation, à l'exception du couvre sol vivant.

Les taux de méthylation semblent similaires dans les zones peu profondes et celles de 10 à 20 mètres de profondeur situées sous la zone photique, mais au-dessus de la thermocline saisonnière. Une fois le Hg méthylé dans les sols inondés, aucune déméthylation nette ne peut être observée durant les mois d'hiver ou au fil des ans jusqu'à 15 ans après l'inondation. En conséquence, il semble qu'une fois formée, la plus grande partie du MeHg s'accumule dans les sols inondés, probablement en raison d'une grande affinité pour la matière organique terrigène.

Une diffusion passive du Hg total ou du MeHg depuis les sols inondés vers la colonne d'eau, confirmant les observations *in vitro,* a pu être observée *in situ* uniquement au-dessus de sols organiques non érodés dans des zones peu profondes en périphérie des réservoirs, là où le temps de séjour de l'eau est relativement prolongé. Ailleurs dans les réservoirs, la diffusion cumulée sur 6 mois étant estimée à quelques dizaines de ng m^{-2} de Hg total et à quelques ng m^{-2} de MeHg, d'après les expériences *in vitro*, il est impossible de détecter une augmentation de concentration de Hg ou de MeHg dans la colonne d'eau, en raison de l'immense dilution y prévalant. Une partie du Hg libéré dans la colonne d'eau peut aussi être adsorbée sur les matières en suspension, puis transférée à la chaîne alimentaire, contribuant à réduire encore d'avantage la possibilité de détecter la diffusion passive du Hg dissous dans la colonne d'eau. Dans les zones peu profondes où les taux de dilution sont réduits, comme dans le cas de la petite retenue LA-40, un facteur d'augmentation moyen d'environ 5 a été obtenu pour les concentrations de MeHg dissous, comparativement à celles des lacs naturels. Les valeurs moyennes alors mesurées atteignent 0,3 ng L^{-1}, confirmant qu'une certaine fraction du Hg initialement lié au système terrestre est libérée dans la colonne d'eau pendant les premiers mois suivant la mise en eau. Il est à noter que les calculs effectués pour le réservoir Robert-Bourassa, à l'aide de données provenant d'expériences d'inondation *in vitro*, génèrent des valeurs comparables dans la colonne d'eau. En effet, la hausse nette estimée pour les 5 premières années aurait été de l'ordre de 0,2 ng L^{-1} de MeHg. Par ailleurs, des mesures *in situ* montrent que la proportion du Hg dissous total qui est sous la forme méthylée, est en moyenne 4 fois plus élevée dans les réservoirs que dans les lacs naturels (12 % par rapport à 3 %).

Malgré la libération dans la colonne d'eau d'une proportion importante de Hg contenu initialement dans la partie verte de la végétation inondée, la charge globale de Hg contenu dans les sols et la végétation inondés demeure pratiquement intacte au fil des années. Une perte totale probablement inférieure à 5 % de la charge initiale en Hg a suffi à augmenter de façon significative la contamination en Hg de toute la chaîne alimentaire des réservoirs. Ce fait démontre à quel point le Hg, principalement sous la forme de MeHg, est efficacement transféré et bioaccumulé le long de la chaîne alimentaire.

Augmentations de mercure dans les organismes à la base de la chaîne alimentaire aquatique

Les facteurs d'augmentation des concentrations de MeHg obtenues en réservoirs sont relativement constants par rapport à celles mesurées dans les lacs naturels. Ils varient habituellement de 3 à 5 pour la partie dissoute (< 0,45 μm), les particules fines (0,45 à 63 μm), le zooplancton (> 150 ou > 210 μm), ainsi que pour les poissons non piscivores et piscivores de longueur standardisée. Par exemple, les concentrations mesurées chez les invertébrés récoltés dans 5 réservoirs différents du complexe La Grande varient de 45 à 680 ng g^{-1} (poids secs) dans les larves d'insectes et de 350 à 550 ng g^{-1} (poids secs) dans le zooplancton (> 150 μm). Ces valeurs correspondent à un facteur d'augmentation moyen de 3 par rapport à celles enregistrées dans les lacs naturels.

Des facteurs de bioamplification comparables, entre un niveau trophique et un autre, ont été obtenus dans les réservoirs du complexe La Grande et les lacs naturels avoisinants, pour les invertébrés, le zooplancton, les insectes et les poissons. Ainsi, les résultats démontrent qu'une augmentation de la biodisponibilité du Hg à la base de la chaîne alimentaire des réservoirs se répercute tout au long de cette chaîne, jusqu'aux poissons piscivores. De plus, les différences notées entre les lacs naturels et les réservoirs, dans les analyses d'isotopes stables effectuées sur les particules fines et le zooplancton, suggèrent que les détritus organiques d'origine terrestre en suspension dans la colonne d'eau sont ingérés par le zooplancton. Ces détritus riches en Hg contribuent donc au transfert de ce métal aux poissons par l'intermédiaire du zooplancton.

L'importance du rôle joué par les formes dissoutes de Hg inorganique et de MeHg sur la contamination de la chaîne alimentaire aquatique demeure incertaine. En raison de l'absence d'une série adéquate de données concernant les concentrations de Hg dans des échantillons purs de phytoplancton, le rôle joué par ces organismes dans le transfert du Hg dissous aux organismes supérieurs demeure inconnu (bien qu'ils soient vraisemblablement les organismes les plus influencés par la phase dissoute). Cependant, l'absence d'augmentations significatives de MeHg dans la fraction de 63 à 210 μm (qui renferme surtout du phytoplancton et du micro-zooplancton) prélevée dans la retenue LA-40, comparativement à celle des lacs naturels, suggère que le phytoplancton ne jouerait pas un rôle déterminant dans le transfert de Hg de l'eau aux organismes situés plus haut dans la chaîne alimentaire. Cette hypothèse serait corroborée par le calcul des flux de biomasse et de Hg dans le réservoir Robert-Bourassa, réalisés à l'aide des données obtenues du réseau de suivi environnemental (RSE). En effet, les résultats de cet exercice révèlent que la chaîne alimentaire pélagique, constituée de la fraction dissoute, du phytoplancton, du zooplancton et des poissons, n'est pas suffisante pour expliquer les biomasses et les concentrations de Hg obtenues pour les poissons non piscivores de ce réservoir. Les calculs suggèrent que le vecteur le plus important de transfert de Hg aux poissons doit provenir des invertébrés littoraux, c'est-à-dire des insectes aquatiques et du zooplancton. Les études sur le régime alimentaire des poissons, ainsi que les mesures *in situ* de la biomasse et des concentrations de Hg enregistrées dans les réservoirs Robert-Bourassa et Laforge 1, de même que dans la retenue LA-40, corroborent cette hypothèse.

Par ailleurs, il est probable que le Hg dissous libéré directement des sols inondés vers la colonne d'eau se lie rapidement aux particules fines en suspension, le rendant plus disponible pour les organismes filtreurs à la base de la chaîne alimentaire. Bien qu'il ne soit probablement pas dominant, le rôle du zooplancton pélagique dans le transfert de ce Hg, valorisé par la filtration de détritus riches en Hg d'origine terrestre, n'est peut-être pas négligeable. En effet, des études complémentaires au RSE ont révélé une abondance de ciscos de lac (*Coregonus artedii*) dans la zone pélagique du réservoir Robert-Bourassa, lesquels s'alimentent essentiellement de plancton.

Transfert actif additionnel des sols inondés à la chaîne alimentaire aquatique

En plus de la diffusion passive depuis la végétation et les sols inondés vers colonne d'eau, ainsi que par la chaîne alimentaire planctonique, le Hg peut aussi être transféré à la chaîne alimentaire aquatique par des mécanismes actifs de nature biotique et abiotique. Premièrement, puisque le MeHg est accumulé dans les sols inondés, les larves d'insectes, fouissant dans les premiers centimètres de ces sols et s'alimentant de la matière organique partiellement dégradée et riche en Hg, peuvent bioaccumuler des concentrations élevées de ce métal et les transférer aux organismes aquatiques de niveau trophique supérieur. Des estimations calculées à partir de mesures *in situ* réalisées dans les réservoirs du complexe La Grande suggèrent que dans les zones peu profondes, la charge en Hg contenue dans les insectes aquatiques peut être jusqu'à 6 fois supérieure à celle contenue dans le zooplancton. Cependant, dans les secteurs pélagiques, la proportion de la charge totale en Hg contenue dans le zooplancton est probablement supérieure, puisque la densité des organismes benthiques est réduite, la couche de périphyton est absente et les températures plus basses sont moins favorables au processus de méthylation.

Deuxièmement, pendant les premières années suivant la mise en eau des réservoirs du Nord du Québec, la majorité des sols inondés présents dans les zones peu profondes de la zone de marnage, et exposés à l'action combinée des vagues et des glaces, sont progressivement érodés. Cette érosion est particulièrement rapide pour l'horizon organique des sols podzoliques qui est généralement très mince. Un triage rapide des particules de sols érodés se produit alors, maintenant en suspension dans la colonne d'eau, pendant un certain temps, les fines particules organiques riches en Hg. Les organismes aquatiques filtrant ces particules (comme le zooplancton) peuvent constituer une voie de transfert importante du Hg vers les chaînes alimentaires aquatiques des réservoirs. Ces particules peuvent par la suite se déposer à la surface des sols inondés un peu plus profonds, où ils pourront constituer une nourriture riche en MeHg pour les organismes benthiques.

Troisièmement, la libération d'éléments nutritifs résultant de la dégradation bactérienne de la matière organique terrigène inondée stimule la production auto-trophe. La dégradation de cette nouvelle matière organique particulièrement labile, contrairement aux composés ligno-cellulosiques des sols inondés, favorise une méthylation additionnelle. Ce processus peut se révéler important dans les zones peu profondes, où le temps de séjour de l'eau relativement long et l'effet combiné de la pénétration de la lumière et de la minéralisation des éléments nutritifs entraî-nent le développement d'une couche de périphyton. Ce dernier, composé d'algues benthiques et de bactéries, favorise la méthylation du Hg et peut représenter une source importante de nourriture riche en MeHg pour le zooplancton et les larves d'insectes. Le zooplancton recueilli dans ces zones peu profondes continue de présenter des concentrations élevées après 13 ou 15 ans d'inondation, contrairement au zooplancton pélagique dont les concentrations de Hg mesurées après environ 8 ans d'inondation correspondent à celles obtenues dans les lacs naturels avoisinants.

Augmentation de mercure dans les poissons

Réservoirs

Les concentrations de Hg dans toutes les espèces de poissons des réservoirs du complexe La Grande ont rapidement augmenté après la mise en eau, à des niveaux de 3 à 7 fois supérieurs à ceux mesurés dans les lacs naturels avoisinants. Chez les espèces non piscivores, telles que le grand corégone et le meunier rouge, les concentrations maximales ont été atteintes 5 à 9 ans après la mise en eau. Les concentrations maximales atteintes chez le grand corégone de 400 mm varient de 0,4 à 0,5 mg kg^{-1} (Hg total, poids humide) d'un réservoir à l'autre, dépassant légère-ment la norme canadienne de mise en marché de 0,5 mg kg^{-1} uniquement dans le réservoir Robert-Bourassa.

Chez les espèces piscivores, telles que le grand brochet et le doré, les concen-trations maximales ont été atteintes plus tardivement, soit 10 à 13 ans après la mise en eau pour les spécimens de longueur standardisée (400 mm pour le doré et 700 mm pour le grand brochet). Les concentrations maximales enregistrées chez le grand brochet de longueur standardisée, qui varient de 1,7 à 4,2 mg kg^{-1} selon le réservoir, sont de 3 à 8 fois supérieures à la norme canadienne de mise en marché de 0,5 mg kg^{-1}.

Des études de contenus stomacaux ont indiqué que les augmentations de Hg total observées chez les poissons des réservoirs sont davantage reliées à l'augmentation des concentrations de MeHg de leurs proies qu'à un changement dans leurs habitudes alimentaires. Ce fait souligne encore l'importance d'une méthylation accrue à la base de la chaîne alimentaire. Les grands corégones des deux milieux passent progressivement d'une alimentation à base de zooplancton à un régime benthique en fonction de l'augmentation de leur taille. Le cisco et le meunier rouge se nourrissent surtout de zooplancton et de benthos respectivement, peu importe leur taille, qu'ils proviennent de réservoirs ou de lacs naturels. Dans le secteur ouest du complexe La Grande, le petit cisco de lac est la proie la plus fréquemment retrouvée dans les estomacs de grands brochets et de dorés. Le grand brochet de cette région se nourrit aussi d'une variété d'autres espèces de poissons, qu'il s'agisse d'espèces strictement non piscivores ou d'espèces strictement piscivores. En grandissant, le brochet se nourrit moins de corégoninés (cisco et corégones) et davantage de poissons piscivores, tels que de dorés, de lottes et de brochets. Ces dernières espèces peuvent représenter jusqu'à 60 % de la biomasse ingérée par les brochets de plus de 400 mm de longueur. Un tel régime alimentaire les expose à des concentrations élevées de Hg. Dans la partie est du territoire, le régime alimentaire des espèces piscivores se compose en grande partie de grands corégones, un régime qui contribue à de plus faibles concentrations de Hg.

Aval des réservoirs

Le suivi des teneurs en Hg dans les poissons du complexe La Grande a aussi démontré que le Hg est exporté en aval des réservoirs. Des études menées au complexe La Grande révèlent que le Hg est principalement exporté par les débris organiques riches en Hg en suspension dans l'eau du réservoir, ainsi que par le plancton, les insectes aquatiques et les petits poissons. Elles suggèrent également que le potentiel d'augmentation des teneurs en Hg des poissons en aval varie selon l'importance des affluents situés en aval, en raison de leur effet de dilution. La distance en aval sur laquelle cette augmentation se produit serait fonction de la présence de grands plans d'eau (lacs ou autres réservoirs) permettant la sédimentation des débris organiques du réservoir, ou la consommation des organismes en dérive par les poissons de ces plans d'eau.

L'importance de ces grands plans d'eau en aval est corroborée par l'absence d'un effet cumulatif dans les teneurs en Hg des poissons capturés de l'amont vers l'aval, depuis le réservoir Caniapiscau jusqu'au réservoir La Grande 4, en passant par les réservoirs Fontanges et Vincelotte. Par exemple, les teneurs moyennes suivantes ont été obtenues en 1993, pour les grands corégones de longueur standardisée : 0,35 mg kg^{-1} aux réservoirs Caniapiscau, Fontanges et Vincelotte et 0,27 mg kg^{-1} à La Grande 4. Les résultats obtenus à partir de ces réservoirs indiquent bien que dans le cas d'une série de grands et profonds réservoirs se déversant les uns dans les autres, l'effet d'un réservoir sur les teneurs en Hg des poissons en aval se limite au premier réservoir situé immédiatement en aval. De plus, la distribution spatiale des teneurs en Hg des poissons à l'intérieur d'un réservoir récepteur révèle que cet effet, dans le cas des grands réservoirs, se limite à la zone située à proximité des apports en provenance du réservoir en amont.

Durée du phénomène d'augmentation des teneurs en mercure dans les poissons

Après avoir atteintes des valeurs maximales 5 à 9 ans après la mise en eau, les concentrations dans les grands corégones de longueur standardisée ont ensuite diminué de façon significative et graduelle dans tous les réservoirs, rejoignant des niveaux représentatifs de ceux mesurés dans les lacs naturels avoisinants après 10, 11 et 17 ans dans les réservoirs La Grande 4, Caniapiscau et Robert-Bourassa respectivement. Ces données, ainsi que l'évolution semblable des teneurs chez les meuniers rouges, suggèrent que les concentrations de Hg retournent à des niveaux normalement mesurés dans les lacs naturels de la région après 10 à 25 ans pour toutes les espèces non piscivores.

Chez les espèces piscivores, telles que le doré et le grand brochet, les concentrations en Hg ont commencé à baisser de façon significative dans tous les réservoirs du complexe La Grande après 15 ans d'inondation. Les données provenant de plusieurs réservoirs plus anciens du Nord du Québec et du Labrador indiquent que les concentrations chez les brochets sont retournées aux niveaux mesurés dans les lacs naturels environnants entre 20 et 30 ans après l'inondation. Une durée semblable a également été observée en Finlande et dans le Nord du Manitoba. Bien que l'âge des réservoirs du complexe La Grande ne variait qu'entre 10 et 17 ans au moment de la dernière campagne de mesures, l'évolution générale des teneurs en Hg chez les espèces piscivores suit la même tendance temporelle.

Un certain nombre de processus clés responsables de l'augmentation des teneurs en Hg des poissons des réservoirs sont temporaires ou diminuent d'intensité après quelques années d'inondation. Ces processus comprennent: (1) la diffusion passive de Hg dans la colonne d'eau depuis la végétation et les sols inondés à la suite de la décomposition de la matière organique terrigène; (2) la libération d'éléments nutritifs stimulant la production autotrophe, dont les matières organiques résultantes, étant particulièrement labiles, favorisent une méthylation additionnelle de Hg; (3) l'érosion de la matière organique inondée dans la zone de marnage, qui rend disponible de fines particules organiques riches en Hg pour les organismes aquatiques filtreurs; (4) le transfert actif du Hg par les insectes aquatiques fouissant dans les sols inondées riches en MeHg; (5) le développement du périphyton sur les sols et la végétation inondés, qui favorise la méthylation du Hg et son transfert actif par les insectes aquatiques et le zooplancton s'y nourrissant.

Décomposition de la matière organique terrigène et libération d'éléments nutritifs

Le suivi de la qualité de l'eau au complexe La Grande a démontré que la décomposition intensive de la matière organique inondée est éphémère en raison de l'épuisement rapide de la matière organique facilement décomposable. Les changements de qualité de l'eau résultant de cette décomposition, comme la diminution des teneurs en oxygène dissous et l'augmentation des concentrations de CO_2 et de phosphore (libération d'éléments nutritifs), sont à toutes fins utiles terminés après 8 à 14 ans d'inondation selon les caractéristiques hydrauliques et morphologiques des réservoirs considérés. Des expériences d'inondation *in vitro* ont confirmé qu'à des températures typiques des eaux du complexe La Grande, la quasi-totalité du Hg libérable serait relâché en quelques années, dont la majorité dès la première année. Ces expériences démontrent de plus que seule une fraction de la partie verte de la végétation inondée (incluant les couvre-sols) est efficacement décomposée et libère une proportion importante de son contenu initial en Hg. Les autres composantes de la végétation et des sols inondés sont constituées de substances résistantes à la dégradation biochimique et ne libèrent qu'une très faible fraction de leur contenu en Hg. Des analyses réalisées sur des troncs d'épinettes inondés pendant 55 ans dans le réservoir Gouin au Québec ont révélé une perte de moins de 1 % de leur biomasse initiale, démontrant que les composantes ligneuses de la végétation et des sols inondés demeurent à toutes fins utiles intactes même après une très longue période d'inondation.

Érosion

Des observations effectuées sur le terrain aux réservoirs du complexe La Grande ont démontré que l'érosion de la matière organique dans la zone de marnage atteignait une intensité maximale au cours des premières années et était complète après une période de 5 à 10 ans. Ainsi, la disponibilité accrue, pour les organismes aquatiques filtreurs (comme le zooplancton), de fines particules organiques en suspension, riches en Hg, est également éphémère.

Transfert actif par les insectes fouisseurs et le périphyton

Le transfert actif de MeHg depuis les sols organiques vers la colonne d'eau, par les organismes benthiques, est aussi très réduit après quelques années, puisque sur de grandes superficies de la zone de marnage des réservoirs, les sols organiques sont rapidement érodés par l'action des vagues et de la glace. Au complexe La Grande, où les sols podzoliques minces sont très répandus, une grande partie du pourtour des réservoirs est rapidement transformée en sable, gravier et roches, réduisant considérablement le transfert passif et actif du Hg depuis les sols inondés vers la colonne d'eau. En conséquence, le zooplancton et les insectes benthiques recueillis dans l'estomac des meuniers noirs du réservoir Desaulniers, situé à proximité du réservoir Robert-Bourassa, présentait, après 17 ans l'inondation, des concentrations de MeHg équivalentes à celles obtenues en lacs naturels dans les mêmes organismes. Dans la zone pélagique d'un certain nombre de réservoirs du complexe La Grande, les concentrations en Hg dans le zooplancton recueilli 8 à 10 ans après l'inondation présentaient également des niveaux de MeHg équivalents à ceux du zooplancton des lacs naturels avoisinants.

Donc, durant les toutes premières années après l'inondation, le transfert du MeHg depuis les sols et la végétation inondés vers la chaîne alimentaire du réservoir s'effectue de façon intensive par les différents processus énumérés précédemment. Après quelques années, l'intensité de tous ces processus de transfert diminue grandement, de sorte le taux de transfert de Hg de la végétation et des sols inondés vers la chaîne alimentaire aquatique des réservoirs devient équivalent au taux de transfert des sédiments lacustres vers la chaîne alimentaire aquatique des lacs. En conséquence, les concentrations de Hg dans les poissons des réservoirs retournent graduellement à des niveaux typiques de ceux des lacs naturels de la région.

Facteurs morphologiques et hydrologiques influençant l'évolution des teneurs en mercure des poissons dans les réservoirs

Le suivi réalisé au complexe La Grande montre que les concentrations en Hg des poissons de tous les réservoirs ont évolué de la même façon, bien que les teneurs maximales atteintes, de même que les taux d'augmentation et de réduction subséquente, diffèrent quelque peu d'un réservoir à l'autre. Nos résultats suggèrent que ces différences peuvent s'expliquer par un nombre limité de caractéristiques physiques et hydrologiques liées aux principaux processus identifiés comme étant responsables de l'augmentation des teneurs en Hg des poissons. Ceux-ci comprennent la superficie terrestre inondée, le volume d'eau annuel transitant dans le réservoir, la durée du remplissage et la proportion de la superficie terrestre inondée située dans la zone de marnage. Ces caractéristiques peuvent être utilisées pour obtenir une première évaluation du potentiel d'augmentation des teneurs en Hg dans les poissons de réservoirs projetés.

Rapport entre la superficie terrestre inondée et le volume d'eau annuel

Le rapport entre la superficie terrestre inondée (en km^2) et le volume d'eau annuel (en km^3) qui transite dans le réservoir (rapport SI/VA), devrait fournir un bon indice du potentiel d'augmentation des teneurs en Hg dans les poissons. La superficie terrestre inondée est un indicateur de la quantité de matière organique stimulant la méthylation bactérienne, ainsi que de l'intensité de la diffusion passive et du transfert actif du Hg. Le volume annuel d'eau transitant dans un réservoir peut aussi être considéré comme un facteur clé car : (1) il constitue un indicateur de la capacité de dilution du Hg diffusé dans la colonne d'eau; (2) il joue un rôle dans le degré d'épuisement de l'oxygène dissous (on sait que les faibles teneurs en oxygène favorisent la méthylation); (3) il détermine le degré d'exportation du Hg vers l'aval d'un réservoir; (4) il joue un rôle au niveau de l'exportation d'éléments nutritifs vers l'aval d'un réservoir, qui réduit la production primaire autotrophe et, par conséquent, la méthylation additionnelle de Hg provenant de la décomposition des matières résultantes (phytoplancton et périphyton). Ainsi, plus le rapport SI/VA est grand, plus le potentiel d'augmentation des teneurs en Hg dans les poisson devrait être élevé.

Superficie terrestre inondée dans la zone de marnage

La proportion de la superficie terrestre totale inondée située dans la zone de marnage serait un également un bon indicateur de l'ampleur et de la durée du transfert biologique actif de MeHg depuis les sols inondés jusqu'aux poissons. D'une part, elle serait un bon indicateur de l'étendue de l'érosion de la matière organique dans la zone de marnage, augmentant la quantité de fines particules organiques en suspension (riches en Hg) pouvant être filtrées par le zooplancton. D'autre part, elle indiquerait la durée du transfert de Hg par les insectes aquatiques fouissant dans les sols organiques non érodés, ainsi que par le zooplancton s'alimentant du périphyton se développant sur ces sols. Ce transfert biologique pourrait jouer un rôle significatif pendant une période prolongée (au moins 15 ans suivant des mesures *in situ*) dans des zones peu profondes, à l'abri de l'action des vagues, où la matière organique n'a pas été érodée. Par contre, dans les réservoirs du complexe La Grande, les sols inondés sont généralement très minces et rapidement érodés pour ensuite se déposer dans les zones plus profondes et froides, moins propices à la méthylation du Hg. Cette érosion réduit la surface des sols organiques inondés où le transfert biologique par les invertébrés se déroule. Ainsi, pour de tels réservoirs, plus la proportion de la superficie terrestre inondée située dans la zone de marnage est grande, plus la durée des fortes teneurs en Hg des poissons serait réduite, en raison de la courte période de transfert actif du Hg par les invertébrés benthiques.

Durée de la période de remplissage

Le temps requis pour remplir un réservoir est aussi considéré comme un facteur important pour déterminer les teneurs maximales en Hg qui seront atteintes dans les poissons à la suite de la mise en eau. Des études *in vitro* ont démontré que la libération de Hg de la matière organique inondée jusqu'à la colonne d'eau s'effectue très rapidement, la majorité du Hg étant libérée pendant les premiers mois d'inondation. Dans le cas d'un réservoir dont le remplissage s'effectue sur un certain nombre d'années, le Hg est libéré sur une période de temps plus longue, mais à un rythme plus lent. Les changements de qualité de l'eau reliés à la décomposition bactérienne ont atteint une intensité maximale après 2 à 3 ans d'inondation dans les réservoirs remplis en moins d'un an (Robert-Bourassa et Opinaca), mais seulement après 6 à 10 ans dans le réservoir Caniapiscau, lequel a été rempli sur une période de 3 ans. Ainsi, dans les cas où les autres facteurs sont équivalents, plus le temps de remplissage est long, moins les teneurs maximales en Hg des poissons seraient élevées, mais plus la période requise pour retourner à des concentrations typiques des lacs naturels avoisinants serait longue. Le temps de remplissage est aussi un bon indicateur du taux d'érosion des sols organiques inondés pendant le remplissage, puisqu'il détermine le taux d'augmentation des niveaux d'eau. Dans un grand réservoir rempli lentement, tel que le Caniapiscau, le niveau d'eau dans la zone de marnage ne s'élève que de quelques centimètres par jour. Les vagues agissent alors sur un même niveau pendant plusieurs jours, de sorte que l'érosion des sols podzoliques minces peut être complétée pendant la période de remplissage. Pour le réservoir Caniapiscau, il semble que ce facteur ait contribué à réduire, chez le grand corégone, le temps de retour vers des concentrations de Hg typiquement mesurées dans les lacs naturels avoisinants (retour en 10 ans par rapport à 17 ans au réservoir Robert-Bourassa).

Risques pour la faune

Deux espèces animales terrestres, le vison (*Mustela vison*) et le Balbuzard pêcheur (*Pandion haliaetus*), qui se nourrissent occasionnellement ou exclusivement de poissons, ont servi de modèles dans le cadre de l'étude des effets potentiels de l'augmentation des teneurs en Hg des poissons des réservoirs sur la faune.

Expérience sur le vison

Mammifère partiellement piscivore, le vison est une espèce largement répandue dans le Nord du Québec. Des spécimens sauvages capturés par les trappeurs cris et des spécimens semi-domestiques ont été utilisés pour mener des études d'exposition *in vivo* et *in vitro* respectivement.

Dans le cadre d'une expérience d'exposition *in vitro*, trois groupes de visons semi-domestiques femelles ont été exposées à des diètes quotidiennes contenant $0,1$, $0,5$ et $1,0 \ \mu g \ g^{-1}$ de Hg respectivement. Ces diètes étaient préparées à l'aide de poissons de réservoirs, de sorte que le Hg étaient principalement ingéré sous forme de MeHg. Les résultats révèlent qu'aucun effet n'a pu être observé chez les groupes de visons exposés à des diètes quotidiennes de 0,1 et 0,5 µg g^{-1}. Par contre, une diminution du taux de fertilité, ainsi qu'une mortalité importante (touchant le 2/3 des animaux), combinée à des signes de toxicité neurologique, ont été observées après 3 mois ou plus d'exposition à une diète quotidienne de 1,0 µg g^{-1}.

Puisque les concentrations moyennes de Hg enregistrées dans tous les tissus des visons sauvages capturés dans le Nord du Québec sont bien inférieures aux valeurs correspondantes mesurées dans les tissus du groupe de visons semi-domestiques exposés à une diète quotidienne de 0,1 µg g^{-1} de Hg, la diète des visons sauvages de cette région doit également être bien inférieure à cette valeur. Le risque que présente le MeHg pour les populations de visons habitant les milieux naturels du Nord du Québec est donc très faible malgré l'augmentation de la charge de Hg dans l'environnement depuis les 60 dernières années.

Pour les populations de visons vivant à proximité des berges des réservoirs du complexe La Grande, le risque semble également faible puisque ces grands plans d'eau, aux fluctuations de niveau importantes et contraires aux cycles saisonniers naturels, offrent peu de possibilité au vison adulte territorial d'y trouver un habitat permanent convenable. Cependant, les berges le long de la voie de dérivation Boyd-Sakami peuvent offrir au vison un habitat plus approprié, n'étant pas soumises à de fortes variations du niveau des eaux.

Les teneurs moyennes en Hg des poissons capturés le long de cette voie de dérivation ont augmenté par un facteur moyen de 5 par rapport aux teneurs observées dans les poissons des lacs naturels de la région. En assumant que les concentrations dans la diète des visons vivant en bordure de la dérivation augmentent proportionnellement à celles des poissons de ce secteur, les teneurs en Hg dans la diète des visons sauvages demeureraient inférieures à 0,5 µg g^{-1}. Les risques pour les populations de visons sauvages habitant à proximité de la dérivation Boyd-Sakami seraient donc également faibles puisqu'aucun effet n'a été observé chez les visons exposés à une diète quotidienne de 0,5 µg g^{-1} de Hg. De plus, la diète estimée pour les visons sauvages de ce secteur est également bien en deçà de la dose de 1,0 µg g^{-1} de Hg à laquelle des effets ont été observés. Cette évaluation du risque pour les visons sauvages du secteur Boyd-Sakami est conservatrice, car les teneurs en Hg dans les tissus des visons capturés en bordure de la voie de dérivation de la rivière Churchill, au Manitoba, avaient augmenté par un facteur moindre que celui des poissons. De nombreux facteurs biologiques, écologiques et environnementaux auraient contribué à réduire l'exposition des visons sauvages au Hg.

Les poissons des réservoirs qui remontent dans les tronçons accessibles des tributaires de ces réservoirs pour se reproduire, peuvent constituer une source additionnelle d'exposition au MeHg pour les visons habitant à proximité de ces cours d'eau. Dans de telles situations, les poissons qui remontent du réservoir avoisinant seraient dilués parmi la population locale de ces cours d'eau. De plus, les poissons des réservoirs ne séjourneraient dans ces milieux que durant la période de reproduction au printemps ou à l'automne. Le risque pour les populations de visons sauvages associé à de tels mouvements serait donc moindre que celui estimé pour les populations habitant le secteur de la dérivation Boyd-Sakami.

Succès de reproduction des Balbuzards pêcheurs

Le succès de reproduction chez les Balbuzards pêcheurs nichant à proximité des réservoirs du complexe La Grande a aussi fait l'objet d'une étude car cette espèce constitue un excellent modèle pour évaluer les effets potentiels de l'augmentation des teneurs en Hg des poissons sur la faune avienne. En effet, cette espèce essentiellement piscivore doit compter sur une vision et une coordination neuro-motrice extrêmement efficaces pour nourrir sa progéniture, des fonctions particulièrement sensibles à l'intoxication au Hg. À l'exception des œufs, qui sont habituellement pondus lorsque les réservoirs sont encore recouverts de glace, les teneurs en Hg total de tous les tissus prélevés chez les adultes et les aiglons étaient considérablement plus élevées chez les individus nichant près des réservoirs, que chez ceux habitant près des milieux aquatiques naturels. Les plumes des oiseaux adultes et des oisillons recueillies à proximité des réservoirs contenaient 3,5 et 5 fois plus de Hg total respectivement que celles des oiseaux provenant d'habitats naturels. En moyenne, les plumes contenaient 16,5 mg kg^{-1} (poids humide) de Hg total chez les Balbuzards pêcheurs adultes des habitats naturels comparativement à 58,1 mg kg^{-1} chez ceux des réservoirs avoisinants. Une tendance similaire a été observée dans le plumage des aiglons âgés de 35 à 45 jours, alors que des concentrations moyennes de 7,0 mg kg^{-1} ont été obtenues chez ceux élevés près dans les habitats naturels par rapport à 37,4 mg kg^{-1} pour ceux élevés à proximité des réservoirs. Malgré une exposition au Hg total nettement plus élevée chez les Balbuzards pêcheurs qui s'alimentent dans les réservoirs, le nombre d'oeufs pondus, ainsi que le nombre de jeunes élevés jusqu'à l'âge où ils quittent le nid, n'étaient pas statistiquement différents entre les nids situés à proximité des réservoirs et ceux retrouvés près des lacs et rivières naturels. Ces résultats indiquent que l'augmentation des teneurs en Hg des poissons des réservoirs du complexe La Grande n'a pas affecté le potentiel reproducteur des populations de Balbuzards pêcheurs de la région.

Nos résultats indiquent que la croissance des plumes, tant chez les oiseaux adultes en mue que chez les oisillons, constitue un bon mécanisme d'excrétion du Hg. En effet, le plumage des jeunes âgés de 5 à 7 semaines contient environ 86 % de la charge corporelle totale de Hg, excluant les autres tissus qui contiennent de la kératine, comme le bec et les griffes, pour lesquels la teneur en Hg total n'a pas été mesurée. Néanmoins, le succès de reproduction des Balbuzards pêcheurs qui font leur nid à proximité des réservoirs, malgré une forte hausse des teneurs en Hg dans leurs tissus, ne peut s'expliquer uniquement par la mue partielle que subissent les oiseaux adultes durant l'été. Plusieurs études ont démontré que la déméthylation du MeHg constitue une voie de désintoxication importante chez les oiseaux de proie. Ainsi, malgré un niveau d'exposition élevé occasionné par leur position en tant que dernier maillon de la chaîne alimentaire, ces oiseaux tolèrent mieux le MeHg qu'on le croyait auparavant.

Conclusion

Bien que les questions concernant les sources et le devenir du Hg dans les écosystèmes aquatiques naturels et aménagés du Nord du Québec n'aient pas toutes été élucidées, les résultats de plus de 10 années de recherches et de plus de 20 ans de suivi environnemental ont permis de bien comprendre les principaux processus biogéochimiques qui entrent en jeu. Ils ont également permis de bien définir l'ampleur et la durée du phénomène d'augmentation des teneurs en Hg des poissons à la suite de la création de réservoirs hydroélectriques. Bien que le risque associé à cette augmentation semble faible pour le vison et le Balbuzard pêcheur, il demeure peu connu pour les autres espèces fauniques piscivores.

En ce qui concerne les risques à la santé des consommateurs de poissons, l'augmentation des teneurs en Hg des poissons des réservoirs était telle, qu'une consommation régulière de poissons piscivores de ces milieux aurait entraîner des niveaux d'exposition supérieurs à ceux généralement recommandés par les organismes de santé publique. Les aspects de santé publique n'ont pas été traités dans le cadre de cette monographie car, en vertu de la Convention de la Baie James sur le mercure, signée en 1986 par le Gouvernement du Québec, les Cris du Québec, la Société d'énergie de la Baie James et Hydro-Québec, le volet santé était sous la responsabilité du Conseil cri de la santé et des services sociaux de la Baie James. Dans le cadre de cette convention, le niveau d'exposition au Hg des Cris du territoire de la Baie James a fait l'objet d'un suivi et plusieurs mesures d'atténuation ont été mises en œuvre pour permettre aux Cris de poursuivre leurs activités traditionnelles tout en réduisant les risques à la santé. Parmi ces mesures, il faut mentionner le financement de pêches communautaires dans des régions où les teneurs en Hg des poissons sont faibles, ainsi que divers aménagements favorisant la production et la récolte d'espèces fauniques non piscivores, à faible teneur en Hg.

Il est clair que dans le cadre de tout nouveau projet de développement hydroélectrique, on devra favoriser les schémas d'aménagement minimisant la hausse des teneurs en mercure des poissons et prévoir la mise en œuvre d'un programme de gestion du risque à la santé. Mais au delà de cette considération, il semble également évident que pour prévenir tout nouveau risque à la santé relié à la consommation de poissons contaminés au Hg, qu'ils proviennent de lacs naturels ou de jeunes réservoirs, la solution ultime réside dans la réduction des émissions anthropiques de Hg, à cause desquelles de plus en plus de Hg est aéroporté sur de grandes distances et incorporé aux écosystèmes terrestres et aquatiques des régions très éloignées des sources d'émissions.

1   Introduction

1.1
The Mercury Issue in Northern Québec

Although there are no direct industrial or municipal sources of mercury (Hg) pollution, the presence of this heavy metal is ubiquitous throughout the northern Québec environment. This is a cause of concern because of the high Hg concentrations measured in fish and certain other wildlife species and because of its potential toxicity for humans. In this region, and generally all across Canada, the principal source of human exposure to methylmercury (MeHg) occurs through consumption of fresh water fish and, for some communities, of sea mammals. During the past two decades, northern Québec has been the site of large scale hydroelectric developments resulting in the impoundment of several reservoirs, flooding extensive tracts of land areas, causing important additional increases of Hg levels in fish.

As elevated Hg concentrations in fish caught in hydroelectric reservoirs were reported in the scientific literature during the mid 1970s, Hydro-Québec initiated its first research program in 1978, before the impoundment of the first La Grande complex reservoir, with the primary objective of monitoring the evolution of Hg levels in fish. Results obtained as early as in 1981 indeed revealed important increases in Hg concentrations in fish of the Robert-Bourassa and Opinaca reservoirs.

Because of the potential health risks to Cree practising subsistence fishing, the "James Bay Mercury Agreement" was signed in 1986 jointly by the Cree of Québec, the Québec government, the Société d'énergie de la Baie James and Hydro-Québec. The principal goals of this 10-year agreement were to determine the nature and extent of the problem caused by the presence of Hg in the environment of the James Bay region, to improve our knowledge and understanding of the phenomenon of Hg increase in hydroelectric reservoirs and to minimise the potential effects of Hg on the Cree's health, way of life and traditional activities.

In parallel, Hydro-Québec, in collaboration with the Chaire de recherche en environnement HQ-CRSNG-UQAM of the Université du Québec à Montréal, the Université de Sherbrooke, the Faculté de Médecine vétérinaire de l'Université de Montréal and the Service canadien de la faune (Canadian Wildlife Service), has put together comprehensive research programs to determine the source and fate of Hg in natural and modified aquatic ecosystems of northern Québec.

This monograph constitutes a comprehensive presentation of the scientific findings derived from the latter research programs conducted over more than a decade. It addresses the biogeochemical processes associated with the presence of Hg in natural and modified aquatic environments of northern Québec, including the study of the food chain, from invertebrates to fish eating mammals and birds. But it does not specifically address the human health aspects which, under "James Bay Mercury Agreement", were the responsibility of the Cree Board of Health and Social Services of James Bay.

1.2
Geographic Setting

The Hg levels in fish observed in natural lakes in northern Québec, as well as their evolution after the impoundment of hydroelectric reservoirs, are the result of a set of physical, chemical and biological conditions. We will begin by briefly present-ing here the principal characteristics of the region of the La Grande complex, the hydroelectric projects built there, and the principal changes which they have brought about in the biophysical environment.

1.2.1
Drainage Basin and Relief

The drainage basin of the La Grande complex, located around latitude 54°N (1 000 km north of Montréal), occupies the northern portion of the James Bay region (Figure 1.1). It covers approximately 175 000 km^2. The relief of this region, shaped by successive glaciations, comprises, from west to east, a coastal plain (150 km wide, with scattered peat bogs and clay deposits), a hilly central plateau studded with numerous lakes, and, at the eastern end, an area of rougher terrain. About 8 000 years ago, the lower basins of Hudson Bay and James Bay were over-run by the Tyrrell Sea, which stretched inland 200 km or more, to an elevation of 290 m (SEBJ 1987). In the depressions of this marine environment, deposits of silty clay and fine deltaic sands were formed. The entire region rests on Canadian Shield made up of igneous and metamorphic rock with outcrops appearing throughout the area.

Fig. 1.1. Location of the study area, Québec, Canada

1.2.2
Climate and Hydrology

The climate is of the cold continental type, characteristic of the humid subarctic zone. The mean annual temperature is -4°C. The prevailing winds blow from west to east. An annual average of 765 mm of water falls on the area, in the form of rain or snow, with precipitation increasing gradually from west to east, but decreasing from south to north.

The annual hydrologic cycle of the rivers is characterized by heavy spring floods, fed by rainfall and snow melt beginning in May or early June; summer low water periods varying in severity from year to year; fall increases in flow resulting from rainfall and finally a winter low water period that begins in November and lasts until early May. The lakes of the region are usually covered by ice from November to mid-May.

1.2.3
Vegetation and Wildlife

The forests are sparse and mainly composed of stands of black spruce (*Picea mariana*) or jack pine (*Pinus banksiana*), with the presence of larch (*Larix laricina*) and aspen (*Populus tremuloides*). The low density and slow growth of trees renders the forests unsuitable for commercial harvesting. The riparian vegetation is dominated by shrubby willows (*Salix* spp.) and alder (*Aulnus* spp.). Podzolic forest soils, fairly thin and lying on a sandy or little weathered substrate cover about 60% of the territory. Peatlands covering about 10% of the territory are found in poorly drained, low-lying areas, while the remaining 30% of forest soils are organic, with drainage conditions intermediate between those of podzolic soils and peatlands. The forest undergrowth is dominated by mosses and lichens (Poulin-Thériault–Gauthier-Guillemette 1993). A total of 39 species of mammals occupy the territory, including moose (*Alces alces*), caribou (*Rangifer caribou*) and beaver (*Castor canadensis*) which are of economic and sporting interest. For wildlife resources as a whole, densities are lower than in more southerly regions. The eastern coast of James Bay constitutes a zone with major potential for migrating waterfowl.

1.2.4
Water Quality and Fish

Bodies of water of the region are characterized by high transparency ($Z_{SD} = 1.5$ to 4.0 m), high dissolved oxygen levels (80 to 100% saturation), mildly acidic pH (5.9 to 6.9), low buffering capacity (0.6 to 11.0 mg L^{-1} of bicarbonates as HCO_3), low mineral content (conductivity of 8 to 30 µS cm^{-1}) and relatively high organic content. Most lakes are quite oligotrophic with low levels of nutrients (0.004-0.010 mg L^{-1} of total phosphorus) (Schetagne 1994). Some 27 species of fresh water fish have been inventoried, the most common are longnose sucker (*Catostomus catostomus*), white sucker (*Catostomus commersoni*), lake whitefish (*Coregonus clupeaformis*), cisco (*Coregonus artedii*), northern pike (*Esox lucius*), lake trout (*Salvelinus namaycush*), walleye (*Stizostedion vitreum*), brook trout (*Salvelinus fontinalis*), and burbot (*Lota lota*). Fish growth rates are lower than in the southern regions of the province, but longevity is greater, sexual maturity is reached later and reproductive cycles are prolonged.

1.2.5
Human Settlements

The James Bay region is inhabited by native Cree. Their population of around 11 000 (in 1997) mainly live in four villages (Waskaganish, Eastmain, Wemindji, Chisasibi) along the eastern coast of James Bay, as well as in four villages located inland (Nemaska, Waswanipi, Mistissini and Oujé-Bougoumou). The Cree population also includes some 600 residents of the community of Whapmagoostui which lies just outside the northern boundary of the James Bay territory, on the eastern coast of Hudson Bay.

The region also supports a non-native population of about 15 000 (in 1995) concentrated mainly in the southern part of the James Bay region, as well as in Radisson (814 individuals in 1991). A changing population of workers connected with the various hydroelectric developments also occupied a number of construction camps from the early 1970s to late 1990s. The region is also visited by approximately 20 000 tourists during the summer season (in 1990).

1.3
Hydroelectric Developments

The La Grande hydroelectric complex called for the construction of 8 principal reservoirs, 8 generating stations and 2 river diversions, that of the Eastmain and Opinaca rivers in the southwest and that of the rivière Caniapiscau in the east (Figure 1.2).

1.3.1
Reservoirs

As indicated in Table 1.1, the Robert-Bourassa[1] reservoir is the oldest reservoir of the La Grande complex, having been flooded in 1979. Next come, in order, the Opinaca, La Grande 3, Caniapiscau, Fontanges, La Grande 4, Laforge 1 and La Grande 1 reservoirs. The maximum land area flooded varies considerably from one reservoir to another, ranging from 40 km^2 at the La Grande 1 reservoir to 3 430 km^2 at the Caniapiscau reservoir (Table 1.1). Depending on the reservoir,

---

[1] Previously called the La Grande 2 reservoir, this reservoir has been renamed the Robert-Bourassa reservoir. The same applies to the La Grande-2 generating station.

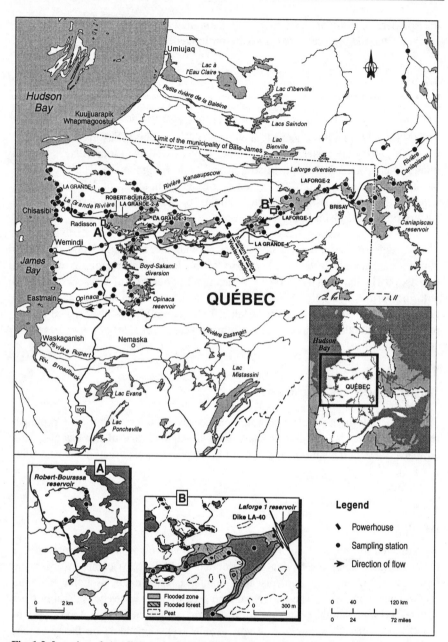

Fig. 1.2. Location of sampling stations in the La Grande complex, Québec, Canada

Table 1.1. Characteristics of the principal reservoirs in the La Grande complex at their maximum levels

Reservoir	Generating station	Mean annual drawdown (m)	Maximum reservoir area (km²)	Maximum land area flooded (km²)	Mean depth (m)	Maximum volume (km³)	Theoretical residence time (months)	Mean annual flow (m³/s)	Filling period
La Grande 1	La Grande-1	1,2 (1.5)*	70	40 (57%)**	18.6	1.3	0.15	3 400	93-10 to 93-11
Robert-Bourassa	Robert-Bourassa, La Grande-2a	3.3 (7.7)	2 835	2 630 (92%)	22.0	61.7	6.9	3 374	78-11 to 79-12
La Grande 3	La Grande-3	5.5 (12.2)	2 420	2 175 (90%)	24.4	60.0	11.0	2 064	81-04 to 84-08
La Grande 4	La Grande-4	8.0 (11.0)	765	700 (89%)	29.4	19.5	4.8	1 534	83-03 to 83-11
Opinaca		3.6 (4.0)	1 040	740 (71%)	8.2	8.4	3.8	845	80-04 to 80-09
Laforge 1	Laforge-1	— (8.0)	1288	923 (72%)	6.2	8.0	3.2	938	93-08 to 93-10
Fontanges	Laforge-2	1.5 (2.0)	286	171 (60%)	6.3	1.8	0.8	804	84-01 to 84-02
Caniapiscau	Brisay	2.1 (12.9)	4 275	3 430 (80%)	16.8	53.8	25.8	790	81-10 to 84-09

* Maximum drawdown indicated in parentheses.

** Percentage of total area consisting of flooded land is given in parentheses.

the percentage of land area flooded relative to the total area varies from 57% (La Grande 1) to 92% (Robert-Bourassa). As will be discussed further, the flooded land area is a major factor in the post-impoundment increase in fish Hg levels. Indeed, Hg becomes bioavailable to the aquatic food chain by its transformation into MeHg by bacteria thriving on the flooded organic matter and by passive or active releases from the flooded terrigenous system to the water column.

The theoretical residence time of the water at its maximum level varies considerably from one reservoir to the next (Table 1.1). It is very short for the La Grande 1 reservoir (4 days), varies from 4 to 11 months for the Robert-Bourassa, La Grande 4, Opinaca and La Grande 3 reservoirs, and reaches 26 months for the Caniapiscau reservoir. The hydrologic characteristics presented in Table 1.1 also play a major role in the Hg issue because: (1) they are indicators of the diluting capacity of the Hg released to the water column, (2) they help determine the extent of oxygen depletion (as anoxic conditions promote Hg methylation), (3) they determine the extent of export of Hg out of a reservoir, and (4) they play a role in the export of nutrients out of a reservoir. Nutrient concentrations control the biological production and thus the additional methylation of Hg brought about by the decomposition of the labile autochthonous matter (phytoplankton and periphyton).

The maximum annual drawdown varies from 1.5 to 12.9 meter from one reservoir to another (Table 1.1). This characteristic is also an important component of the Hg cycle in a reservoir as it influences the areas of flooded land potentially eroded by wave or ice action. As will be presented in the monograph, the erosion of terrigenous organic matter can either enhance the transfer of the contaminant to the aquatic organisms or reduce it through the destruction of suitable habitats for burrowing invertebrates. The time required to fill each reservoir (Table 1.1.) is also important when studying the fate of Hg because it corresponds to the rate of erosion of the flooded organic soils during the impoundment, a slow water level increase rate allowing more erosion of the soils of the area. For the La Grande complex reservoirs, the filling time varied from 1 to 35 months.

In addition, two small reservoirs (total area of less than 10 km^2), the Desaulniers which adjoins the Robert-Bourassa reservoir in the western part of the complex and the LA-40 located next to the Laforge 1 reservoir in the eastern part, were created by the construction of dikes to limit the land area flooded by the larger reservoirs. Their fluctuations in water level are different from those in the large reservoirs in the complex, not being related to hydroelectric generation.

1.3.2
Other Modified Environments

The creation of the Opinaca reservoir required the cut-off of the Petite Opinaca river, in July 1979, and then that of the Opinaca and Eastmain rivers, in April and July 1980, respectively. This resulted in a 90% decrease in mean flow at the mouth of the Eastmain river.

Since its cut-off in October 1981 to impound the Caniapiscau reservoir, the Caniapiscau river has lost 40% of its initial flow at its confluence with the Koksoak river. However, during most of the 1984 ice-free season (from June to October) as well as during the months of June and July 1985, spillages from the Caniapiscau reservoir led to flows approximately equivalent to those recorded before the cut-off.

The development of the La Grande hydroelectric complex also called for the diversion of rivers, creating areas characterized by series of lakes and rivers with increased flow, such as the 140-km long Boyd-Sakami diversion or the 250-km long Laforge diversion (Figure 1.2). These diversions increased and regulated the flow of the La Grande river below the Robert-Bourassa reservoir. From 1979 to 1986, the mean annual flow of this section of river increased gradually from $1\ 700\ m^3\ s^{-1}$, when Robert-Bourassa generating station was commissioned in 1979, to around $3\ 400\ m^3\ s^{-1}$.

As will be shown, the monitoring of fish Hg levels in these modified environments enabled us to better understand the export of Hg downstream from reservoirs.

1.4
Evolution of the Biophysical Environment

The biophysical characteristics of the Robert-Bourassa, Opinaca and Caniapiscau reservoirs were monitored for nearly 20 years, as part of the "Réseau de Suivi Environnemental" (RSE) of Hydro-Québec, an environmental effects monitoring program which followed the changes occurring in water quality, plankton, benthos and fish in different modified environments of the La Grande complex. As demonstrated in the monograph, these long term monitoring studies have been extremely useful in developing our understanding of the biogeochemical cycle of Hg in northern reservoirs.

1.4.1
Water Quality

The evolution of the principal water quality parameters measured in the photic zone of these 3 reservoirs, during the ice-free period, is shown in Figure 1.3. The results show that pH, CO_2 and oxygen values always remained adequate for most aquatic organisms. Temporary water quality changes due to the leaching and the decomposition of flooded organic matter have been weak in the photic layers of the reservoirs (Schetagne 1994). This decomposition caused a consumption of dissolved oxygen, an increase of CO_2, a drop of pH and a release of nutrients such as phosphorus (Figure 1.3). For the Robert-Bourassa and Opinaca reservoirs overall, the maximum water variations of these variables were reached quickly, 2 or 3 years after the start of impoundment. For all intents and purposes, the temporary water quality changes related to the decomposition of flooded organic matter, as shown in Figure 1.3, were over by 1988, 9 to 10 years after the start of impoundment (Schetagne 1994). In the Caniapiscau reservoir, the changes measured for these parameters were comparable to those recorded at the other reservoirs. However, the maximum modifications for total phosphorus, chlorophyll *a* and silica were reached later, between 6 and 10 years after the start of impoundment, and the return to values representative of natural conditions was nearly complete after 14 years (Chartrand et al. 1994).

The greatest variations in oxygen, major ions and nutrients were measured in the bottom layers of the reservoirs where redox potential drops permitted better exchanges between the flooded substrates and the overlying waters (for example, see Table 1.2). However, these restricted areas had very little effect on the overall water quality of these large and deep reservoirs (Schetagne 1994).

1.4.2
Plankton and Benthos

Phytoplanktonic biomass, measured by chlorophyll *a*, also increased for a number of years inducing a depletion of silica, as diatoms are an important phytoplankton group in this region (Figure 1.4). Increased water residence time and nutrient enrichment contributed to temporary increases in zooplankton and benthos biomass in most of the modified bodies of water. For example, figure 1.4 shows the changes in the density and biomass of zooplankton and benthos observed in the Robert-Bourassa reservoir.

1.4.3
Fish

For the first year, reservoir impoundment caused a dilution in fish populations. After that, populations in the three large reservoirs sampled showed increases in

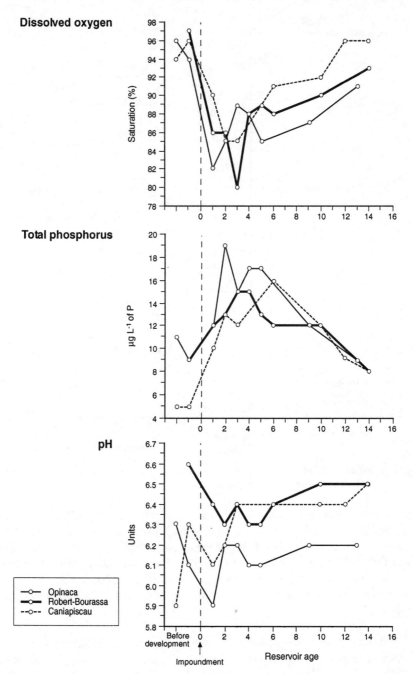

Fig. 1.3. Evolution of some water quality variables as a function of the age of the Opinaca, Robert-Bourassa and Caniapiscau reservoirs

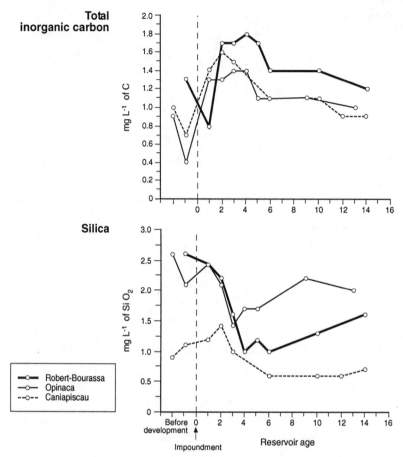

Fig. 1.3. Evolution of some water quality variables as a function of the age of the Opinaca, Robert-Bourassa and Caniapiscau reservoirs (continued)

overall yield resulting from the water's enrichment. In the Robert-Bourassa reservoir, the fish catch per unit of effort doubled after 4 years of impoundment and has been gradually returning towards initial values since then. This increase was rapid in the Robert-Bourassa and Opinaca reservoirs, but more gradual in the Caniapiscau reservoir, where it peaked in 1991, 9 years after impoundment. Figure 1.5 illustrates the evolution of the fish catch per unit of effort for the Robert-Bourassa reservoir.

Northern pike and lake whitefish are the species that benefited most from the creation of large reservoirs. Reproduction rates for these species suggest that much of their increase in yield is attributable to better recruitment, associated with a better survival rate in the first years after impoundment (Deslandes et al. 1995).

Table 1.2. Values of main water quality parameters measured in near-surface and one-meter-from-bottom samples of the Robert-Bourassa reservoir in 1982, 3 years after impoundment

Parameter	Deep zone (one meter from bottom)		Surface
	End of winter	After spring turnover	After spring turnover
Dissolved oxygen (% saturation)	0-6	66-91	73-82
pH (units)	5.8-6.2	6.0-6.3	6.3-6.6
Total inorganic carbon (mg L^{-1} of C)	8.8-15.8	2.1-3.3	1.0-1.8
Total phosphorus (μg L^{-1} of P)	92-178	13-28	14-33
Total Fe (mg L^{-1})	1.9-7.2	0.05-0.90	0.05-0.75
Manganese (mg L^{-1})	0.2-1.0	0.02-0.23	< 0.02
Ammonia (mg L^{-1})	0.07-0.39	< 0.02-0.10	$\leq$ 0.02

The growth rate of the principal species in the Robert-Bourassa reservoir increased markedly after impoundment (Deslandes et al. 1995). Like growth, the condition factor of almost all the species increased after impoundment. This increase was especially pronounced for walleye, longnose sucker and cisco, for which it increased by 18.8%, 20% and 21.3%, respectively (Deslandes et al. 1994). After several years, the fish's condition factor gradually returns to normal values.

1.5
Contents and Rationale

In order to fulfill the objective of determining the source and fate of Hg in natural and modified aquatic ecosystems of northern Québec, methodological improvements were required, especially for the determination of total and MeHg in surface water and pore water samples, with concentrations at or below the ppt level, as well as for minute samples of organic debris, plankton and invertebrates. These developments, as well as all other methodological aspects, are presented in Chapter 2.0.

As different biophysical inventories related to future development schemes in natural areas north and south of the La Grande drainage basin revealed high

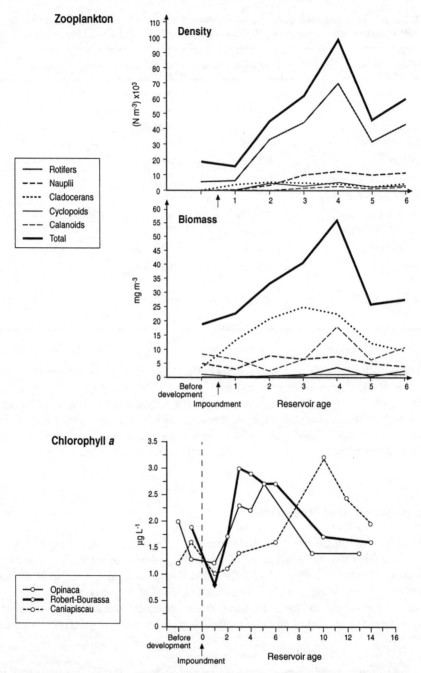

Fig. 1.4. Trends in density and biomass of plankton and macro-invertebrates in the Robert-Bourassa reservoir

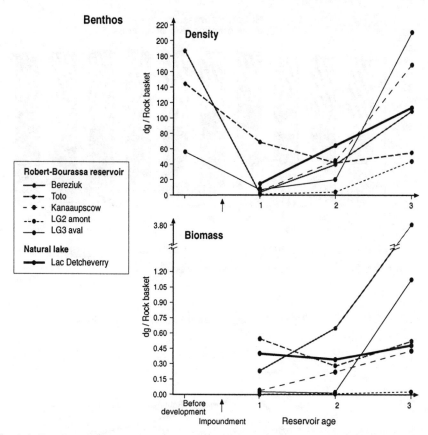

Fig. 1.4. Trends in density and biomass of plankton and macro-invertebrates in the Robert-Bourassa reservoir (continued)

concentrations in fish and wildlife (Chapters 5.0 and 6.0), it became important to document the history and importance of Hg contamination in the region (Chapter 3.0). A priority was set on the study of the processes by which inorganic Hg is accumulated in organic components of the terrestrial ecosystems, transported to the aquatic environments where it is methylated and transferred up the aquatic food chain (Chapters 3.0, 4.0 and 5.0).

In the reservoirs, emphasis was put on the processes by which the Hg contained in the flooded soils and vegetation is methylated, passively and actively released to the water column and transferred up the aquatic food chain (Chapters 7.0, 8.0 and 9.0). Monitoring of Hg levels in fish of modified and natural environments continued throughout all these activities and were enhanced by complementary studies investigating the diet of the principal species and the Hg burden in most common prey organisms (Chapters 9.0, 10.0 and 11.0).

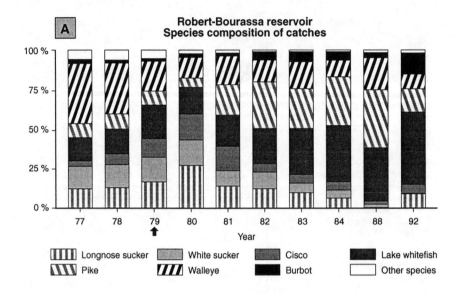

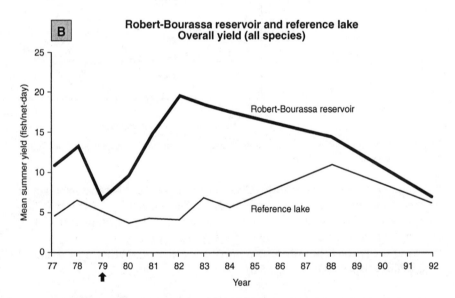

Evolution of species composition of catches, overall yield (all species combined) and mean annual yields of the main fish species in the Robert-Bourassa reservoir and its reference lake (lac Detcheverry; station SB400).

↑ Impoundment.

Fig. 1.5. Evolution of fish biomass and capture per unit of effort in Robert-Bourassa reservoir and a natural lake

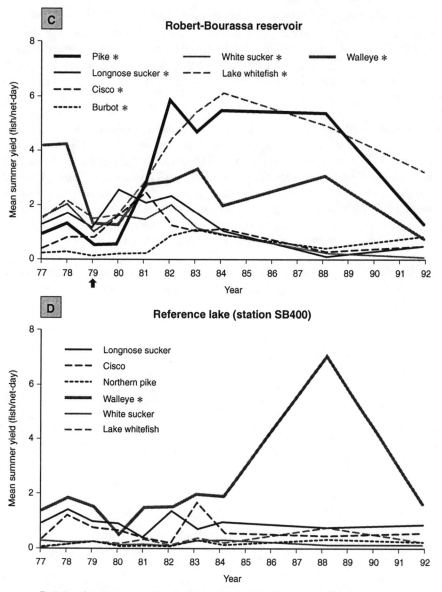

Evolution of species composition of catches, overall yield (all species combined) and mean annual yields of the main fish species in the Robert-Bourassa reservoir and its reference lake (lac Detcheverry; station SB400).

∗ Significant variations over time detected using single-classification ANOVA (reference) or repeated-measures ANOVA (reservoir).

⬆ Impoundment.

Fig. 1.5. Evolution of fish biomass and capture per unit of effort in Robert-Bourassa reservoir and a natural lake (continued)

As a preliminary investigation to verify the feasibility of developing a mechanistic model to predict Hg levels in reservoir fish, with a secondary objective of
identifying missing information, results obtained from the environmental effects
monitoring program (RSE), and from the monitoring of fish Hg levels, were used
to estimate biomass and Hg fluxes in fish of the Robert-Bourassa reservoir (Chapter 12.0).

In addition to the efforts allocated to the aquatic environment, two terrestrial
animals, feeding occasionally or exclusively on fish, were selected as models for
the study of potential effects of increasing Hg levels in reservoir fish on terrestrial
animals. First, an *in vitro* exposure experiment was carried out on mink (*Mustela
vison*) (Chapter 13.0). Second, the breeding success of Osprey (*Pandion haliaetus*)
nesting near La Grande complex reservoirs was also studied (Chapter 14.0). These
top predatory birds rely on highly efficient sight and neuro-motor co-ordination to
feed their young; functions particularly sensitive to Hg intoxication.

The findings of all these studies are presented in 14 separate articles, each
forming a distinct chapter. In addition, two final chapters present a general synthesis and future prospects.

2 Analysis of Total Mercury and Methylmercury in Environmental Samples

Pierre Pichet, Ken Morrison, Isabelle Rheault and Alain Tremblay

Abstract

During a ten year collaborative project between Hydro-Québec, the Université de Sherbrooke and the Université du Québec à Montréal, analytical methods to measure total mercury (total Hg) and methylmercury (MeHg) concentrations in different compartments of hydroelectric reservoir and natural lake ecosystems were developed and adapted. As a result of improvements made to our analytical procedures for measuring dissolved Hg species techniques (i.e., gold amalgamation with cold vapor atomic absorption spectrophotometry and direct measurement with atomic fluorescence spectrophotometry) both the sample volumes needed (from 1-10 L to 10-50 mL) and the detection limits (from 5 ng L^{-1} to 0.1 ng L^{-1}) have decreased. Similarly, in biological samples, we reduced the required sample size from 300 mg dry weight (dw) to < 1 mg dw. By allowing us to determine Hg concentrations in small environmental samples, improvements to analytical techniques represent a key link in our understanding of Hg cycling in reservoir and natural lake ecosystems.

2.1
Introduction

Numerous researchers working in remote regions have reported elevated concentrations of total Hg in fish of lakes and hydroelectric reservoirs (Håkanson et al. 1990a; Wren et al. 1991; Verdon et al. 1991; Bodaly et al. 1993; Haines and Brumbaugh 1994; Driscoll et al. 1995; Chapter 11.0). The high levels of this contaminant in fish inhabiting remote lakes or reservoirs with low ambient Hg concentrations have led to an increased interest in the biomagnification of this element in the lower food web (see Chapters 4.0 and 9.0). Until recently, studies of the water column and biological compartments (excluding fish) were restricted by difficulties associated with analysis of subnanogram levels of Hg, thus limiting our understanding of Hg cycling in aquatic ecosystems. Analytical advances have, however, improved detection limits while the increasing awareness of the necessity for ultra-clean techniques during sample collection has considerably reduced problems associated with contamination (Bloom and Crecelius 1983; Bloom and Fitzgerald 1988).

In order to investigate the dynamics of Hg in natural lakes and hydroelectric reservoirs we needed to be able to measure the concentrations in many environmental compartments. To meet these needs we developed and adapted analytical methods for a variety of environmental samples (e.g., water, sediments, forest soils, plankton, benthos, suspended particles, pore water). For the analysis of total Hg and MeHg we used two techniques for measuring elemental Hg, cold vapor atomic absorption spectrophotometry (CVAAS) with a detection limit on the order of 10^{-9} g g^{-1} and atomic fluorescence spectrophotometry (AFS) with a detection limit of approximately 10^{-12} g g^{-1}.

2.2
Total Mercury Measurements

2.2.1
Total Mercury by Cold Vapor Atomic Absorption Spectrophotometry

2.2.1.1
Organisms and Sediment Samples

At the onset of the project, we used the simple, low-cost technique of CVAAS for fish, bird, mammal or plant samples having relatively high concentrations of total Hg. The method corresponds to NAQUADAT procedure #80801 (Environment Canada 1979) and comprised a hot sulfuric acid/nitric acid digestion at 85°C for 2 hours, followed by an overnight permanganate/persulfate oxidation, pre-reduction with hydroxylamine, adjustment of the volume with distilled de-ionized water (DDI), and injections of fixed volumes into a bubbler containing a $SnCl_2$ reducing solution, and finally, detection by CVAAS. Quantification was achieved using standard calibration curves. The used of ultra-pure acids and the purification of the permanganate and persulfate solutions with iron hydroxide yielded very low blanks (< 1 ng) with corresponding detection limits of 0.05 mg kg^{-1} wet weight (ww). The accuracy of the methods were tested by analyzing three different standards of the National Research Council of Canada (Table 2.1). These measurements yielded mean values of 1.84 ± 0.15 mg kg^{-1}, 0.78 ± 0.04 mg kg^{-1} and 0.33 ± 0.05 for DOLT-2, DORM-1 and Lobster hepatopancreas, respectively, as compared to the certified values of 1.99 ± 0.10 mg kg^{-1}, 0.79 ± 0.07 mg kg^{-1} and 0.33 ± 0.06. This method, however, required large amounts of material per analysis (0.5-1.0 g dw) so we developed a technique using atomic fluorescence that allowed for the determination of total Hg with as little as 1 mg of material.

Table 2.1. Total mercury concentrations of reference materials analyzed by CVAAS (accepted and measured values, mean ± standard deviation, ng g^{-1} dry weight)

Standards	Type of material	Accepted values (ng g^{-1} dw)	Measured values (ng g^{-1} dw)
DOLT-2	Dogfish liver	1 990 ± 100	1 840 ± 150
DORM-1	Dogfish muscle	790 ± 70	780 ± 40
TORT-1	Lobster hepatopancreas	330 ± 60	330 ± 50
Intercalibration[*]	Fish flesh	1 210 ± 110	1 240 ± 80
		680 ± 80	686 ± 60
		250 ± 20	247 ± 18
		543 ± 50	540 ± 30

[*] Fisheries and Oceans Canada, Winnipeg, 1996.

2.2.1.2
Water Samples

To determine dissolved total Hg concentrations in pore water and in the water column, we used two different techniques: one based on CVAAS (water column only) and the other on AFS (pore water and water column). The CVAAS used was a modification of the method of Robertson et al. (1987). Briefly, a BrCl oxidation was followed by pre-reduction with hydroxylamine, reduction with $SnCl_2$, sparging of elemental Hg to gold amalgamation traps, thermal desorption of the Hg followed by CVAAS detection using a LDC Hg Monitor. The basic setup is shown in Figure 2.1. With regards to modifications made to the original technique, for highly colored waters we found it preferable to add the BrCl reagent at 1% rather than 0.5%. Some of the water samples from *in vitro* experiments had very high organic carbon levels (> 1000 ppm) and caused severe foaming problems, so these samples were oxidized using the sulfuric acid/nitric acid/perman-ganate/persulfate digestion of APHA-AWWA-WPCF (1989). For samples that caused minor foaming problems, a 15 mL bubbler with anti-foaming agent placed immediately downstream from the main bubbler proved suitable for trapping foam without interfering with the Hg determination. The direct addition of anti-foaming agent to samples was found to cause extreme interference with the $SnCl_2$ reduc-tion. Since the reagent water had residual total Hg concentrations of about 0.5 ng L^{-1}, reagent blanks were determined using previously stripped samples as discussed by Blake (1985). Using BrCl oxidation, the detection limit was 0.1 ng L^{-1}, while for the more complete oxidation procedure it was 5 ng L^{-1}.

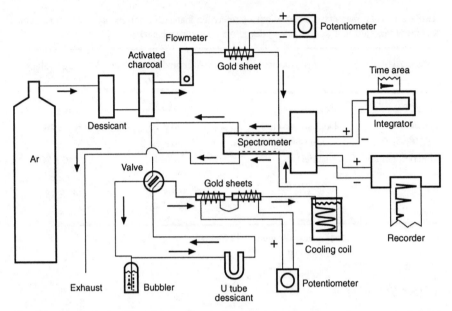

Fig. 2.1. Schematic drawing of the cold vapor atomic absorption spectrophotometry technique. The direction of flow of the carrier gas is indicated by the arrows

2.2.2
Total Mercury by Atomic Fluorescence Spectrophotometry

The major obstacle to be overcomed at the onset of our project was the lack of a commercially available atomic fluorescence apparatus for the detection of Hg in a diverse assortment of sample types covering a large range of concentrations (from ng L^{-1} to mg g^{-1}). As a consequence we set out to manufacture our own. An atomic fluorometer is easily constructed and, with various models having been tested over the years, their sensitivity is as good or better than the commercial models now available.

Atomic fluorescence is obtained by the emission of a wavelength at 254 nm which illuminates Hg vapor. The Hg atoms are excited and, as they return to a more stable state, they emit energy as fluorescence which can be captured by a detector. This technique is usually 100 times more sensitive than atomic absorption since the interference caused by other gases is greatly reduced and enables the analysis of small samples (< 1 mg) or samples with low Hg concentrations. There are, however, a few drawbacks. First, since the emission and excitation wavelength is the same, reflection or scattering by solid particles or fluorescence of aromatic organic molecules can affect the signal. Therefore, precautions must be taken to reduce the influence of these factors. This was done by coloring the cell walls black and by adding a 254 nm filter between the cell and the detector. Stabilizing the wavelength emission, by keeping the Hg lamp at a constant temperature, also helps to reduce interference with the fluorescence signal. Second, atomic Hg fluorescence is very sensitive to quenching by molecules containing double bonds (e.g., N_2 or O_2). For example, 1 ppm of O_2 reduces the fluorescence signal by 50%. Thus, to prevent the quenching effect, all measurements were performed in the presence of a noble gas (Ar or He).

2.2.2.1
Organisms and Sediment Samples

To avoid risks of Hg contamination, all labware was acid washed in 10% HCl, triple-rinsed with NANOpure® water and then heated to 300°C for 16 hours. The analytical method was modified from Bloom (1989). Briefly, 100-1 000 mg (dw) of homogenized soil or lake sediments were transferred to glass tubes and digested in 10 mL of a 10:1 16N HNO_3 / 6N HCl mixture for 6 hours at 121°C. For small samples (< 100 mg) of insects, phytoplankton, zooplankton, periphyton, suspended particles and plant debris, total Hg was analyzed following the digestion of either 5-15 mg (dw) of material in a small glass tube with 1 mL of the acid mixture or the digestion of 1-2 mg of material in Reacti-vials with 40 μL of 16N HNO_3 and 5 μL of 6N HCl for 4 hours at 121°C. The solution was then diluted to 0.5-10 mL with NANOpure® water.

For analysis, 200 to 1 000 μL of the digestion solution were injected into a reaction vessel (a 50 mL Teflon® centrifuge tube) containing 3 mL of a 10% $SnCl_2$ solution which was continuously bubbled by a stream of Hg-free argon gas. Following the reduction of the Hg to elemental mercury vapor (Hg^0), the gas stream, at an optimal flow of 150 mL min^{-1}, carried the Hg^0 to a fluorescence cell where detection occurred. The concentration of total Hg was calculated from the area of the fluorescence peak. The detection limit, three times the standard deviation of the procedural blanks, was 5 picograms Hg, which corresponds to 5 ppb for a typical 1 mg sample. During the evaluation period, the accuracy of the methods was tested by analyzing different standards (Table 2.2) which yielded mean concentrations comparable to the certified values.

2.2.2.2
Water Samples

Due to volume requirements of the CVAAS technique (1 L or more) for the determination of dissolved total Hg, we opted for the AFS method following a modification of that described by Bloom and Fitzgerald (1988). Samples were

Table 2.2. Total mercury concentrations of reference materials analyzed by AFS (accepted and measured values, mean ± standard deviation, ng g^{-1} dry weight)

Standards	Type of material	Accepted values (ng g^{-1} dw)	Measured values (ng g^{-1} dw)
TORT-1[a]	Lobster hepatopancreas	325 ± 30	310 ± 30
DORM-1[a]	Dogfish muscle	798 ± 74	795 ± 37
BEST-1[a] or MESS-2[a]	Marine sediment	92 ± 9	91 ± 5
SRM 1575[b]	Pine needles	150 ± 50	116 ± 12
Intercalibration	Hair[c]	11 700 ± 900	12 100 ± 300
		8 800 ± 600	8 900 ± 200
		6 800 ± 800	7 500 ± 100
	Mussel[d]	128 ± 13.7	120 ± 2
	Sea plant[e]	38 ± 6.0	42 ± 1.8

[a] National Research Council of Canada.
[b] National Bureau of Standards, USA.
[c] Intercalibration with L. Bigras, Health and Welfare Canada.
[d] Intercalibration with M. Horvat, IAEA-142, Monaco.
[e] Intercalibration with M. Horvat, IAEA-140, Monaco.

pre-treated using a photochemical digestion prior to analysis in order to liberate and oxidize Hg^{2+} associated with stable organic complexes. Two 10 mL aliquots of each sample were transferred to pre-cleaned quartz tubes, to which were added 100 µL of an oxidizing agent, a 50 g L^{-1} potassium persulfate solution. The tubes were sealed with Parafilm®, stirred and the solution subjected to ultraviolet oxidation for 20 minutes in a photochemical reactor. The reaction vessel was prepared by injecting 2 mL of the 10% SnCl$_2$ solution, 0.5 mL of 6N distilled HCl, and 1.0 mL of NANOpure® water. A 5 mL aliquot was injected into the reactor and, in accordance with the method of Bloom and Fitzgerald (1988), dissolved Hg^{2+} was reduced to Hg° and then analyzed by AFS. Our system, however, involved direct volatilization and therefore did not include a preconcentration step with a gold trap (Figure 2.2). Instead, Hg° was stripped out of the reactor and carried directly to the fluorescence cell as previously described. Figure 2.3 presents the signals from both direct volatilization and gold-amalgamation. Although the direct volatilization signal was lower, the detection limit in pg was on the same order of magnitude as that for gold amalgamation, since the blank did not need to be subtracted from the signal (Table 2.3). For sample sizes larger than 50 mL, gold amalgamation had a higher precision, however, when the volumes are on the order of 10 mL, both methods yielded comparable results. The detection limit for the direct volatilization method was 0.3 ng L^{-1}, with procedural blanks of 0.5 ng L^{-1}. In view of the usefulness and simplicity of direct volatilization, in addition to a reduction in sampling effort and cost, most of the field water samples were analyzed with this technique.

A Direct volatilization measurements

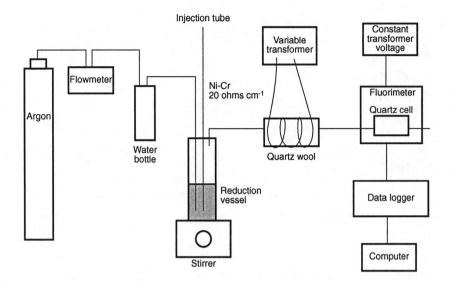

B Gold amalgamation

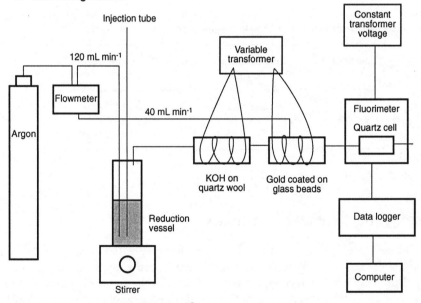

Note: All tubing and connectors are in Teflon ®

Fig. 2.2. Setup for the determination of total mercury by atomic fluorescence. A. Direct volatilization measurement. B. Gold amalgamation. All tubing and connectors are in Teflon®

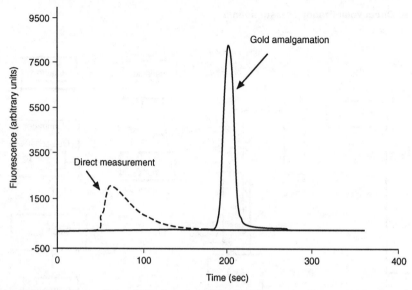

Fig. 2. 3. Comparison of experimental total mercury fluorescence for direct volatilization versus gold amalgamation. Experimental conditions: Ar flow 40 mL min^{-1}, 10 mL reduction vessel, Hg 20 pg

Table 2.3. Total mercury concentration in water sample (ng L^{-1}) analyzed by gold amalgamation and direct measurement with 5 mL samples (mean±standard deviation (n))

Sample number	Gold amalgamation (ng L^{-1})	Direct measurement (ng L^{-1})
1	1.64 ± 0.25 (6)	1.52 ± 0.42 (5)
2	1.89 ± 0.60 (6)	1.64 ± 0.39 (5)
3	1.59 ± 0.16 (6)	1.62 ± 0.23 (5)

2.3
Methylmercury Measurements

In the past, MeHg was measured by gas chromatography (GC) using an electron capture detector. The detector, however, is not selective to Hg resulting in the need for a complex extraction procedure to confirm the signal obtained. The selectivity of atomic fluorescence eliminates this complex procedure. The MeHg halide or the organomercury derivatives (dimethylmercury, methyethylmercury) are separated on a chromatographic column, decomposed in a quartz tube at high temperature (> 500°C) and then detected as elemental Hg. The critical step in measuring MeHg concentrations in environmental samples is its extraction. Here we present methods used to extract MeHg from samples of varying size and matrix.

2.3.1
Organisms and Sediment Samples

Using subsamples of 0.3 to 1.0 g, MeHg bound in sediments or flooded soils was released with a $CuSO_4$-HCl mixture and then extracted with toluene (Westoo 1967). The MeHg chloride content of the toluene extract was separated by GC on a 10% free fatty acid phase. The MeHg concentration was measured as total Hg by atomic fluorescence following thermal decomposition on silica (400°C) with an argon flow of 60 mL min^{-1} (Bloom 1989). Accuracy was monitored by standard addition measurements which gave a 90 ± 15% recovery rate. The detection limit for a 1 mg (dw) sample was 5 ng Hg g^{-1} and the coefficient of variation (< 3%) was verified on a routine basis using standards.

Because various organic substances extracted with the organic solvent were decomposing in the chromatograph injector and catalyzing the decomposition of MeHgBr, thus leading to low analytical results, and because of the large samples required (300-500 mg dw) for these analyses, a distillation process was used to measure MeHg contents in samples of insects, plankton, periphyton and plant debris. This analytical procedure, modified from that of Horvat et al. (1993a), allowed as little as 3 mg dw of material to be analyzed. A mixture of 0.7 mL water, 0.2 mL 2M H_2SO_4 and 0.1 mL 4M KBr was added to a 3-10 mg sample. Distillation was carried out under a 15 mL min^{-1} nitrogen gas flow at a constant temperature of 90°C and lasted 90 minutes. This procedure released MeHg as volatile CH_3HgBr which was then recovered in a quartz tube filled with 1 mL of NANO-pure® water and kept on ice. The tube was then sealed with Parafilm® and stirred. To check for the presence of inorganic Hg which may have co-distilled and contaminated the solution, a 100 μL subsample was taken and analyzed for the total Hg by AFS as described above. If inorganic Hg contamination was greater than 15% of the MeHg concentration of the sample, analysis was repeated on a replicate of the sample. To each tube, 100 μL of a 50 g L^{-1} potassium persulfate solution were added as an oxidizing agent. The tubes were resealed with Parafilm® and stirred before subjecting the solution to ultra-violet oxidation for 20 minutes to oxidize CH_3HgBr to Hg°. The resulting total Hg was then analyzed by AFS as described above. The MeHg recovery rate of the distillation procedure as determined on standard insect samples was 99 ± 11%. The accuracy of the method was verified by analyzing TORT-1, DORM-1 and BEST-1 standards (Table 2.4).

The distillation technique is, however, time consuming allowing 5 to 10 analyses per day and required a certain amount of material, which was often not available for biological samples. Thus we developed a simple saponification technique,

Table 2.4. Methylmercury concentrations of reference materials (accepted and measured values, mean ± standard deviation)

Standards	Type of material	Accepted values (ng g^{-1} dw)	Measured values (ng g^{-1} dw)		
			Distillation	KOH-MeOH extraction	Toluene extraction
TORT-1[a]	Lobster hepatopancreas	128 ± 14	128 ± 14	126 ± 5	119 ±10
DORM-1[a]	Dogfish muscle	731 ± 60	793 ± 47	759 ± 38	712 ± 61
Intercalibration	Mussel[b]	47.7 ± 4.3	-----	46 ± 2	-----
	Sea plant[c]	0.62 ± 0.11	-----	< 1.0	-----

[a] National Research Council of Canada.
[b] Intercalibration with M. Horvat, IAEA-142, Monaco.
[c] Intercalibration with M. Horvat, IAEA-140, Monaco.

a modification of Bloom (1989). The digestion of 0.5-5 mg (dw) was performed in 0.5 mL of a KOH/MeOH (1 g/4 mL) solution during 8 hours at 68°C. After its extraction, MeHg was converted to MeEtHg with sodium tetraethylborate in a buffered solution at pH 4.5. MeEtHg was then trapped on a column, separated by gas chromatography and quantified using AFS. In the original tetraethylborate technique (Bloom 1989), the organomercury compounds were trapped on a Carbotrap® column and desorbed on a cryogenic chromatographic column. With this method, there was a partial decomposition of the organomercury derivative during the desorption step. After testing other absorbents, the absorbent was changed to Tenax®, a polymer used for air analysis, since no decomposition was measured during its use. We also found that typical chromatographic conditions (Se-30 on chromosorb at 80°C) gave satisfactory results without resorting to a cryogenic chromatographic column. The detection limit was about 0.6 picogram of MeHg, which corresponds to 0.3 ppb for a typical 2 mg sample and the accuracy of the method was tested by analyzing different standards (Table 2.4).

2.3.2
Water Samples

As was the case for total Hg in water, we used the CVAAS and AFS techniques to measure dissolved MeHg concentrations. With the CVAAS method and up until 1991, preconcentration on sulfhydryl cotton fiber (SCF) was used (Lee 1987; Lee and Mowrer 1989). Samples of 4 L (*in vitro* experiments) or 10 L (field samples) were passed through columns containing SCF absorbent at low flow rates (< 20 mL min^{-1}). For each analysis a spiked sample of the same water was also passed through a column. The collected MeHg was desorbed with a HCl-NaCl solution and extracted into benzene (C_6H_6), subsamples of which were injected into a GC with an ECD. The GC had a 50 m DB-5 capillary column, and temperature programming was used. This technique was slow and laborious, requiring large sample volumes, and produced chromatograms with large numbers of spurious peaks due to the sensitivity of the ECD to any halogen-containing compounds, not just CH_3HgCl. This method was replaced in 1992 with the CH_2Cl_2 extraction-aqueous phase ethylation method proposed by Bloom (1989) and Brooks-Rand (1990). In this method 60 mL of acidified sample were extracted with 40 mL of CH_2Cl_2, the phases being subsequently separated and the CH_2Cl_2 phase placed in a borosilicate BOD bottle with 10 mL of DDI water and Teflon® boiling stones. The CH_2Cl_2 was boiled off at 80°C. It was preferable to put only 10 mL of water rather than 100 mL as described in the references, as well as to use the boiling stones to have a more-controlled boil. Afterwards, 100 mL of DDI water were added to the bottle. This water was then buffered, the ethylating agent was added, and the sample was sparged to a trap containing graphitised carbon black (Carbotrap®). The trap was subsequently desorbed onto a 1 m packed column held at 90°C. At the exit from the column the gas went through a pyrolysis unit (~900°C) and into an AFS detector. Quantification was accomplished via standard additions. This method was much quicker and yielded much cleaner chromatograms than the previous technique, but the limits of detection were similar, ~50 pg L^{-1}. Extraction of MeHg to CH_2Cl_2 has been found to be comparable to distillation for water samples, giving insignificantly lower results (Horvat et al. 1993b; Bloom et al. 1995).

With the AFS technique, we used a distillation-ethylation technique, similar to the one used for biological samples. Briefly, 27-32 mL of water was weighed into a Teflon® tube, and 0.2 mL of a 1% ammonium pyrrolidine dithiocarbamate, 0.5 mL of 4M KBr and 0.5 mL of 2M H_2SO_4 were added. The solution was then distilled at 110°C in the presence of nitrogen for 4 to 5 hours, transferred to an ethylation vial and ethylated with sodium tetraethylborate in a buffered solution at pH 4.5. As for the biological samples, MeEtHg was trapped on a Tenax® column, separated by gas chromatography and quantified using AFS. The detection limit was about 1 pg MeHg L^{-1}. The coefficient of variation (< 6%) was routinely verified using standards.

2.4
Conclusions

During the past ten years of collaboration between Hydro-Québec and the universities, we have greatly improved our analytical techniques by lowering the detection limits and the amount of material needed to perform the analyses. We are now able to determine, on a routine basis, concentrations of total Hg and MeHg in the water column, suspended particulate matter, individual benthic invertebrates or single species of plankton and periphyton. All these compartments were out of reach with the previously available techniques. Moreover, these analytical improvements enable us to explore these compartments in terms of their role in the Hg cycle, thus representing a key link in our understanding of the biogeochemistry of Hg in the ecosystems of hydroelectric reservoirs and natural lakes.

Acknowledgements

This research was supported by a grant from Hydro-Québec (HQ), the Conseil de recherches en sciences naturelles et en génie (CRSNG) du Canada, and the Université du Québec à Montréal (UQAM) to the Chaire de recherche en environnement (HQ/CRSNG/UQAM). This research was also supported by a grant from Hydro-Québec to the Université de Sherbrooke. The participation of the Corporation des services analytiques Philip in this study is also gratefully acknowledged.

Mercury and Methylmercury in Natural Ecosystems of Northern Québec

3 Mercury in Natural Lakes and Unperturbed Terrestrial Ecosystems of Northern Québec

Marc Lucotte, Shelagh Montgomery, Bernard Caron and Martin Kainz

Abstract

Mercury (Hg) in natural lakes and unperturbed terrestrial systems is ubiquitous throughout northern Québec, at sites situated 200 to 1 400 km away from the closest industrial centers. The diagenetic stability of Hg in recent lacustrine sediments makes it possible to give historic interpretations of the deposition of this heavy metal in remote regions. Accelerated anthropogenic activities over the last century are responsible for the 2- to 3-fold increase in the atmospheric Hg deposition rate in all lakes studied between latitudes 45°N and 56°N in northern Québec. Hg concentrations in lake sediments are proportional to the amounts of terrestrial organic carbon brought from the catchment area. Once incorporated into the organic horizons of the soils of the boreal forest, the airborne Hg efficiently binds itself to humic matter. Natural Hg has been accumulating in these horizons since the beginning of soil formation at the end of the ice age. The recent additional inputs of anthropogenic Hg to these soils mixes with the existing natural Hg pool and represents only a small fraction of the total Hg burden. The cumulative burden of Hg in the humic horizon of podzolic soils averages 2 mg m^{-2}, with less than 25% of it being attributable to anthropogenic Hg deposition, while peatlands accumulate 4 to 5 times more Hg in their thick peat layer. In boreal forests a substantial fraction of newly deposited atmospheric Hg is readily transported to lacustrine systems by snow melt in the spring or by surficial (i.e., not fully penetrating the humic horizons) runoff during summer storms. This explains the paradox between the 2- to 3-fold increase in Hg deposition rates in recent sediments, with the Hg concentration present in lake sediments being of terrestrial origin in spite of the small fraction of anthropogenic Hg in soils. Using a combined set of environmental indicators, a fairly uniform pattern of Hg atmospheric deposition, with no detectable gradient decreasing away from the Great Lakes industrial belt, can be drawn for the entire domain of the Québec boreal forest south of latitude 54°N. These include Hg concentrations in snow, in yearly shoots of spruce or in epiphytic lichens, Hg burdens normalized to carbon in humic layers of podzols, and normalized sedimentary Hg enrichment since the onset of the industrial era. Particulate matter at the bottom of a lake or in suspension in the water column plays a key role in the transfer of Hg to the base of the aquatic food chain. It has not yet been quantified to which extent the substantial increase of anthropogenic Hg in lake sediments relates to an equivalent increase in Hg bioaccumulation in aquatic organisms.

3.1
Introduction

Numerous researchers working in remote, temperate regions have reported elevated concentrations of Hg in fish of lakes (Björklund et al. 1984; Swain and Helwig 1989; Håkanson et al. 1990a; Wren et al. 1991; Chapter 5.0). In the New England states and Eastern Canadian provinces, several fish consumption advisories have been issued in the past years to protect public health. It is now widely accepted that atmospheric deposition is the major source for Hg and that both wet and dry processes are of importance in the forested catchments (Jackson 1997). Although it is well known that municipal waste combustors and utility and non-utility boilers are major contributors of anthropogenic Hg emissions to the atmosphere of northeastern America (Pirrone et al. 1996; Jackson 1997), it is still unclear to which levels of intensity and to which spatial extent these local pollution sources of atmospheric Hg are responsible for the contamination of regions, such as northern Québec, far away from industrial centers.

This chapter presents a synthesized image of the presence of natural and anthropogenic Hg in the various environmental compartments (excluding living organisms which are described in Chapters 4.0, 5.0 and 6.0) of both the aquatic and terrestrial ecosystems of northern Québec. To do this synthesis, we compiled the Hg data generated by the team of the Chaire de recherche en environnement HQ/CRSNG/UQAM since 1990 and compared them to other Hg data gathered for the same region by other research groups. First, we propose an historical reconstruction of the contribution of recent anthropogenic Hg emissions relative to the ubiquitous presence of the heavy metal in that remote region based on a careful interpretation of sedimentary records taken from a series of natural lakes. Second, we discuss the ability of the various types of boreal forest soils to retain direct (precipitation) or indirect (litter fall) atmospheric Hg fallout in an attempt to link the presence of Hg in lakes to weathering processes on their catchments. Finally, using a combination of multiple lacustrine and terrestrial indices, we present a regional pattern for the long range atmospheric deposition of Hg over the vast domain of the boreal forest of Québec.

3.2
Materials and Methods

The study area included in this chapter covers the Laurentian forest and the boreal forest ecozones of western Québec between latitudes 45°N and 55°N. The bedrock is composed of Precambrian-age igneous and metamorphic rocks. Numerous terrestrial and aquatic samples have been collected over the past 7 years during several multidisciplinary field campaigns (Lucotte et al. 1995a; Grondin et al. 1995; Montgomery et al. 1995; Caron 1997).

3.2.1
Aquatic Systems

The sampled lakes were chosen for their pristine environment, with difficult or no direct road access. All lakes were small in size (less than 6 km^2), with a maximum depth of 6 to 13 m, a slightly acidic to circumneutral pH (6.0 to 7.3) and oligotrophic characteristics. Recent sediments were collected with a 15-cm in diameter core tube manually inserted by divers in the deepest part of a series of lakes distributed along a 10° north/south transect in western Québec. In the field, the retrieved cores were immediately subsampled at 1 cm intervals, placed in clean scintillation vials and kept frozen until analyses. Redox potential (Eh) profiles in the sediments were obtained by inserting a platinum-Ag/AgCl combination electrode at each depth prior to subsampling.

The collection of water samples for the dissolved fraction was conducted following an ultra-clean procedure (Montgomery et al. 1995, 1996). Briefly, water was collected into Teflon$^®$ bottles using a manually operated peristaltic pump and a short length of Masterflex$^®$ silicone tubing to which was attached Teflon$^®$ tubing and, at the outlet, a 47 mm Millipore$^®$ in-line filter holder assembly. For each sample, two filters, a glass fiber (GF) filter and a 0.45 μm GN6 mixed cellulose ester Gelman$^®$ filter, were placed one atop the other in the filter unit. At the time of sampling, the filters were rinsed with approximately 250 mL of sample water and the bottles were rinsed three times.

For the particulate fraction, 200 L of water collected with an electrical pump and prefiltered on-line with 64 μm and 210 μm Nytex$^®$ filters were stored in 50 L carboys for the subsequent concentration of the fine particulate fraction. The > 210 μm and 64-210 μm fractions were rinsed from the filters and stored in Sardstedt$^®$ tubes while the fine particulate matter < 64 μm (FPM) was concentrated the same day by continuous flow centrifugation at a speed of 10 000 rpm or 8 500 G. Previous laboratory tests have indicated that about 72 to 85% of particles retained by 0.45 μm GN-6 filters are recovered by centrifugation (Mucci et al. 1995). Following centrifugation, the FPM was stored in Teflon$^®$ tubes and all the samples were stored frozen.

3.2.2
Terrestrial Systems

Beginning in 1993, samples of various species of terrestrial plants of the boreal domain in northern Québec were collected between latitudes 47°N and 56°N. Needles of the last annual growth of white spruce (*Picea glauca*) and black spruce (*Picea mariana*) were manually picked wearing clean vinyl gloves and using clean tweezers. The needles were generally collected at a height varying between 1 and 3 meters above the ground, on trees of similar size (about 10 cm at breast height). Epiphytic lichens growing on the trees were collected the same way as spruce needles. In our sampling, we systematically separated the two most common species of lichens (*Evernia mesomorpha* and *Bryoria* sp.) growing in the study area.

A bank of soil samples representing the two major pedologic features of the Québec boreal forest, well drained podzolic soils and ombrotrophic peatlands, exists since 1992. The same technique for soil sampling has been used since that time. Briefly, a PVC tube of 15 cm in diameter and 30 to 40 cm long was manually inserted into the soils. The roots were progressively cut with a knife on the external side of the tube. The soil samples were then extracted from their tube, sliced at one centimeter intervals with the subsamples being placed in plastic bags. In all cores retrieved we always took a subsample of the living ground cover, predominantly *Cladina* spp. lichen for forest soils and *Sphagnum* spp. moss for peatlands. All plant and soil samples were kept frozen until freeze dried.

Snow accumulated over the winter season was sampled at the end of March 1994, at 9 sites spread on a latitudinal transect between latitudes 47°N and 56°N. At that period of the year, little sublimation or melting had occurred. The samples were obtained by inserting a Teflon® tube 10 cm in diameter and 60 cm long into the snow accumulated on the bank of a lake or on the edge of a forest, at a minimum distance of 100 m from the nearest road. The samples were then immediately placed to thaw over a Teflon® bottle under a dust free atmosphere. The melted snow was then filtered on a pre-cleaned 0.45 μm GN6 filter and acidified to pH 2 with ultra pure HCl. The analyses of dissolved Hg content were then run on these samples within a few days.

3.2.3
Analyses

As detailed procedures for the Hg analyses in the various environmental compartments are given in Chapter 2.0, we only present here a brief summary of the various analytical techniques followed. Total Hg in wet sediments or soil samples was extracted by a digestion in a 10:1 solution of nitric and hydrochloric acids at 121°C for 6 hours. Total dissolved Hg was liberated as Hg(II) from the complexes it forms with organic matter using a photochemical digestion. A tin chloride solution was then added to all extracts containing dissolved Hg(II) in order to reduce this form of Hg to elemental Hg vapor, Hg°, which was then stripped out of the reactor and carried in an Ar gas stream. Gaseous Hg concentrations were then measured by atomic fluorescence. Our system involved direct volatilization and therefore did not include a preconcentration step on a gold trap, even for the least concentrated extracts.

The levels of C and N were measured using a CN Carlo Erba® elemental analyzer with a reproducibility on the order of 1%. The burdens (B) of organic C and Hg were evaluated for given soil horizons by summing, for each centimeter, the concentration of C or Hg, multiplied by the dry density of that specific layer. Operationally defined reactive iron and manganese oxyhydroxides, extracted using a citrate-dithionite-bicarbonate buffer (Fe_{cdb} and Mn_{cdb}) were analyzed by atomic absorption (Lucotte and d'Anglejan 1985).

3.3
Results and Discussion

3.3.1
Mercury in Natural Lakes of Northern Québec

Bottom sediments are the main repository of Hg in lacustrine ecosystems. In northern Québec, lake sediments usually bear Hg concentrations ranging from 50 to 300 ng g^{-1}, which is 4 to 5 orders of magnitude higher than total Hg (dissolved plus particulate) concentrations in the water column, usually below 10 ng L^{-1} (Lucotte et al. 1995a; Montgomery et al. 1995). Even though the highest Hg concentrations in lakes, recorded in their sediments, are of the same magnitude as those in soils, aquatic organisms will bioaccumulate much more Hg than their equivalent counterparts (in terms of trophic position in the food chain) in terrestrial ecosystems. This is due to the much higher bioavailability of that heavy metal in aquatic environments, principally as the inorganic forms of Hg are methylated via microbial activity at various levels of the aquatic system. Methylmercury (MeHg) is readily accumulated in the organisms as a result of both its strong affinity for proteins and slow elimination rates (Lindqvist 1991).

3.3.1.1
Presence and Diagenesis of Mercury in Lake Sediments

Just about all sedimentary profiles of Hg concentrations published for cores retrieved from the bottom of natural lakes in Europe (e.g., Johansson 1985; Verta et al. 1990) and North America (e.g., Rada et al. 1989; Swain et al. 1992; Lockhart et al. 1995) exhibit the same pattern; that is, a sharp exponential-like increase in the top centimeters above background concentrations. Hg concentration profiles reported for the sediments of over twenty five lakes in northern Québec follow the same rule (Lucotte et al. 1995a). A typical example of such a profile is given for Lake 154 (54° 14' N, 72° 52' W) near the Laforge 1 reservoir in Figure 3.1.

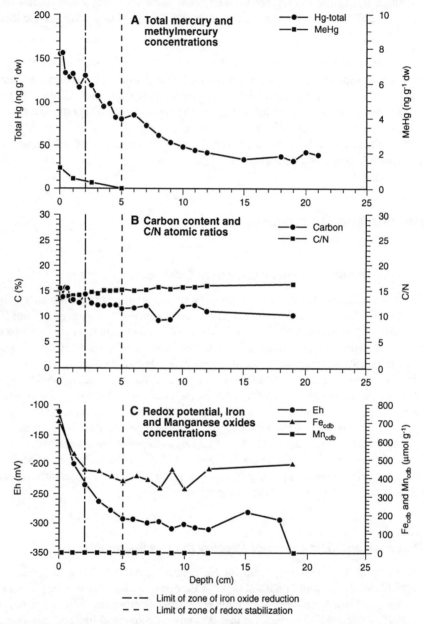

Fig. 3.1. Typical sedimentary profiles of a natural lake in northern Québec near the Laforge 1 reservoir (F_{ecdb} and Mn_{cdb})

Like most lakes in northern Québec, Lake 154 is oligotrophic. Thus, little autochthonous organic matter is produced in the water column and subsequently deposited to the bottom of the lake. Indeed, the high atomic carbon/nitrogen ratios of organic matter reaching the bottom of the lake indicates the relatively high contribution of terrigenous organic matter to the lake sediments. More generally speaking, the terrigenous nature of the sedimentary organic matter seems to be a characteristic of boreal lakes in Québec, as indicated by stable isotope studies (Lucotte et al. 1995b). Only a small fraction of this organic matter appears readily biodegradable (< 25%). In the example of Lake 154, the degradation of the labile organic matter takes place in the top 5 cm of the core and the demand in oxygen it requires is reflected in the sharp drop in the redox potential measured over that active layer. Along with this reaction, the Fe_{cdb} are totally reduced in the top 2 cm of the core. Much more sensitive to reducing conditions, all Mn_{cdb} have been released right at the sediment-water interface. The Hg profile appears totally decoupled from these marked diagenetic reactions (Figure 3.1). Hg concentrations smoothly decrease in the top 12 cm of the core, without being influenced by the biodegradation of labile C and concomitant drop in the redox potential, nor by the sharp reduction of Fe_{cdb}.

The limited mobility of Hg in lake sediments is often attributed to the high content of organic matter offering numerous and strong binding sites for the heavy metal (Linqvist 1991; Dmytriw et al. 1995; EPRI 1996). Several other indices suggest that Hg accumulates in lake sediments without major diagenetic remobilization. In the sediments of lakes Gardsjön and Härsvatten in Sweden, Munthe et al. (1995) found a sharp decrease in the Hg accumulation rates following a sharp decrease in the Hg atmospheric deposition in north eastern Europe.

3.3.1.2
Historic Interpretation of the Presence of Mercury in Sediments

In a collective review paper, Fitzgerald et al. (1998) recently reaffirmed that since Hg accumulated in lake sediments over time remains stable, it is possible to assess the temporal trends in Hg deposition to the bottom of the lakes. In all dated sediment cores across North America and Europe the sharp increase in the deposition rate of Hg in the studied lakes has been shown to correspond to the onset of the industrial era over the last century. Lakes of northern Québec are no exception to this trend, although the increase in anthropogenic Hg deposition seems to have started some 70 years ago, which is several decades later than in the vicinity of industrial centers (Lucotte et al. 1995a). The ratio of the surface to baseline Hg concentrations (anthropogenic sedimentary enrichment factor, ASEF) in Lake 154 (Figure 3.1) reaches 3.5, a value somewhat higher than what has been found for other northern Québec lakes (Lucotte et al. 1995a). It, above all, emphasizes that significant amounts of anthropogenic Hg are transported by the atmosphere over long distances and subsequently deposited on remote regions of northern Québec.

3.3.1.3
Sedimentary Mercury and Terrigenous Organic Matter

In northern Québec lakes, Landers et al. (1997) calculated that the average sedimentary Hg accumulation rates ranged from 5 to 50 $\mu g \, m^{-2} \, y^{-1}$. These values do not represent the average atmospheric deposition rates for the nearby region of the lakes, as they are strongly modulated by variable effects of focussing characterizing the sedimentary accumulation in the different parts of the lake. As a matter of fact, the quantity of Hg accumulated at a given spot at the bottom of a northern Québec lake appears directly related to the amount of organic C deposited at that location (Lucotte et al. 1995a). The example of Lake 154 (Figure 3.1) exactly corresponds to the linear relationship found between the C content and the Hg content for 16 other lake cores by Lucotte et al. (1995a). As most of the C found at the bottom of boreal lakes has a mainly terrigenous source, one may conclude that most of the Hg found in the northern Québec lake sediments must have been brought by the washout of terrestrial organic matter from the drainage basin.

Along with the biodegradation of the labile fraction of the most recently sedimented organic matter at the bottom of Lake 154 (Figure 3.1), one can outline the appearance of significant concentrations of MeHg. The transient building of MeHg concentrations in surface sediments is described in the literature as a faster rate of methylation than demethylation of the Hg already present in the sediments (e.g., Ramlal et al. 1986; Rudd 1995). In an ongoing study we see that the rate of Hg methylation that accompanies early diagenetic processes in a sediment is modulated by environmental parameters, including the nature of the sedimentary organic matter and the depth at which the sediments lie. In surface sediments of northern Québec lakes, methylation rates appear to be positively correlated to the presence of organic matter leached from nearby peatlands and enhanced in the photic zone. These findings corroborate those of St. Louis et al. (1996) which show that wetlands could be an important source of MeHg to downstream ecosystems.

3.3.1.4
Mercury in the Water Column of Natural Lakes

Until recently, water column studies of Hg have been hampered by difficulties associated with the sampling and analysis of subnanogram levels of total Hg. Analytical advances (Chapter 2.0) have improved detection limits for the determination of total and MeHg in water, while increased awareness of the necessity for ultra-clean techniques during sample collection has considerably reduced problems associated with contamination. In the scope of an investigation begun in 1990, designed to study the sources and fate of Hg in hydroelectric reservoirs, numerous natural systems were sampled for determination of both the dissolved and particulate concentrations of Hg. While water column sampling began in 1993, the results presented in this chapter are from five sampling campaigns conducted between 1994 and 1996. A detailed description of the 1993 results for total dissolved Hg (Hg_{T-D}) has been previously published (Montgomery et al. 1995) and since that time improvements to the analytical and sample handling techniques made it possible to increase the sensitivity of analyses. With improved blanks, less variability in replicates and the ability to perform dissolved $MeHg_D$ analyses, we are more confident in our recent data (see Table 3.1 for a comparison of 1993 and 1994-1996 results).

Table 3.1. General results of total dissolved mercury and dissolved methylmercury analyses for the natural systems sampled in 1993 and between 1994-1996. The units for the average, standard deviation and max./min. are ng Hg L^{-1}

	Hg_{T-D}	$MeHg_D$
1993		
Average	2.39	n.a.
Standard deviation	0.15	n.a.
Max./min.	3.67/0.77	n.a.
Number of analyses	53	n.a.
1994-1996		
Average	1.51	0.049
Standard deviation	0.06	0.004
Max./min.	2.6/0.4	0.115/0.018
Number of analyses	70	30

The overall results for Hg_{T-D} and $MeHg_D$ in several natural lakes of northern Québec are shown in Figure 3.2. With these two sets of data, we find no significant variation in concentrations among lakes and from one season to the next. Furthermore, we observe that the percentage of Hg_{T-D} that is methylated averages about 3% and does not exceed 8%. The importance of these findings emerges when one considers the ongoing question concerning the transfer of MeHg to aquatic biota. That is, in neighboring lakes, Hg concentrations in fish of the same species and size have been found to vary significantly (Chapter 5.0). While volumetrically the dissolved phase represents a considerable pool, in view of the low proportion of Hg_{T-D} that is methylated and the small range of Hg_{T-D} and $MeHg_D$ concentrations measured, the potential contribution of MeHg associated to the dissolved fraction to the biota is probably less important than that of MeHg linked to the particulate phase. In that respect, we examined, using an inverted microscope, the three fractions of suspended particulate matter sampled from the natural systems: $> 210\ \mu m$, 64-$210\ \mu m$, and FPM ($< 64\ \mu m$). The $> 210\ \mu m$ samples were predominantly composed of zooplankton (e.g., *Daphnia* spp.) and some terrestrial

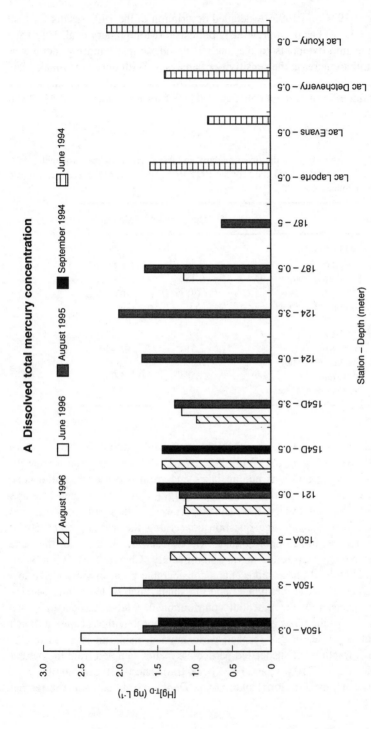

Fig. 3.2. Summary of concentrations of total and methylmercury in the water column of natural systems sampled in northern Québec between 1994 and 1996. The sampling depth (in meters) is indicated along the x-axis. The mean water temperatures were 20°C in August 1995, 12°C in September 1994, June and August 1996, and 3°C in June 1994

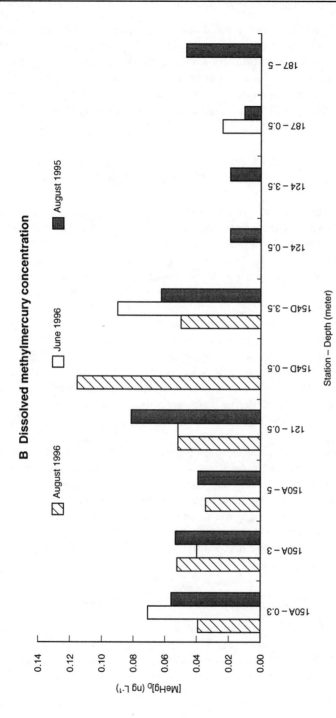

Fig. 3.2. Summary of concentrations of total and methylmercury in the water column of natural systems sampled in northern Québec between 1994 and 1996. The sampling depth (in meters) is indicated along the x-axis. The mean water temperatures were 20°C in August 1995 and 12°C in June and August 1996 (continued)

debris. The 64-210 μm samples contained mostly phytoplankton with some micro-zooplankton, and the FPM was essentially made of terrestrial organic matter with some algae (e.g., *Dinobryon* spp.). As a means of assessing the importance of these three solid fractions in the accumulation of Hg, the partitioning between the dissolved and particulate phases was examined. Partitioning can be described by a simple, linear partition coefficient, Kd, defined here under steady state conditions as:

$$K d = \frac{[\text{Hg}]\text{particulate (or [MeHg]particulate) (ng g}^{-1})}{[\text{Hg}]\text{dissolved (or [MeHg]dissolved) (ng mL}^{-1})}$$

and representing one of the most widely used parameters for estimating the crucial partitioning of metals between dissolved and particulate phases (Benoît 1995). Generally applied only to < 0.45 μm (filtrate) and > 0.45 μm (filter-retained) fractions, we have expanded its use to describe Hg accumulation. A plot of the log (Kd) values for total Hg and MeHg is shown in Figure 3.3. The first striking feature of this plot is the separation of the > 210 μm fraction along the vertical axis (log (Kd) MeHg), with values significantly higher than those for the 64-210 μm and FPM fractions ($p = 8.9 \times 10^{-5}$ and $p = 1.2 \times 10^{-6}$, respectively), suggesting an accumulation of MeHg in the zooplankton fraction. Secondly, the log (Kd) values for MeHg of the 64-210 μm fraction are slightly higher than those of the FPM ($p = 0.04$). If we explain the slight difference as being caused by the presence of microzooplankton in the coarser phase, we may conclude that these two fractions, the living organisms of which derive their energy from dissolved nutrients (i.e., algae), accumulate little MeHg.

While the observed accumulation of MeHg with respect to the > 210 μm samples does not fully explain the transfer pathway for MeHg to aquatic biota, considering the dissolved phase as a less important potential direct route does suggest that the ingestion of organic particles by zooplankton may represent one link in the food chain. Aspects concerning the structure of the food web are discussed in Chapters 4.0 and 5.0, while the importance of fine particulates as a source of food for invertebrates is discussed in Chapter 4.0.

3.3.1.5
Mercury in Aquatic Macrophytes and in Riparian Plants

We present in Table 3.2 Hg concentrations in common and ubiquitous aquatic macrophytes growing in shallow areas of a series of eight natural lakes lying between latitudes 46°N and 52°N in Québec. Considerable differences in Hg concentrations are observed for the various plant species growing at the same location of a given lake. This observation indicates that the plant physiology plays

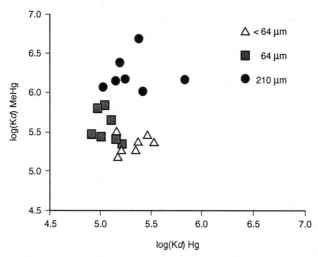

Fig. 3.3. The partition coefficients for methylmercury (log (Kd) MeHg) and total mercury (log (Kd) Hg$_T$) for the three particulate fractions; fine particulate matter (FPM), 64 µm and > 210 µm

an essential role in its Hg uptake. Most of these macrophytes (*Alisma* spp., *Carex* spp., *Cassandra* spp., *Myrique* spp., *Nymphoides* spp., *Potamogeton* spp. and *Sparganium angustifolium*) frequently exhibit maximum Hg concentrations in their root system. This situation for Hg parallels the well known association of various metals with the roots of macrophytes, either externally on their oxidized microlayer, or after penetration through the cell wall (Booserman 1985). Inversely, *Nuphar variegatum* accumulates as much Hg in its leaves as its roots, and much less than any other plant species growing nearby (Table 3.2). This observation suggests that *Nuphar variegatum* principally takes up Hg from the water column through its leaves and stems, with limited transfer from the sediments to the rhizosphere (Campbell et al. 1985; Grondin 1994). Indeed, Hg concentrations in any part of *Nuphar variegatum* are always one order of magnitude lower than those of the sediments in which they grow, contrary to *Sparganium angustifolium* which bears Hg concentrations in its roots equivalent to those of its substrate (Grondin 1994).

3.3.2
Mercury in Terrestrial Systems of Northern Québec

Gaseous Hg of both natural and anthropogenic sources released into the atmosphere is transported downwind over long distances. Terrestrial ecosystems, being the dominant feature of the northern Québec landscape, obviously represent the main receptors of the airborne contaminant. In this section, we present Hg accumulation in above ground and ground cover vegetation of the Québec boreal

Table 3.2. Mercury concentrations (Mean ± SD) in common aquatic macrophytes of lakes of northern Québec (adapted from Grondin 1994) (L: Leaves, R: Roots, Rh: Rhizome)

Latitude (° North)	46° 41'	47° 17' Cabonga			47° 12'	47° 17'	53° 29'	53° 29'
Lake	Lusignan	Site #1	Site #2	Site #3	Laporte	Beaver 2	Duncan	Beaver 3
Alismaplantago aquatica						L: 9 ± 9 Rh: 105 ± 15		
Carex sp.						L: 56 ± 9 R: 59 ± 7		
Cassandra calyculata	R: 110			L: 9 R: 189		L: 17 ± 3 R: 226 ± 55		L: 31 R: 257
Nuphar variegatum	L: 48 ± 4 Rh: 15 ± 15		L: 12 ± 1 Rh: 14 ± 9	L: 13 ± 6 Rh: 5 ± 5	L: 53 ± 3 Rh: 21 ± 20		L: 24 ± 0 Rh: 12 ± 13	
Nymphaeaceae	R: 661			L: 38 R:113				L: 52 ± 44
Potamogeton epihydrus	L: 55 Rh: 228	L: 88 ± 7 Rh: 149 ± 6	L: 172 ± 52 Rh: 57 ± 49					
Sparganium angusifolia	L: 95 ± 29 R: 360±136	L: 161 ± 3 R: 179 ± 57	L: 54±20 R: 137 ± 0	L: 255±33 R: 206 ± 25	L: 92±16 R: 139 ± 37			L: 57 ± 19 R: 41 ± 2

domain. We then examine the ability of the various types of soils, namely forested podzols and peatlands, to retain and accumulate Hg, deposited either through direct precipitation plus litter fall, or leached from the upland catchment. Relatively little Hg is transferred from the soils or the vegetation to the living organisms. Nevertheless, terrestrial ecosystems indirectly play an important role in the presence of Hg in aquatic organisms as variable fractions of the heavy metal deposited on the soils are weathered and transported to lacustrine environments.

3.3.2.1
Mercury in Terrestrial Vegetation of Northern Québec

For a given species of spruce, Hg contents in the last year of growth appear to be fairly comparable among the various sampled sites (Figure 3.4). In most cases, the variations in the Hg concentrations from one site to the next are of the same order of magnitude, considering the spread of concentrations observed for the replicates of a same station. Similarly, no significant variation in Hg concentrations was observed among distinct samples taken from the same tree, nor among samples taken from trees of different ages. The average Hg concentration in the young needles of black spruce, the dominant tree species in Québec, of latitude 48°N is 8.4 ± 1.1 ng g^{-1}. For white spruce, found growing in more southern regions, the Hg concentrations in yearly needles are significantly higher than those of the black spruce, with an average value of 12.5 ± 0.5 ng g^{-1}. The latter average concentration is either quite similar or somewhat smaller than the ones reported for young needles of the same species growing in Ontario (Rasmussen et al. 1991; Moore et al. 1995).

The difference between black and white spruce does not seem to be related to the geographical proximity of industrial centers since within a given region, both species growing side by side still exhibit the same differences in Hg concentrations in their young needles (Figure 3.4). Although further testing is needed, the higher Hg concentrations in white spruce needles as compared to those of black spruce could simply be explained by the different physiology of the two species.

Generally speaking, the Hg concentrations in young spruce needles appear to be very low when compared to the Hg concentrations in the organic horizon of the soils on which they grow (see below). As indicated by several studies, there seems to be little translocation of Hg from the humic layer to vascular plants (Lindqvist 1991; Steinnes 1995). This fact appears to be particularly true for spruce as it has been estimated that 99% of the Hg adsorbed by their roots does not concentrate in the upper portions of the tree (Godbold and Hüttermann 1988).

On the other hand, the loss by volatilization, of Hg from the soils seems to directly influence the Hg concentrations in the leaves of the trees (Cocking et al. 1995; Lindberg 1996).

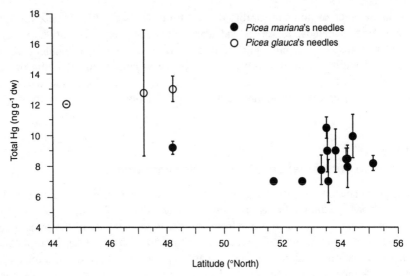

Fig. 3.4. Mercury concentrations in spruce needles as a function of the latitude. Each point represents several samples collected at the same site (adapted from Caron 1997)

Cladina spp. lichens represents the most ubiquitous ground cover of northern Québec uplands (podzolic soils), whereas *Sphagnum* spp. moss dominates the covering of poorly drained soils (wetlands). In contrast to spruce needles, Hg concentrations in these non vascular plants are fairly variable within replicates of a same species collected on a single site (Figure 3.5). The ranges of Hg concentrations in *Cladina* spp. lichens (22 to 126 ng g^{-1}, median of 50 ng g^{-1}) and *Sphagnum* spp. moss (39 to 98 ng g^{-1}, median of 53 ng g^{-1}) are significantly higher than the reported Hg concentrations in spruce needles. These high and variable Hg concentrations are nevertheless not exceptional, as similar results have been found for lichens and moss in Ontario (Rasmussen et al. 1991; Moore et al. 1995), as well as the Grande Baleine region of northern Québec (SOMER Inc. 1993). Since the variations in Hg concentrations for a given ground cover species occur within a 100 m radius, it is unlikely that the Hg concentrations in lichens and moss derive from the influence of the local geological substrate, as previously suggested by Rasmussen et al. (1991). Rather, they most probably result from micro-scale features such as the drainage conditions, the growth rate of the plants and the exposure to atmospheric deposition (Berthelsen et al. 1995; Landers et al. 1995; Evans and Hutchinson 1996).

Hg concentrations in epiphytic lichens are considerably higher than those found for the spruce needles or the ground covers. They range from 150 to 1 170 ng g^{-1} for *Evernia mesomorpha* and 170 to 3 310 ng g^{-1} for *Bryoria* sp. (Figure 3.6). Paradoxically, the lowest concentrations for both species have been found in

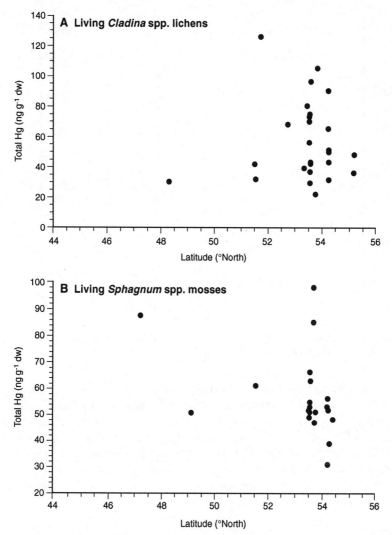

Fig. 3.5. Mercury concentrations in ground cover: (A) lichens of podzolic soils and (B) mosses of wetlands as a function of the latitude (adapted from Caron 1997)

samples taken in southern Québec and the highest in samples coming from the northernmost sites. These plants had first been chosen in our study as they potentially represent interesting geographical indicators of atmospheric deposition of Hg, being rootless and solely depending on precipitation for their growth (e.g., Henderson 1992). But the paradox of finding the lichens furthest away from industrial centers with the highest Hg concentrations in their tissues led us to modify the latter assumption. We first found that at a given site, lichens growing on bigger

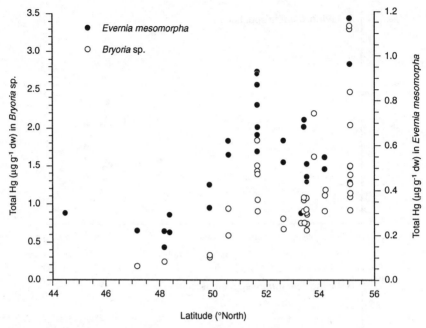

Fig. 3.6. Mercury concentrations in epiphytic lichens as a function of latitude. Each point represents a distinct sample (adapted from Caron 1997)

trees systematically exhibited elevated Hg concentrations. This finding means that Hg concentrations in lichens are dependent upon the time they had to accumulate atmospheric Hg. In order to avoid that bias, we chose to sample lichens on spruce of comparable size throughout the studied area, however, this was met with difficulty since trees at latitude 55°N grow more slowly than those at latitude 46°N. Consequently, the higher Hg concentrations in lichens growing on the northernmost trees reflect their older age and thus do not necessarily reflect greater atmospheric Hg deposition in remote regions.

3.3.2.2
Distribution of Mercury in Forest Soils and Peatlands of the Boreal Domain

Undulating relief and subarctic conditions are key factors in the evolution and composition of northern Québec soils. Forest soils are usually podzolic, being fairly thin and lying on a sandy or little altered substrate. They predominate in the 10° north/south transect covered by this study, representing about 60% of the territory (Poulin-Thériault–Gauthier-Guillemette 1993). They are characterized by a superficial organic horizon (average C content > 40%) less than 15 cm thick. The transition to the mineral horizon (average C content < 17%) is abrupt. Accumulation of Fe_{cdb} in the illuviated B horizon is typical of these soils. While podzols are found in upland areas where the terrain is relatively well drained, peatlands predominate in poorly drained, low-lying areas. These soils cover approximately 10% of the La Grande watershed (Poulin-Thériault–Gauthier-Guillemette 1993). Their organic horizon is usually more than 30 cm thick. The majority of peatlands in northern Québec are ombrotrophic as they only receive minimal drainage inputs and are thus qualified as "bogs" (Fleurbec 1987). The remaining 30% of the soil cover is represented by organic forest soils, with poor drainage conditions intermediate between those of podzolic soils and peatlands (Poulin-Thériault–Gauthier-Guillemette 1993). Soils in northern Québec have been developing since the retreat of the glaciers, about 8 000 years ago, and since then they have been accumulating Hg originating from the alteration of the mineral substrate and from wet and dry atmospheric deposition.

Two representative Hg profiles for the well drained forest soils of northern Québec are presented in Figures 3.7a and 3.7b. Respectively, they represent about three quarters and one quarter of the 20 podzols randomly sampled throughout Québec (Grondin et al. 1995; Caron 1997). The most common distribution of Hg in these soils is characterized by elevated Hg concentrations ranging from 100 to 350 ng g^{-1} and limited to the base of the ground cover layer (LF horizon) and to the thin humic layer (H horizon) (Figure 3.7a). In these soils, little Hg is found in the mineral horizons. The efficient retention of Hg in the humic layer of the podzols has been explained by the fact that Hg has a strong affinity for the OH^- ions of the organic matter (Aastrup et al. 1991; Johansson et al. 1991; Schuster 1991; Steinnes 1995; Dmytriw et al. 1995). The stable hydrated Hg species in soils, Hg (II), can also be complexed by organic ligands possessing the S^- radical (Schuster 1991) and/or with carboxyl groups of humic acids (Allard and Arsenie 1991). For the second group of podzols, in addition to the presence of Hg in the surface organic horizons, fairly high Hg concentrations (around 50 ng g^{-1}) are also found in the illuviated horizon (Figure 3.7b). The presence of Hg in these mineral horizons is principally associated with the zone of iron oxyhydroxide reprecipitation known for its adsorptive properties of other metals (Barrow and Cox 1992; Dmytriw et al. 1995). In terms of cumulative burden of Hg per unit of surface, the total amount of Hg that is present in the illuviated horizon may become as high as the one in the humic layer. It is interesting to note that in a similar way to that of Hg, Pb is retained in the humic horizon of podzolic soils, but does not reabsorb onto iron oxides in the Bf horizon as efficiently as Hg (Figures 3.7a and 3.7b).

Forest fires periodically affect boreal forests. While the surface organic layer of extremely well drained soils is completely burned during such events, the C content of the humic horizon is little affected, as illustrated in the C profile of a recently burned podzolic soil in northern Québec (Figure 3.8). Along with the limited perturbation of the mor layer, the Hg content remains intact, and no apparent volatilization of Hg is observed. Only the fraction of Hg directly associated with the burned fraction of the above ground vegetation (twigs and branches) is readily released to the atmosphere. Intensified weathering of the exposed soils after burning could, nevertheless, promote an increased transport of Hg still bound to the humic matter towards the peatlands or the lakes situated downslope.

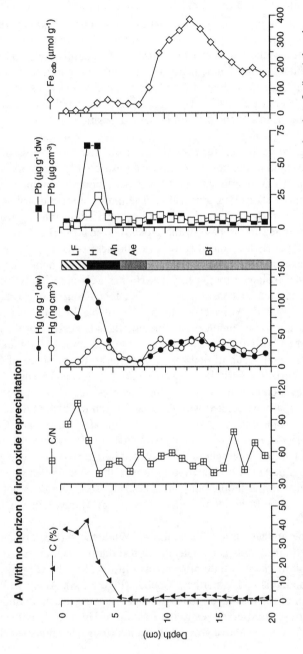

Fig. 3.7. Typical concentration profiles of carbon, mercury, lead, iron oxides, carbon/nitrogen ratios in two podzols of northern Québec (adapted from Caron 1997)

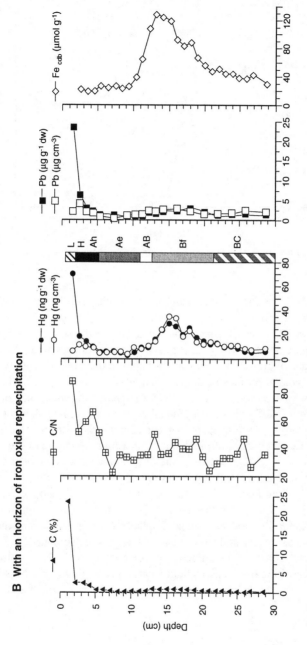

Fig. 3.7. Typical concentration profiles of carbon, mercury, lead, iron oxides, carbon/nitrogen ratios in two podzols of northern Québec (adapted from Caron 1997) (continued)

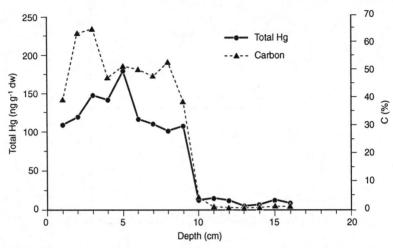

Fig. 3.8. Profiles of mercury and carbon concentrations in a recently burned podzolic soil, region of the La Grande 4 reservoir

Of the 20 cores randomly taken from ombrotrophic peatlands throughout the studied area, the distribution of Hg in these thick organic soils can be described in one of two ways: (1) peak Hg concentrations appear at the surface, just below the living peat (Figure 3.9a) or, (2) peak Hg concentrations appear at depth, above the mineral substrate (Figure 3.9b). Pb usually behaves the same way as Hg, as evidenced by the parallel profiles of the two heavy metals in most of peatland cores analyzed (Grondin et al. 1995; Caron 1997). In contrast to forest soils, one cannot simply attribute the differential heavy metal accumulation in peatlands to distinct pedologic horizons characterized by a particular geochemistry. The only factor that seems to effectively control the depth of maximum retention of Hg and Pb in peatlands is the state of degradation of the organic matter accumulated over time. Hg and Pb accumulation attains a maximum in zones where the organic matter is most intensively humified, as indicated by the lowest C/N atomic ratios (Figures 3.9a and 3.9b). Moreover, the seasonal fluctuations of the level of the water table may be responsible for the vertical displacement of organic gels, themselves being efficient adsorbents of heavy metals (Arakel and Hongjun 1992). Consequently, peatland cores cannot directly be used for historical reconstruction of atmospheric deposition of metals without a careful screening of the possible internal movements that may have occurred in these water saturated environments (Shotyk 1996).

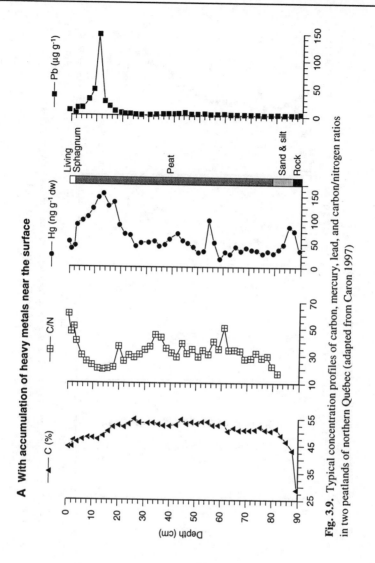

Fig. 3.9. Typical concentration profiles of carbon, mercury, lead, and carbon/nitrogen ratios in two peatlands of northern Québec (adapted from Caron 1997)

3.3.2.3
Anthropogenic Fraction of Mercury in Soils of Northern Québec

The origin of Hg present in forest soils located far from the direct influence of industrial point sources may be two-fold, either being a product of alteration of the mineral substrate (rocks, till, clays) (Rasmussen 1994) or deriving from the slow accumulation of atmospheric inputs (Johansson et al. 1995). In the region of

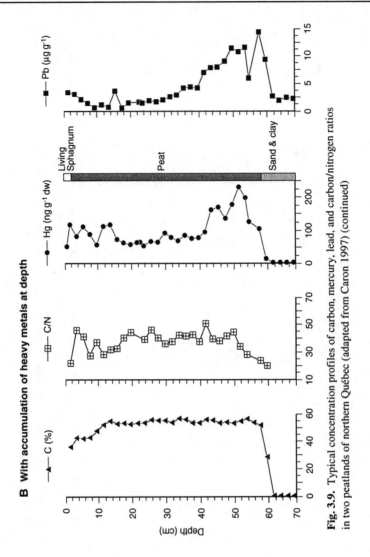

Fig. 3.9. Typical concentration profiles of carbon, mercury, lead, and carbon/nitrogen ratios in two peatlands of northern Québec (adapted from Caron 1997) (continued)

northern Québec that we studied, no geological anomaly of Hg has been described (Caron 1997). The influence of the mineral substrate should therefore be quite uniform over the entire range of forest soils that we sampled. On the other hand, vascular plants transfer little Hg from the soil humus to their root system, and thus to their above ground parts (Lindqvist 1991; Steinnes 1995). Once a podzolic soil is well developed, the composition of its humic layer mainly results from the humification of the litter fall of the vascular plants it bears and from direct atmospheric fall out. Consequently, Hg found in the mor horizon of a boreal forest soil should principally originate from wet and dry atmospheric deposition, either directly to the soils, or through litter fall for the fraction of atmospheric Hg that has been intercepted by the above ground vegetation (Johansson et al. 1991, 1995). It should also be noted that in this process, a small fraction of the Hg found on the cuticle of the leaves may be recycled Hg which is lost by volatilization from the soils (Poissant and Casimir 1997).

We calculate the cumulative burden of Hg per unit of surface in specific horizons of boreal soils by summing, over its thickness, the product of the Hg concentration by the dry density. For podzolic soils of northern Québec, cumulative Hg burdens in the mor horizon range from approximately 1 to 3 mg m^{-2} (average of 2.1 mg m^{-2}), and they occasionally reach 9 mg m^{-2} in organic soils having thicker humic horizons (Figure 3.10a). These Hg burdens are comparable to those already reported for the humic layer of soils in Québec (Grondin et al. 1995), in Sweden (Aastrup et al. 1991; Lindqvist 1991) and in the Great Lakes states in the USA (Nater and Grigal 1992). In order to get a first evaluation of the anthropogenic fraction of Hg accumulated in the mor horizon of these soils, we hypothesized that all Hg reaching the soils as direct atmospheric deposition or along with the litter fall would be retained for at least one century in the humus. This gross approximation is quite justified by the very high retention capacity of Hg(II) by humified organic matter. By extension of the work of Swain et al. (1992) in mid continental North America, we suppose that the anthropogenic Hg deposition rate steadily increased from 0 to 12.5 µg m^{-2} y^{-1} over the past 70 years in Québec (Lucotte et al. 1995a).

This gives an average anthropogenic deposition of Hg of about 0.45 mg m^{-2} since the onset of the industrial era. Thus, in spite of the present atmospheric pollution, it appears that only a small fraction (in general less than 25%) of the Hg present in the mor horizon of a typical podzolic soil of northern Québec is anthropogenic. We may then conclude that most of the Hg in the mor horizon of boreal forest soils has a natural origin, and has been slowly accumulating since the formation of the soils after the last glaciation.

The geochemical influence of the mineral substrate on the composition of the thick peat accumulation in a bog is very limited. The Hg present in the organic layer of a peatland then originates, for the most part, from the local atmospheric deposition or from the drainage basin. Calculations of Hg burdens based on the same principle as the one described for the mor horizon of the forest soils gave fairly high cumulative values over the entire peat accumulation. Typical Hg burdens are of the order of 8.5 mg m^{-2} over a 80 cm thick peat accumulation. While these values are significantly higher than the ones recorded for the humic horizon of podzolic soils, they indicate that peatlands efficiently accumulate Hg leached from uplands, and do not promote the process of loss of Hg by volatilization. Considering only the direct atmospheric deposition of Hg, the anthropogenic

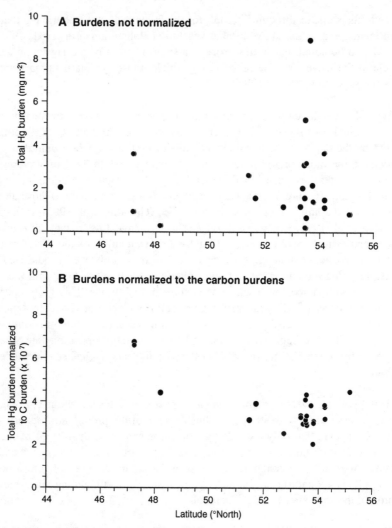

Fig. 3.10. Mercury burdens in the mor horizon of podzolic soils of northern Québec as a function of the latitude (adapted from Caron 1997)

fraction of Hg accumulated in the peat horizon would be even smaller (less than 10%) than the one estimated above for forest soils. At first glance, it is difficult to correlate the 2- to 3-fold increase in Hg concentrations in remote lake sediments, attributed to long range air pollution, with the facts that: 1) Hg is mainly transported to the lakes along with the runoff of terrigenous organic matter, and 2) anthropogenic Hg only constitutes a small fraction of the Hg present in the soils. In order to explain this apparent paradox, one must argue that substantial fractions of Hg deposited on the soils are entrained to the downslope lacustrine systems without fully penetrating the organic horizons of the soils. This may happen with the snow melt on still partially frozen ground in the spring and with the surficial runoff during storm events in the summer or fall (Aastrup et al. 1991).

3.3.3
Patterns of Atmospheric Mercury Deposition throughout Northern Québec

A governmental program for the monitoring of atmospheric Hg deposition across the USA has been implemented since 1995. It clearly reveals spatial differences in wet deposition with a 3-fold difference between sites in the northeastern region of the USA and Florida. Such monitoring of long range atmospheric deposition of Hg is not yet fully in place throughout all northern regions of Canada. In the absence of spatially distributed and long term measurements of Hg deposition, we used a series of direct and indirect indicators, taken from both terrestrial and aquatic environments, to establish a preliminary regional pattern of long range atmospheric deposition of Hg in Québec. As is the case for wet sulphate (Ouellet and Jones 1982) or for other airborne pollutants such as Pb (Sturges and Barrie 1989), we expected to find a decreasing intensity in the long range atmospheric transport of Hg from the southern-most sites of our regional study, closest to industrial centers, to remote sites as far as latitude 55°N.

Both epigeous and ground cover plants could *a priori* constitute good indicators of the intensity of Hg precipitation over the last few months for a given region, as little translocation of Hg from the soils takes place with vascular or non vascular plants (Lindqvist 1991; Steinnes 1995). However, local variations in Hg concentrations for each species of plant studied, yearly spruce needles, *Cladina* spp. lichens and *Sphagnum* spp. moss, have proven to be systematically higher than their respective ranges over the 10° of latitude covered by our study (Figures 3.4 and 3.5). In fact, it appears that Hg concentrations in these plants depend upon Hg volatilization from the soils, and micropedological features such as drainage conditions (Caron 1997). We are thus forced to conclude that the study of the spatial distribution of Hg concentrations in terrestrial plants is of little use in resolving the pattern of present atmospheric Hg deposition in northern Québec. This nevertheless suggests that if a decreasing gradient in Hg atmospheric deposition does exist, this gradient is fairly moderate, and is not recorded as such by living terrestrial plants.

We used the Hg concentrations in various snow samples as a direct indicator of the pattern of Hg deposition. In contrast to rain, the filtration of samples of snow accumulated during the winter period gives an integrated value of the wet Hg deposition over several months. The Hg concentrations determined for the 9 samples collected in March 1994 along a 9° north/south transect ranged from 0.7 to 2.3 ng L^{-1} (Figure 3.11). These concentrations are very low, being comparable to those measured in recent snow and older ice of the Antarctic but one order of magnitude lower than those reported in unfiltered precipitation of the Great Lakes region (Hoyer et al. 1995). The marked variations in Hg concentrations from one snow sample to its closest (Figure 3.11) suggests that there might have been, in certain cases, some losses of Hg either due to partial melt events or to volatilization under UV exposure. Consequently, Hg concentrations in snow samples do not enable us to discern any clear image for winter Hg deposition between the meridional and septentrional sites of Québec.

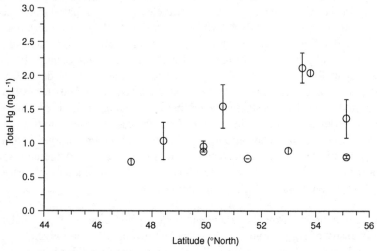

Fig. 3.11. Mercury concentrations in snow as a function of the latitude (adapted from Caron 1997)

The absorption of atmospheric Hg by epiphytic lichens depends upon several biotic and abiotic factors such as climate, age and specific growth rate. Thus, their absolute Hg concentrations yield little information that can be used to deduce a pattern of atmospheric Hg deposition in remote regions (Figure 3.6). On the other hand, these rootless plants may be used as reliable indicators of the origin of the airborne pollutants deposited. To achieve this, it has been suggested that their Hg concentrations be normalized to aluminum (Al) and silicium (Si) contents, two abundant and insoluble elements typically representing aerosol deposition originating from the common rocks of the region studied (Caron 1997). Between latitudes 44°N and 54°N in Québec, all lichens samples belonging to the same species present relatively stable Hg/Al and Hg/Si ratios (Figure 3.12). In particular, contrary to what has been reported for Pb, no influence of the mining sector of Abitibi (49°N) could be detected (Caron 1997). The fairly constant Hg/Al and Hg/Si ratios reported in lichens growing to the south of latitude 54°N suggest that this entire region is subjected to a uniform deposition of airborne Hg, most probably originating from the industries of the Great Lakes area. North of latitude 54°N, atmospheric Hg appears to originate from an additional source, as indicated by contrasting Hg/Al and Hg/Si ratios. In fact, in the absence of a direct source of pollution, this remote region may be under the influence of Hg either volatilized from the nearby Hudson Bay or transported across the pole from Eurasia.

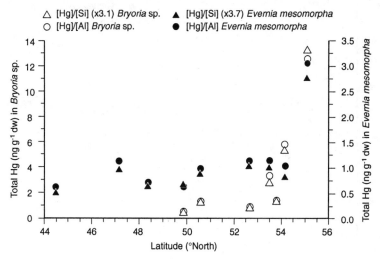

Fig. 3.12. Mercury concentrations normalized to silicium or aluminum contents in epiphytic lichens as a function of the latitude (adapted from Caron 1997)

The geographical pattern of atmospheric Hg deposition in northern Québec could also be addressed using environmental indicators integrating several years. In that sense, while the absolute Hg fluxes at the bottom of a given lake cannot be used directly as an indication of the local Hg atmospheric deposition rate, lacustrine sedimentation being heavily modulated by in-lake depositional processes, lake sediment records may provide time-averaged information on Hg deposition. Sedimentary enrichment factors representing the ratios between modern and pre-industrial Hg inputs help to locally estimate the relative importance of anthropogenic Hg deposition with respect to the natural conditions. Lucotte et al. (1995a) and Landers et al. (1997) concluded that human activities contributed to a uniform 2- to 3-fold increase in Hg deposition within the entire region of Québec situated between latitudes 45°N and 55°N. Since the majority of Hg transported to a lake is done so with the washout of terrestrial organic matter from the drainage basin, we normalized the Hg concentrations of the surficial anthropogenic enrichment (EHg) to the organic C content in order to get an absolute picture of the long range atmospheric transport of Hg in a given region (Figure 3.13). When expressed with respect to the latitude, the plot of the EHg/C ratios for all lakes sampled in Québec over the past years is remarkably invariable. This finding suggests that non particulate atmospheric Hg deposition is independent of the latitude over the boreal forest domain of Québec. The uniform atmospheric Hg contamination contrasts with that of Pb, which decreases towards the north over the same latitudinal span, away from the industrial centers of the Great Lakes (Lucotte et al. 1995a).

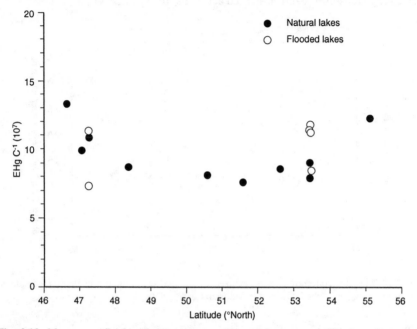

Fig. 3.13. Mercury surficial sediment anthropogenic enrichments normalized to the organic carbon content as a function of the latitude (completed from Lucotte et al. 1995a)

The retention of Hg in the mor horizon of podzolic soils is facilitated by the great affinity of humified organic matter for that heavy metal (Aastrup et al. 1991; Johansson et al. 1991; Arakel and Hongjun 1992; Steinnes 1995). The cumulative burden of Hg in the mor of a forest soil has commonly been used as an indicator of the local atmospheric Hg deposition (Nater and Grigal 1992; Johansson et al. 1991, 1995). But as that burden appears to be heavily dependant upon the C content of the humic layer (Caron 1997), we suggest here to normalize the Hg burdens to the C burdens in the mor layer in order to more rigorously distinguish any geographical trend in the Hg deposition. The latitudinal distribution of the normalized Hg burdens appears to be reduced by half between latitudes 44°N and 54°N (Figure 3.10b). As this sharp drop may correspond to the gradual transition from the Laurentian forest, with fairly thick organic soils, to the boreal forest, dominated by black spruce and characterized by much thinner humic layers, no clear pattern of atmospheric Hg deposition can be drawn over the entire study area.

3.4
Conclusions

Hg is a ubiquitous constituent in all compartments of natural ecosystems of remote regions of northern Québec. In terrestrial environments, the humic horizon of podzolic soils and the thick peat accumulation of ombrotrophic peatlands represent the largest reservoirs of Hg. Airborne Hg of natural origin has progressively been accumulating in these organic layers since the beginning of soil formation following the last deglaciation. In addition to this natural Hg, for the past century, anthropogenic Hg has also started to accumulate in remote soils of northern Québec. Given the large amounts of natural Hg accumulated over several thousands of years, overall anthropogenic Hg constitutes only a small fraction of the total Hg burden in the organic layers of soils of the boreal forest. This observation implies that the Hg responsible for the transitory contamination of the aquatic food chain in newly created reservoirs (Chapters 9.0 and 11.0) would originate from the release of mainly natural Hg accumulated in soils since their initial stage of development, and only marginally from the anthropogenic Hg accumulated over the last century. In the humic horizon of unperturbed soils, Hg appears to remain fairly stable, being strongly attached to humified organic matter. A fraction, however, of the Hg currently deposited with precipitation or litter fall from plants could be transported downslope during spring melt or during storm events via surficial runoff.

In lacustrine ecosystems, sediments constitute the main reservoir of Hg. The quantity of Hg brought to a lake appears to be directly proportional to the amount of C leached from the surrounding soils in the drainage basin. As for Hg in the humic horizon of soils, Hg is very stable in lacustrine sediments, being strongly bound to the organic matter (mostly of terrigenous origin), and little affected by diagenetic remobilization once it is sedimented. Consequently, contrary to soils, the newly deposited Hg in sediments is not incorporated into the already existing Hg pool, but progressively accumulates on top of older layers. This particular behavior thus makes it possible to give historic interpretations of Hg archives. In all lakes sampled between latitudes 45°N and 56°N in northern Québec, a doubling to tripling in accumulation rates of Hg has been observed since about 1940 with respect to natural background rates. Anthropogenic Hg then represents the major fraction of total Hg in surficial sediment layers. The apparent paradox between the small fraction of anthropogenic Hg in soils and the dominant contribution of anthropogenic Hg to surface sediments is explained by the combined effects of surficial weathering of newly deposited atmospheric Hg on the soils and the diagenetic stability of Hg in lake sediments.

In the absence of an integrated program monitoring atmospheric pollution throughout the vast domain of the boreal forest of Québec, several environmental indicators may represent proxies for mapping the geographical pattern of atmospheric Hg deposition rates, although their specific usage may bear some restrictions. The expected decreasing gradient from the southern industrialized belt to the northern pristine regions of western Québec appears too subtle to be observed by the mapping of Hg concentrations in living plants (too dependent upon Hg volatilization from the soils and micropedological features) or in snow (as losses of Hg may occur with sublimation or partial melt). On the other hand, the presence of Hg in epiphytic lichens, once normalized to Al or Si (representing the local influence of the bedrock), is quite illustrative of the source of atmospheric Hg. The entire region between latitudes 44°N and 54°N in western Québec appears to lie under the influence of a diffuse but homogeneous source of Hg, probably identifiable as the industrial area of the Great Lakes region. It is only north of latitude 54°N that the industrial source of Hg may be overshadowed by Hudson Bay or transpolar influence of Hg. Finally, the cumulative Hg burdens normalized to C burdens in the humic layer of podzols or the sedimentary enrichment of Hg normalized to organic content in sediments deposited since the onset of the industrial era suggest that the distribution pattern of long range Hg contamination is independent of latitude over the boreal forest domain of Québec. This pattern contrasts with the clear decreasing gradient in Pb or sulphate deposition over the same region and suggests that the residence time of Hg in the atmosphere is fairly long, enabling its transport over longer distances.

Hg bioaccumulates in the aquatic food chain after inorganic Hg is transformed into bioavailable MeHg, principally at the surface of the sediments, where organic matter is also degraded. In these processes, particulate organic matter and its coating of bacteria plays a particular role, being a food source to invertebrate organisms. It is not clear to what extent the present doubling to tripling in Hg concentrations in surface sediments corresponds to an equivalent increase in Hg in aquatic organisms, as there exist no long term studies of Hg accumulation in aquatic organisms over the last century. Nevertheless, one may argue that the recent increase in Hg inputs to lakes most certainly has direct effects in increased Hg accumulation in organisms living in sensitive aquatic environments, that is, where methylation processes are facilitated.

Acknowledgements

This research was supported by a grant from Hydro-Québec, the Conseil de recherches en sciences naturelles et en génie (CRSNG) du Canada and the Université du Québec à Montréal (UQAM) to the Chaire de recherche en environnement HQ/CRSNG/UQAM. Additional financial support, in the form of student scholarships, was supplied by the fund for the Formation de Chercheurs et l'Aide à la Recherche (Fonds FCAR), the Programme de Formation scientifique dans le Nord (PFSN) and CRSNG. The laboratory and field work for this project could not have been accomplished without the assistance of many motivated people, particularly,

P. Pichet, P. Ferland and I. Rheault, as well as, M. Bégin, R. Canuel, L. Cournoyer, É. Duchemin, B. Fortin, A. Grondin, C. Guignard, D. Lebeau, Y. Plourde, L.-F. Richard, S. Tran, and A. Tremblay.

4 Bioaccumulation of Mercury and Methylmercury in Invertebrates from Natural Boreal Lakes

Alain Tremblay

Abstract

Plankton was sampled in 13 natural lakes using conical nets of different mesh size, aquatic insect larvae were sampled in 11 lakes by dragging 250 μm mesh framed nylon bags and adult insects were collected by emergence floating traps. Total mercury (Hg) concentrations in insects (larvae and adults) and zooplankton ranged from 31 ng g^{-1} dry weight (dw) to 793 ng g^{-1} dw. Methylmercury (MeHg) levels were lower but showed similar variation from 25 ng g^{-1} to 575 ng g^{-1} dw. The mean proportion of MeHg to total Hg concentrations depended on the feeding behavior of the animals, increasing from 35-45% in detritivore insect larvae (dipterans, ephemeropterans, trichopterans) to 70-85% in predators (heteropterans, coleopterans, odonates). Similarly, the mean proportion of MeHg increased in the planktonic food web from 20-25% in the 20-75 μm mesh plankton to 60-70% in the 225 μm mesh plankton. These differences are attributed to the biomagnification of MeHg in the food web with the biomagnification factor between two adjacent trophic levels being about 3. Two principal factors control the contamination of predators at the top of the food chain: (1) the bioavailability of Hg to lower trophic levels (influenced by the quality and the Hg concentration of the ingested food) and (2) the feeding behavior of the animals.

4.1
Introduction

Uptake from food is the predominant pathway for Hg accumulation in many organisms and, in combination with slow elimination from their tissues, leads to the bioaccumulation of Hg (e.g., Jackson 1991; Meili 1991a; Cabana et al. 1994; Hall et al. 1997). This may present a serious problem for northern countries such as Canada and the United States, where thousands of lakes contain predatory fish with Hg concentrations greater than the national standard of 0.5 μg g^{-1} (wet weight) for the commercial sale of fish (Health and Welfare Canada 1985). In the province of Québec, over 90% of the lakes have northern pike (> 700 mm) with total Hg concentrations surpassing this standard (MEF-MESS 1995). Such statistics have important socio-economic consequences for sport and commercial fisheries.

While many studies have examined relationships between environmental factors and Hg accumulation in fish, little is known about the influence of these factors on the accumulation of MeHg at lower trophic levels. Since diet is an important source of Hg to fish, and given that MeHg is the predominant form of Hg at higher trophic levels, an understanding of the factors influencing the availability of MeHg at the base of the food chain would provide important insight into the accumulation of Hg in fish. Invertebrates such as insects and zooplankton play a key role in Hg cycling because they constitute a major food source for many fish species (Chapter 10.0; Scott and Crossman 1974; Dumont and Fortin 1978; Doyon et al. 1996).

This chapter presents results from a five year collaborative project between the Université du Québec à Montréal and Hydro-Québec on the cycling of Hg in aquatic insects and plankton. In the first part of this chapter, we define what is meant by the bioavailability of MeHg to invertebrates at low trophic levels and its implications for the food chain. This is then linked, in the second part of the chapter, to the role of the food web structure in the accumulation of Hg by aquatic organisms.

4.2
Materials and Methods

4.2.1
Study Areas

Sixteen Québec lakes were sampled during the ice free periods of 1990-1995. Eleven lakes (Duncan, Detcheverry, Des Vœux, Hazeur, Ivry, Jobert, Lake 136, Lake 150, Plamondon, Robertson, Rond-de-Poêle) were visited during three seasons (spring, summer and fall), while five others (Evans area, Laporte, Matagami, Opinaca, and Koury) were sampled only once in the summer of 1994 (Figure 4.1). The catchments of all of these mostly oligotrophic lakes are dominated by coniferous forest, shallow podzolic and peat soils, and igneous bedrock. The morphological and physico-chemical characteristics of the water and sediments are presented in Table 4.1. Additional information describing the study area can be found in Tremblay and Lucotte (1997) and Grondin et al. (1995).

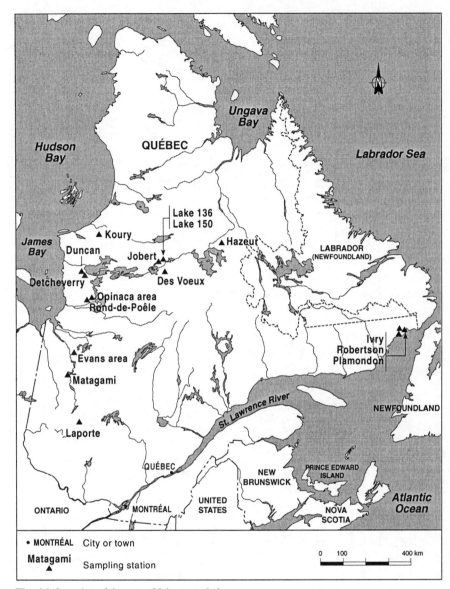

Fig. 4.1. Location of the natural lakes sampled

Table 4.1. Lake characteristics, water and sediment chemistry and sediment total mercury and methylmercury concentrations in the natural lakes sampled in 1990-95.

Name	Type of organisms sampled	Lake		Water column			Sediments			
		Area (km²)	Watershed area (km²)	pH	Colour (Pt mg L⁻¹)	Cond (µs cm⁻¹)	C (%dw)	C/N (g g⁻¹)	Total Hg (ng g⁻¹ dw)	MeHg (ng Hg g⁻¹ dw)
Des Vœux	Insects-plankton	29.3	90.0	6.9-7.1	18	12	8-9	15-17	75-88	0.41-1.20
Detcheverry	Insects-plankton	8.3	53.7	6.8-7.3	49	37	8-9	17-19	88-102	0.14-0.39
Duncan	Insects-plankton	100.0	281.5	6.8-7.1	68	30	3-4	15-17	52-59	0.36-0.64
Evans area	Insects-plankton	0.03	0.3	6.8-7.1	74	50	17-18	20-22	175-196	0.65-1.04
Hazeur	Plankton	3.9		5.9-6.8	23	11				
Ivry	Plankton			6.0	35-41	13				
Jobert	Insects-plankton	15.1	58.8	6.8-7.1	46	14	12-13	15-16	154-160	0.24-0.75
Koury	Insects-plankton	1.0	3.7	6.4-6.8	52	32	2-3	16-17	45-56	0.23-0.40
Lake 136	Insects-plankton	0.03	0.2	6.9-7.3	42	15	6-10	16-18	53-77	0.45-1.58
Lake 150	Insects-plankton	4.7	13.2	7.0-7.3	37	13	3-6	15-17	40-59	0.30-0.69
Laporte	Insects	0.6	2.3	6.3-6.6	54	30	17-18	20-22	294-351	0.63-1.44
Matagami	Insects	228.8	1 086.8	6.9-7.1	68	30	8-10*	14-15*	73-83*	
Opinaca area	Insects	0.08	0.3	6.2-6.5	73	21	7-8	21-22	115-142	0.49-1.37
Plamondon	Plankton			6.2	39	14				
Robertson	Plankton			6.0-6.2	39-43	12-13				
Rond-de-Poêle	Plankton			5.8-7.0	49	23				

* Obtained from a neighboring lake.

Sediments were sampled in the pelagic zone. The values for pH and water colour represent the productive surface water (1-2 meters)

4.2.2
Sampling

4.2.2.1
Insects

Insect larvae were collected from the littoral zones (< 2 m of water) of eleven lakes by dragging a 250 µm mesh framed nylon bag. From the surface sediments collected with the net, all insects visible to the naked eye were picked using tweezers with stainless steel tips and sorted by genus or species.

Adult insects were collected from the littoral zone of Lake 136 using 1 m^2 emergence floating traps with a 250 µm mesh size net. Emerging adults were retained in a flask containing a 1:1 solution of ethylene glycol and distilled water. Upon return to the field laboratory, the insect samples were rinsed with both tap and distilled water to remove the excess ethylene glycol and preserved in 85% ethanol until further analysis. Using a binocular and tweezers with stainless steel tips, adult insects were sorted by species.

The abundance of adult insects (number of individuals m^{-2} year^{-1}) was calculated using the mean total number of individuals of two emergence traps for the entire sampling season from June 20 to September 15, 1994. The traps were emptied every 2 or 3 weeks during that period. The biomass of insects (mg m^{-2} year^{-1} dw) was estimated from the abundance and from the mean dry weight of 10 individuals of the most common taxa. Mean dry weight of *Agrypnia* spp. (11 mg), *Polycentropus* spp. (1 mg), other trichopterans (5 mg), and dipterans (0.05 mg) were used to evaluate the biomass. More details describing the manipulation of insects are available in Tremblay et al. (1998b).

4.2.2.2
Plankton

Plankton sampling was carried out in surface waters during the day using conical nets of either 20, 75, 150 or 225 µm. Suspended particulate matter (SPM, < 40 µm) was collected by continuous flow centrifugation of about 50 L of water. The samples were kept frozen until analysis. Additional information describing sampling techniques are available in Plourde et al. (1997) and Doyon (1995ab).

Zooplankton abundance (number of individuals L^{-1}) was determined for 50 L water samples in four lakes (Duncan, Detcheverry, Lake 150 and Des Vœux) twice during summer-fall 1994. Ten individuals of the most common taxa were measured and their dry weight was calculated from length-weight regressions (Malley et al. 1989). Zooplankton biomass (mg L^{-1} dw) was obtained by multiplying the abundance by the mean dry weight.

4.2.3
Analyses

Pooled insect samples consisting of at least 10 individuals and bulk plankton samples were freeze-dried for 3 days and then homogenized. In order to reduce manipulation and the risk of Hg contamination, samples were ground to a powder directly in their storage containers using a glass rod. The use of homogenized samples was decided upon since this procedure allows all analyses (total Hg, MeHg, C/N) to be conducted on the same sample. The concentration of total Hg and MeHg extracted by a distillation technique were determined by atomic fluorescence spectrophotometry as described in Chapter 2.0.

The total Hg and MeHg concentrations are expressed per dry weight. To evaluate the homogeneity of the samples, every tenth sample was analyzed in triplicate. The coefficient of variation between analyses of the same sample was always less then 6%. The preservative solutions, ethylene glycol and 85% ethanol, were analyzed for trace of Hg and MeHg and in both cases the concentrations were below the detection limits.

4.3
Results and Discussion

4.3.1
Total Mercury and Methylmercury Concentrations in Invertebrates

Our studies showed a range in total Hg concentrations from 31 ng g^{-1} to 793 ng g^{-1} dw in the insect larvae and from 25 ng g^{-1} to 575 ng g^{-1} in plankton (Table 4.2a; Tremblay et al. 1996ab; Tremblay et al. 1998a). MeHg concentrations were lower but showed similar variations for both the insects and the plankton with a range from 25 ng g^{-1} to 510 ng g^{-1}. Our results were comparable to those reported by Parkman and Meili (1993) for different species of insect larvae of eight Swedish lakes and by Rask et al. (1994) on a Finnish lake, as well as those reported for plankton by Back et al. (1995) and Westcott and Kalff (1996) (Table 4.2).

From these results and those presented in Table 4.2, one can see that there is a wide variation (by a factor of 5) in invertebrate total Hg and MeHg concentrations among lakes for a given taxon (see also Figures 4.2 and 4.3). Although these differences may partly be due to differences in the Hg loading to the aquatic ecosystem (Chapter 3.0), they probably reflect differences in the sensitivity of the lakes in terms of bioavailability and biological uptake.

Table 4.2. Data from the literature on mercury concentrations (ng g^{-1} dw, mean ± standard deviation or range) and the proportion of methylmercury to total mercury (%) in invertebrates of natural aquatic ecosystems

Taxa	Mean Hg (ng g^{-1}, dw) (variations)	Prop. of MeHg (%)	Site location	References
INSECTA				
EPHEMEROPTERA				
Leptophlebia vespertina	270-560 (ww)		Idrijca river, ex-Yougoslavia	Nuorteva et al. 1980
	50-66 (ww)			Huckabee et al. 1979
Leptophlebia spp.	140-190		Lake Nimetön, Finland	Rask et al. 1994
	50-216	40-60	12 lakes, Québec	Tremblay et al. 1996b
ODONATA				
	48-210			Huckabee et al. 1979
	93-409		8 lakes, Sweden	Parkman and Meili 1993
	240 (70-420)	76	Finland	Verta et al. 1986a
	50		L979 and L632, Ontario	Hall et al. 1998
SOMATOCHLORA AND	152-501	68-84	8 lakes Sweden	Tremblay et al. 1996b
CORDULIA	107-419	80-85	12 lakes, Québec	Tremblay et al. 1996b
Aeschna grandis	110-200		Lake Nimetön, Finland	Rask et al. 1994
Aeschna juncea	160-200			
Agrion hastulatum	140-310			
Cordulia aenea	110-310			
Leucorrhinia sp.	180-350			

Table 4.2. Data from the literature on mercury concentrations (ng g^{-1} dw, mean ± standard deviation or range) and the proportion of methylmercury to total mercury (%) in invertebrates of natural aquatic ecosystems (continued)

Taxa	Mean Hg (ng g^{-1}, dw) (variations)	Prop. of MeHg (%)	Site location	References
HETEROPTERA				
NOTONECTIDAE	75		L979 and L632, Ontario	Hall et al. 1998
Notonecta glauca	460-580		Lake Nimetön, Finland	Rask et al. 1994
Gerris spp. and	130-600	75-85	12 lakes, Québec	Tremblay et al. 1996b
Sigara spp.				
TRICHOPTERA	390		Rivière Qu'Appelle, Saskatchewan	Hammer et al. 1982
	240 (40-760)	66	Finland	Verta et al. 1986b
	35		L979 and L632, Ontario	Hall et al. 1998
LIMNEPHILIDAE	20-54 (ww)			Huckabee et al. 1979
	50-255	25-50	12 lakes, Québec	Tremblay et al. 1996b
	25-299	15-55	8 lakes, Sweden	Tremblay et al. 1996b
Limnephilus lunatus	40-140		Lake Nimetön, Finland	Rask et al. 1994
COLEOPTERA				
DYSTISCIDAE	174		L979 and L632, Ontario	Hall et al. 1998
Dysticus sp.	320-590		Lake Nimetön, Finland	Rask et al. 1994
GYRINIDAE	80		L979 and L632, Ontario	Hall et al. 1998
Gyrinus spp.	75-550	65-75	12 lakes, Québec	Tremblay et al. 1996b

Table 4.2. Data from the literature on mercury concentrations (ng g^{-1} dw, mean ± standard deviation or range) and the proportion of methylmercury to total mercury (%) in invertebrates of natural aquatic ecosystems (continued)

Taxa	Mean Hg (ng g^{-1}, dw) (variations)	Prop. of MeHg (%)	Site location	References
DIPTERA				
Chironomus sp.	10-50		North of Manitoba Qu'Appelle	Jackson 1988b
	340-1100		river, Saskatchewan	Hammer et al. 1982
	115-1349		8 lakes, Sweden	Parkman and Meili, 1993
	70-200	25-50	12 lakes, Québec	Tremblay et al. 1996b
Sergentia sp.	33-667	6-36	8 lakes, Sweden	
Procladius sp.	71-4600		8 lakes, Sweden	Parkman and Meili 1993
Chaoborus sp.	42-332		8 lakes, Sweden	Tremblay et al. 1996b
	60-225	27-43	8 lakes, Sweden	
Benthos (mixed...)	55 (8-155) (ww)	80	Pihlajevesi lake, Finland	Surma-Aho et al. 1986
(Trichoptera and Odonata)	45-(3-112) (ww)	85	Seinäjärvi lake, Finland	
CRUSTACEA				
MALACOSTRACA				
Aellus aquaticus	330-360		Lake Nimetön, Finland	Rask et al. 1994
	63-394		8 lakes, Sweden	Parkman and Meili, 1993
	228-446	45-66	8 lakes, Sweden	Tremblay et al. 1996b
ISOPODA				
AMPHIPODA				
Hyallela azteca	25		L979 and L632, Ontario	Hall et al. 1998
	95-266	32-52	12 lakes, Québec	Tremblay et al. 1996b

Table 4.2. Data from the literature on mercury concentrations (ng g^{-1} dw, mean ± standard deviation or range) and the proportion of methylmercury to total mercury (%) in invertebrates of natural aquatic ecosystems (continued)

Taxa	Mean Hg (ng g^{-1}, dw) (variations)	Prop. of MeHg (%)	Site location	References
Zooplankton (bulk samples)	88±49		North of Minnesota	Sorensen et al. 1990
	80-500	40-80	8 lakes in Sweden	Meili 1991a
	250 (20-550)	70	Finland	Verta et al. 1986b
	59 (27-150)		St. Louis river, Minnesota	Glass et al. 1990
	220 (120-390)	40-70	Lake Pihlajevesi, Finland	Surma-Aho et al. 1986
	280 (110-550)	40-75	Lake Seinäjärvi, Finland	
	20-450 (MeHg)		24 lakes in Ontario	Westcott and Kalff 1996
	108 (26-378)		24 lakes in northern Ontario	Tremblay et al. 1995
Copepodes	20-350		Lake Nimêton, Finland	Rask et al. 1994
Cladocerans	120-600			
Cyclopoides	74-123 (MeHg)	45-77	Mouse lake, Ontario	Tsalkitzis 1995
Calanoids	106-133 (MeHg)	45-77		

Table 4.2. Data from the literature on mercury concentrations ($ng\ g^{-1}$ dw, mean ± standard deviation or range) and the proportion of methylmercury to total mercury (%) in invertebrates of natural aquatic ecosystems (continued)

Taxa	Mean Hg ($ng\ g^{-1}$, dw) (variations)	Prop. of MeHg (%)	Site location	References
Daphnia + Bosmina spp.	82-244 (MeHg)		3 lakes in Québec	Plourde et al. 1997
Daphnia spp.	1-211 (MeHg)		Mud lake, Wisconsin	Back et al. 1995
Bosmina spp.	69 (MeHg)		2 lakes in Québec	This study
	17-63(MeHg)	8-26	Mouse lake, Ontario	Tsalkitzis 1995
Diaptomus spp.	22-66 (MeHg)		Mud lake, Wisconsin	Back et al. 1995
Hydracarina spp.	251 (MeHg)			
Holopedium spp.	40-419 (MeHg)	50-75	2 lakes in Québec	This study
	148-164		Mouse lake, Ontario	Tsalkitzis 1995
	176-274 (MeHg)	67-109	2 lakes in Québec	This study
	82-111 (MeHg)			
Cyclops spp.	24-30 (MeHg)	50-70	Mud lake, Wisconsin	Back et al. 1995
Mesocyclops spp.	80-245 (MeHg)		3 lakes in Québec	Plourde et al. 1997
Senecella spp.	142-286		2 lakes in Québec	This study
Leptodora spp.	76-196 (MeHg)		3 lakes in Québec	Plourde et al. 1997

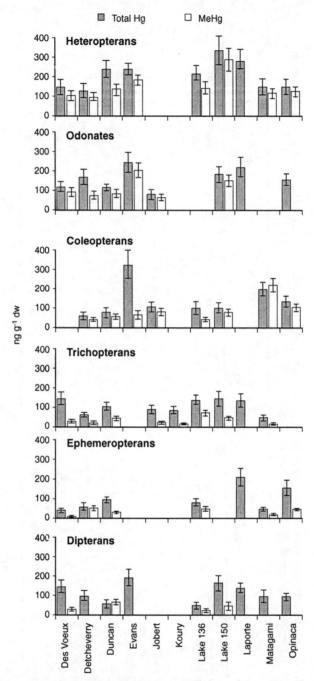

Fig. 4.2. Total mercury and methylmercury concentrations (ng g^{-1} dw, mean ± standard deviation) in various insect orders from different lakes in Québec

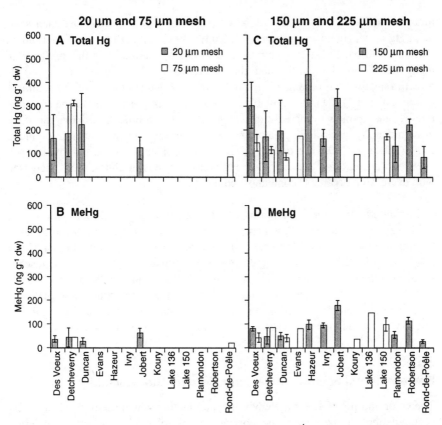

Fig. 4.3. Total mercury and methylmercury concentrations (ng g^{-1} dw, mean ± standard deviation) in various plankton mesh sizes from different lakes in Québec (modified from Tremblay et al. 1998a)

4.3.2
Bioavailability of Total Mercury and Methylmercury for Insect Larvae

Only a small fraction of the total Hg burden in lake sediments is transformed (principally by microbial activity) into MeHg, the form accumulated by the organisms (e.g., Xun et al. 1987; Korthals and Winfrey 1987). Environmental factors that modify microbial activity and MeHg production (e.g., temperature, humic substances and the organic content of sediments) likely play key roles in the bioavailability of MeHg to organisms at the base of the food web. This concept of bioavailability of MeHg, while not clearly defined, is essentially the quantity of MeHg that can be taken up by the organisms.

The generally weak relationships observed between total Hg or MeHg concentrations in the aquatic insect larvae and those in sediments indicate that Hg accumulation by these organisms may be determined by parameters other than sediment Hg or MeHg concentrations (this study; Bissonnette 1975; Jackson 1988ab; Parkman and Meili 1993; Tremblay et al. 1996ab). Sediment is the substrate of many detritivores-herbivores and the quality of available food, measured as growth rates of such organisms, is related to the organic content of the sediment (Mattingly et al. 1981). It is possible that a low carbon (C) content of the substrate requires the animals to increase their ingestion rates in order to achieve the same growth thereby resulting in greater exposure to Hg for a given concentration in the food (Parkman 1993; Tremblay et al. 1996ab). This is supported by Jackson (1988b) who reported that inorganic Hg concentrations in chironomids increased with decreasing organic C content of sediment.

Further evidence for the importance of the nature of ingested food with respect to Hg exposure is provided by a comparative study of Québec and Swedish lakes (Tremblay et al. 1996b). We found both the highest total Hg and the lowest MeHg levels in profundal zone detritivores insects as compared to shallow zone insects, suggesting that the accumulation of inorganic Hg by profundal detritivores resulted from their high exposure to inorganic Hg in their food and a poor quality of C. Since the C present in sediments of the profundal zone is generally highly refractory to degradation, it is usually related with a poor nutritive value.

4.3.3
Environmental Factors Influencing Mercury Bioavailability

The color and pH of water are environmental factors which seem to affect MeHg accumulation in the planktonic community of relatively pristine lakes since Watras and Bloom (1992), Meili (1991a) and Westcott and Kalff (1996) reported significant correlations between these environmental factors and the MeHg concentrations in plankton.

Several mechanisms have been proposed to explain the effects of low pH on the bioavailability of MeHg and its uptake by the biota. They include, increased production of MeHg in the water column and surface sediments (Furutani and Rudd 1980; Xun et al. 1987), decreased loss of volatile Hg from water (Rada et al. 1987) and increased uptake of both organic and inorganic Hg by aquatic organisms (Rodgers and Beamish 1981; Ponce and Bloom 1991).

Organic matter decomposition decreases the pH and produces humic substances. These may increase the uptake of MeHg by organisms as they carry Hg from the drainage basin to the lake (Lee and Hultberg 1990; Lee and Iverfeldt 1991; Mierle 1990; Mierle and Ingram 1991), stimulate the within-lake production of MeHg by providing methylating bacteria with a substrate for growth (McMurtry et al. 1989; Wren et al. 1991; Miskimmin et al. 1992), and promote the abiotic methylation of inorganic Hg in highly organic soils thereby influencing the supply of MeHg to lakes (Lee et al. 1985; Lee and Hultberg 1990).

A study done on 24 natural Canadian lakes by Westcott and Kalff (1996) have shown that the effect of the humic substances outweighs the pH effect. This suggests that water color influences MeHg transfer to the food chain primarily via its role in the export of Hg from the drainage basin. This hypothesis is supported by the positive relationships observed between MeHg levels in the biota and drainage ratio as well as the percent wetland in the catchment (Swain et al. 1992; St. Louis et al. 1994; Westcott and Kalff 1996; Chapter 3.0). According to Driscoll et al. (1994), dissolved organic carbon (DOC) released from the wetlands within the drainage basin plays a complicated role in transport and regulating concentrations of both total Hg and MeHg, and ultimately the supply of Hg to the food web. This is another argument in favor of the transfer of MeHg to the food chain since humic substances are ingested by bacteria and possibly zooplankton (Tranvik 1992; Hessen 1992; Jones 1992).

In our study, we found a significant correlation between water temperature and MeHg concentrations in chironomids (t-test, $r = 0.99$, $p < 0.07$) and odonates (t-test, $r = 0.64$, $p < 0.006$). Total Hg concentrations in insect larvae increased from spring to fall, from 60-80 ng g^{-1} to 125-175 ng g^{-1} dw in detritivore-grazer taxa, and from 88-120 ng g^{-1} to 180-200 ng g^{-1} in predator taxa (Figure 4.4). The patterns are similar for MeHg concentrations, but the proportion of MeHg to total Hg remained constant from spring to fall in the detritivores-grazers, and increased slightly in the predators.

For plankton, seasonal patterns of total Hg accumulation were not as well defined, since an increase over the season was only observed in some lakes (Figure 4.4). The SPM MeHg concentrations however were higher in August than in June (Plourde et al. 1997). This indicates that microbial and/or algal activity was stimulated by the warmer temperatures during the summer months in midnorthern Québec and suggests a delayed effect of temperature, as the time to build up microbial populations and rates of MeHg accumulation are expected to vary more slowly than water temperatures (Kelly et al. 1997). Indeed, high temperatures are known to stimulate microbial activity responsible for the methylation of Hg (e.g., Korthals and Winfrey 1987; Bodaly et al. 1993). Similarly, Parks et al. (1989) reported a positive correlation between water temperature and MeHg concentrations in the water column of the English-Wabigoon river system. Jackson (1988a), however, found no relationship between the temperature and the accumulation of total Hg in invertebrates and suggested that the increase of total Hg in invertebrates over the season could be related to the nature of ingested food rather than to the temperature itself. Similar conclusions were also reported by Parkman and Meili (1993). Although the processes of methylation are poorly

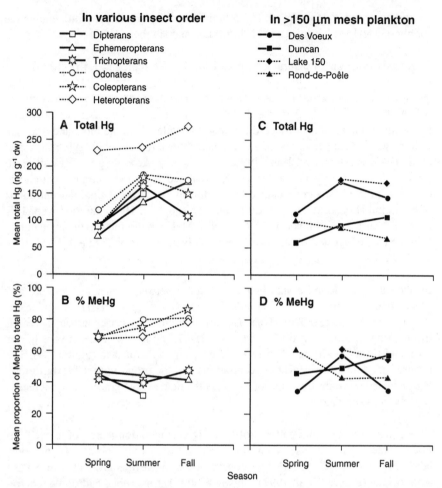

Fig. 4.4. Seasonnal variation (spring, summer, fall) in mean total mercury (ng g^{-1} dw) and proportion of methylmercury to total mercury in various insect order and in > 150 μm mesh plankton

understood and merit further research, environmental factors are crucial for the uptake of MeHg by the biota since they influence bioavailability of MeHg to organisms.

4.3.4
Biomagnification of Methylmercury along the Food Web

Two distinct trophic levels of the benthic invertebrate food web structure have been defined by Tremblay et al. (1996b) using the C content of insect larvae of 11 lakes, their MeHg concentration, as well as their MeHg to total Hg ratios. The first trophic level includes organisms with C content between 38-50%, MeHg concentrations lower than 100 ng g^{-1} dw and MeHg/total Hg ratios lower than 60%. These organisms are detritivores-grazers and correspond to the general feeding habits of the insects reported by Merritt and Cummins (1985). The second level includes organisms having C contents always > 48%, MeHg levels > 100 ng Hg g^{-1} dw and MeHg/total Hg ratios > 60% and are herbivores-predators as classified by Merritt and Cummins (1985).

The results of our studies indicated a biomagnification of MeHg along the invertebrate food web, since the MeHg concentrations increased when moving from the detritivore-herbivore insects to the herbivore-predator insects (Figure 4.5, Tremblay et al. 1996ab). The increased proportion of MeHg from 30-50% in detritivores to 70-95% in the predator insects was comparable to those reported in the literature for benthic invertebrates (Table 4.2). Increases in MeHg concentration with trophic level have also been reported in the fish community with piscivorous species containing the highest Hg levels (e.g., Chapters 5.0 and 11.0; Cabana et al. 1994).

Similarly, based on a study of four lakes, four trophic levels for the pelagic food web structure were defined by Tremblay et al. (1998a) using the mesh-size of the collecting plankton nets. The SPM smaller than 40 μm corresponds to the first trophic level. The material collected by the 20-75 μm-mesh contained many groups of small organisms such as rotifers and large phytoplankton, but excluded the nanoplankton and corresponds to the second trophic level. The material collected with 150 μm-mesh and 225 μm-mesh nets contained mainly large zooplankton but also large phytoplankton and possibly organic debris, and correspond, respectively, to the third and the fourth trophic levels.

We observed a similar pattern for plankton with our data showing an increase in the MeHg:total Hg ratio of 5-7% in the SPM, 20-30% in the 20-75 μm mesh plankton, 35-45% in the > 150 μm mesh plankton and 50-60% in the > 225 μm mesh plankton (Figure 4.5a; Plourde et al. 1997; Tremblay et al. 1998a). The 20-75 μm represents mainly small organisms and phytoplankton and the > 150-225 μm are dominated by larger animals such as *Daphnia* spp. or *Holopedium* spp. (see Table 4.5). Furthermore, in three of our lakes we separated the phytoplankton from the zooplankton and observed a doubling of the MeHg:total Hg ratio from phytoplankton to zooplankton (Doyon et al. 1996). These results indicate that the increase of the MeHg:total Hg ratio with the plankton size is a biomagnification of MeHg along the plankton food web. Similar results were reported by Watras and Bloom (1992) for phytoplankton (13% of MeHg) and zooplankton (29%) in Little Rock lake.

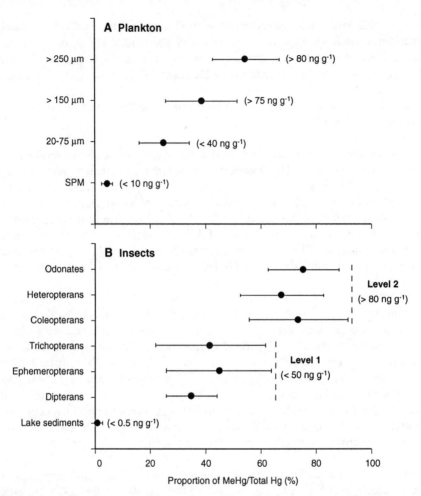

Fig. 4.5. Trophic levels of plankton (A) and insects (B) with the proportion of methylmercury to total mercury (%, mean ± standard deviation) in the organisms. For the plankton, trophic levels refer to suspended particulate matter and various plankton mesh size (see text). For the insects, trophic levels refer to detritivore-herbivore organisms (Level 1) and herbivore-predator organisms (Level 2) (see text, modified from Tremblay et al. 1996a; Tremblay et al. 1998a). The numbers in parentheses represent the mean methylmercury concentration (ng g^{-1} dw)

Following the calculation of biomagnification factors (BMF, concentration of total Hg or MeHg in one trophic level divided by the concentration in the level below), our data showed negative BMF (in log units) for the total Hg of the insects and plankton and positive BMF for the MeHg. These observations support the hypothesis of Watras and Bloom (1992) that total Hg becomes more diluted while MeHg is concentrated as it is moved up the lower food web. Indeed, the mean ratio of MeHg concentrations in both the insects and the plankton, is about three between two adjacent trophic levels (Figure 4.5). The threefold difference in

MeHg concentrations between trophic levels has been documented elsewhere. For example, in a comprehensive food chain study of Swedish lakes which included the compartments of SPM to predatory fish (Meili 1991b), as well as a study of phytoplankton and zooplankton in Little Rock lake (Watras and Bloom 1992).

In our study, we found no relationship between size and MeHg concentrations of 32 samples of small, medium or large odonates (Somatochlora and Cordulia, t-test, all $p > 0.05$, Figure 4.6). Similarly, Parkman and Meili (1993), in a study of eight Swedish lakes, found no significant relationship between total Hg content and body size of predatory invertebrates, despite a 100-fold variation in body weight. The absence of a size or weight effect on MeHg accumulation in insects confirms that the increase up the food chain of the proportion of MeHg to total Hg in insects, observed in our study, is a food web effect (Tremblay et al. 1996ab). This conclusion agrees with that of Cabana et al. (1994) who compared lakes with and without *Mysis* sp. and forage fish and showed that Hg accumulation in lake trout increases with the length of the food chain.

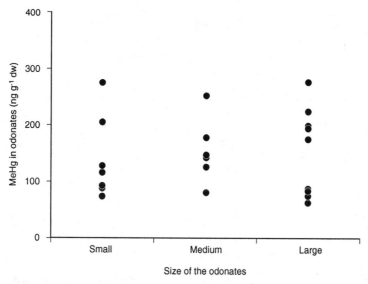

Fig. 4.6. The relationship of methylmercury concentration (ng g^{-1} dw) with size in odonates

The total Hg and MeHg concentrations in a given taxon (e.g., dipterans) vary as much as threefold between lakes (Figures 4.2; Tremblay et al. 1996ab). A comparable pattern in a study of eight Swedish lakes was observed for various insect species (Parkman and Meili 1993) and for crustacean zooplankton (Meili 1991a). In particular lakes (e.g., Lake 150, Figure 4.2), when MeHg concentrations are higher in a given taxa, they are usually also higher for all taxa. This is also true for plankton (Figure 4.3) and fish communities (Chapter 5.0). For instance, both lakes Hazeur and Jobert showed some of the highest MeHg levels in plankton (150 and

225 µm mesh, Figure 4.3) while Doyon (1995b) reported the same trend for white-fish. Similarly, Parkman and Meili (1993) and Meili (1991a) studied the food web of eight Swedish lakes and reported that the highest Hg levels in the different groups of organisms (insects, plankton and fish) were from the same two acidic-dystrophic lakes (Bottentjärn and Skårhultssjön).

Another argument in support of the similarity of the biomagnification of MeHg among ecosystems is that we observed constant ratios of mean total Hg or MeHg concentrations between given taxa in the insect community (e.g., odonates/heter-opterans, dipterans/odonates, dipterans/heteropterans, etc) (Table 4.3; Tremblay et al. 1996a). These ratios give an idea of the relative position of a taxon within the food chain and it seems that, while the MeHg concentrations may differ by a fac-tor of 2-3 (Tremblay et al. 1996b), the relationships between given taxa are the same for different lakes. This is a quite general pattern in boreal lakes since very similar ratios in the insect community were observed in our comparative study of Québec and Swedish lakes (Tremblay et al. 1996b; Chapter 9.0).

Table 4.3. Ratio of total mercury concentrations between different aquatic insect taxa from different natural lakes (modified from Tremblay 1996)

Name of the lakes	Ratio between Hg in Heteropterans and Hg in Odonates	Ratio between Hg in Dipterans and Hg in Odonates	Ratio between Hg in Dipterans and Hg in Heteropterans
Des Vœux	0.94	1.35	0.97
Detcheverry	0.97		0.89
Duncan	0.97	1.01	1.01
Evans area	0.71	1.10	1.02
Jobert	1.29	0.75	0.99
Koury	1.14		0.89
Lake 136	1.10	0.88	0.98
Lake 150	1.10	1.06	0.90
Laporte	0.96	1.26	0.88
Matagami	1.32		0.89
Opinaca area	0.72	1.14	1.14
Mean	**1.02**	**1.07**	**0.96**
Standard deviation	**0.20**	**0.19**	**0.08**

This is further supported by Meili (1991a), who reported a constant ratio of five between the total Hg level in zooplankton and in pike for eight Swedish lakes, while the total Hg concentrations varied 2-3 fold. Thus the bioavailability of MeHg, for organisms at the base of the food chain, plays a key role in determining MeHg concentrations for the whole community. It is reasonable to think that abi-otic factors could be responsible for 2-3 fold differences in bioavailablity of MeHg among lakes. These differences would lead to a 2-3 fold difference in the accumulation of MeHg in invertebrates constituing the first trophic level (i.e., the detritivores). Once MeHg has entered the food chain, our results showed that it will be transferred and amplified from one trophic level to the next and that 2-3 fold differences among lakes will still remain in top predators (Chapters 5.0 and 11.0). This tends to show that the patterns of MeHg accumulation in food chains are similar in all lakes and that the uptake of MeHg by organisms at the base of the food chain will determine the contamination in the uppermost. However, part of the variability of MeHg concentrations in invertebrates among lakes could also be due to the structure of the community (Chapter 10.0). For a given uptake at the base, top predators in long food chains will have higher MeHg levels than those in short food chains.

4.3.5
Diversity, Biomass and Mercury Burdens

In this section, we examine the adult insect community from Lake 136 and the zooplankton communities from lakes Duncan, Detcheverry, Des Vœux and 150 and we evaluate the relative proportion of MeHg, in terms of biomass and of MeHg burdens, in these groups of organisms for a generic water body with an area of 1 km^2 and a depth of 5 meters. To our knowledge, this is done for the first time, since no estimation of MeHg burden in invertebrate communities (both zooplankton and zoobenthos) are available in the literature.

A total of 24 taxa of adult insects were identified in the littoral zone of Lake 136 (Table 4.4). Chironomini dominated with 46% of the total number of taxa. *Agrypnia straminea*, Orthocladiinae, *Psectrocladius* spp., and *Procladius denticulatus* were the most abundant taxa with *Agrypnia straminea* making up 42% of the total number of individuals. In Lake 136, the mean number of emerging adult chironomids was about 75-80 Ind. m^{-2} yr^{-1}, which corresponds to the abundances reported by Vallières and Gilbert (1992) for insect larvae collected in Bob-Grant lake in eastern Québec and to those of Magnin (1977) for seven natural lakes of northern Québec.

With respect to zooplankton, the communities were dominated by *Senecella* spp., *Daphnia* spp. and *Bosmina* spp., both in biomass and in MeHg burden. For Lake 150, *Holopedium* spp. was also important. In a general characterisation of zooplankton species of 46 lakes of northern Québec, Magnin (1977) identified 46 different species and reported that *Bosmina longirostris*, *Holopedium gibberum* and *Daphnia longiremis* dominated the cladocerans. Our abundances varied from 0.7 to 11 Ind. L^{-1} (Table 4.5) and are comparable to those measured by Magnin (1977) in Nathalie and Hélène lakes in northern Québec.

Table 4.4. Abundance, biomass and mercury burden in insects collected from the littoral zone of Lake 136 during summer 1994

Taxon	Abundance (Ind. m^{-2} year^{-1})	Biomass (mg m^{-2} year^{-1} dw)	Total Hg burden (ng m^{-2} year^{-1})	MeHg burden (ng m^{-2} year^{-1})
TRICHOPTERA				
LIMNEPHILIDAE				
Asynarchus curtus	1	5	0.8	0.45
Limnephilus extractus	1	5	0.8	0.45
MOLANNIDAE				
Molanna flavicornis	2	36	5.6	3.2
POLYCENTROPODIDAE				
Agrypnia colorata	5	55	8.5	4.9
Agrypnia straminea	74	814	126.0	73.0
Neureclipsis validus?	2	10	1.6	0.9
Polycentropus cinereus	7	7	1.1	0.6
Polycentropus flavus	1	18	2.8	1.6
Polycentropus sp.	1	1	0.2	0.1
DIPTERA				
Cryptochironomus fulvus	1			
Procladius denticulatus	10	0.1	0.01	0.001
ORTHOCLADIINAE	30	1.5	0.1	0.05
Parakiefferiella spp.	11	0.2	0.01	0.001
Psectrocladius spp.	17	0.9	0.05	0.03
Others species	16			
TOTAL	179	954.0	176.05	85.3

Abundance is expressed as a mean integrated total emergence over the entire sampling season (Ind. m^{-2} year^{-1}) and the mean biomass (mg m^{-2} year^{-1} dw) was estimated by the mean dry weight of individuals (see text). We used a mean concentration of 155 ng g^{-1} dw for total mercury and 90 ng g^{-1} dw for methylmercury to calculate the mean annual burden in the the adult insects (modified from Tremblay et al. 1998b).

Based on the literature values, average larval biomass lost by emergence is estimated to be 30-35% and the average mortality due to predation is about 50-60%, leaving 5-15% for other causes of mortality (Miller 1941; Hayne and Ball 1956; Jonasson 1965). The biomass of adult insects can therefore be multiplied by 2 to estimate the burden of MeHg of the ingested insects. Zooplankton being organisms which have a short life cycle (3-4 weeks); we have thus considered five generations over the summer (Wetzel 1983) and multiplied the biomass by the same number. Considering that the measured biomass are representative of

Table 4.5. Abundance, biomass and mercury burden of zooplankton collected with a 225 μm mesh in an integrated water column (0-10 m) during fall 1994

	Duncan	Detcheverry	Lake 150	Des Vœux
Abundance (Ind. 50 L^{-1})				
Bosmina spp.	182	45	53	17
Cyclops spp.	85	3	58	30
Daphnia spp.	70	18	27	13
Diacyclops spp.			111·	3
Epischura spp.			13	
Holopedium spp.			41	
Leptodora spp.	7	5		
Senecella spp.	355	23	325	91
Others species	8	6	1	9
Total	707	100	629	154
Biomass (ng L^{-1})				
Bosmina spp.	1 300	321	379	121
Cyclops spp.	385	14	263	136
Daphnia spp.	4 682	1 204	1 806	870
Diacyclops spp.			285	8
Holopedium spp.			1 788	
Senecella spp.	3 835	248	3 511	983
Total	10 202	1 787	8 032	2 118
Total Hg burden (femto g L^{-1})				
Mean levels in the zooplankton	84	118	171	147
Bosmina spp.	10.9	3.8	6.5	1.8
Cyclops spp.	3.2	0.2	4.5	2.0
Daphnia spp.	39.3	14.2	30.9	12.8
Diacyclops spp.			4.9	0.1
Holopedium spp.			30.6	
Senecella spp.	32.2	2.9	60.0	14.5
Total	85.6	21.1	137.4	31.2
MeHg burden (femto g L^{-1})				
Mean levels in the zooplankton	43	86	99	63
Bosmina spp.	5.6	2.8	3.8	0.8
Cyclops spp.	1.7	0.1	2.6	0.9
Daphnia spp.	20.1	10.4	17.9	5.5
Diacyclops spp.			2.8	0.1
Holopedium spp.			17.7	
Senecella spp.	16.5	2.1	34.8	6.2
Total	43.9	15.4	79.6	13.5

Abundance represents the number of individuals in 50 L of water and the biomasses were determined by measuring the length of individuals and the dry weight was estimated by weight-length regressions (see text).

the shallow area of natural lakes, and despite the spatial heterogeneity, we have estimated the annual MeHg burdens of both adult insects and zooplankton for a water body with an area of 1 km^2 and a depth of 5 meters. With respect to MeHg, the insects and zooplankton would represent 853 mg km^{-2} and 48 mg km^{-2} respectively, and the insects would constitute 95% of the MeHg burden of the invertebrate community of this generic lake (Tremblay 1996; Tremblay et al. 1998ab).

Fish feeding on the benthos (whitefish or suckers) will ingest organisms at the base of the benthic food chain, mainly dipterans and trichopterans (Chapter 10.0; Doyon et al. 1996), which usually contain less than 50 ng g^{-1} dw of MeHg. On the other hand, fish feeding on the zooplankton (cisco) will ingest the bigger organisms (> 150 μm-mesh, see Chapter 10.0), at the top of the planktonic food web, which usually have MeHg levels higher than 75 ng g^{-1} dw. This indicates that fish feeding on zooplankton may be exposed to higher levels of MeHg than those feeding on aquatic insects, considering they have similar bioenergetics (Chapters 5.0 and 11.0), they would therefore show higher MeHg concentrations. This is supported by the higher total Hg concentrations observed in cisco compared to suckers or whitefish collected from natural lakes (Chapter 10.0).

4.4
Conclusions

The proportion of MeHg to total Hg in aquatic invertebrates is controlled by their feeding behavior, and increases from detritivores to predators. This effect is due to biomagnification through food web rather than to body size. The MeHg concentrations of predatory animals are partly governed by trophic structure and dynamics of the food web, the latter being quite similar among aquatic ecosystems. However, MeHg concentrations in aquatic invertebrates at low trophic levels are more dependent upon abiotic factors (such as temperature, humic substances and C content of sediments) that vary between systems, thus controlling the bioavailability of the whole food chain. The nutritive value of the ingested food, measured as growth rates of organisms, may represent an important factor influencing uptake of Hg by invertebrates at low trophic levels. Indirectly, temperature affects biological uptake of MeHg, by stimulating microbial activity responsible for the methylation of Hg. Since the biomagnification of MeHg from one trophic level to the next tends to be around 3, a higher bioavailability at the base of the food chain would lead to higher levels in predators.

Acknowledgements

I am thankful to Maxime Bégin, Bernard Caron, Louise Cournoyer, Hugo Poirier, Isabelle Rheault and Louis-Filip Richard from Université du Québec à Montréal for their assistance in the field and laboratory. The participation of the Groupe-conseil Génivar in this study is also gratefully acknowledged. This research was supported by a grant from Hydro-Québec (HQ), the Conseil de recherches en sciences naturelles et en génie (CRSNG) du Canada, and the Université du Québec à Montréal (UQAM) to the Chaire de recherche en environnement (HQ/CRSNG/UQAM).

5 Mercury in Fish of Natural Lakes of Northern Québec

Roger Schetagne and Richard Verdon

Abstract

Mercury (Hg) concentrations in fish have been documented for over 180 sampling stations located in natural lakes and rivers of northern Québec. Mean Hg concentrations were estimated for fish of standardized length and of the following species: longnose sucker (*Catostomus catostomus*), white sucker (*Catostomus commersoni*), lake whitefish (*Coregonus clupeaformis*), northern pike (*Esox lucius*), walleye (*Stizostedion vitreum*), and lake trout (*Salvelinus namaycush*). Mean concentrations for 400-mm non-piscivorous fish of various lakes, ranging from 0.05 to 0.30 mg kg^{-1}, were always well below the Canadian marketing standard of 0.5 mg kg^{-1} wet weight. For piscivorous species, concentrations often exceeded this standard, with mean levels ranging from 0.30 to 1.41 mg kg^{-1} for standardized lengths of 400 to 700 mm (depending on the species). Inter-lake variability within the same region is important for all fish species, as estimated mean concentrations often vary by factors of 3 to 4 for neighboring bodies of water. In the Nottaway-Broadback-Rupert region, where lakes and rivers display a wide range of physical and chemical properties, higher fish Hg concentrations were usually found in bodies of water with high organic content as described by color and concentrations of tannins, as well as total and dissolved organic carbon.

5.1
Introduction

As a result of the development of hydroelectric resources, numerous environmental studies have been carried out in remote regions of northern Québec since the early 1970s. Water quality and Hg levels in fish have been studied with particular emphasis; the former being for environmental impact assessment and monitoring purposes and the latter due to potential health risks to native Cree practising traditional hunting and fishing activities. Results of such studies have been documented for over 100 natural rivers and lakes lying between latitudes 49°N and 56°N.

In most of northern Québec, as there are no direct industrial sources, Hg reaches pristine rivers and lakes essentially by long range atmospheric transport. Lucotte et al. (1995a) showed that, during the last 50 to 70 years, anthropogenic atmospheric inputs have contributed to increase Hg concentrations in natural lake sediments of the area by an average factor of 2.3. Previous papers have shown that Hg concentrations in fish of pristine bodies of water of northern Québec often exceed the Canadian marketing standard of 0.5 mg kg^{-1} (Verdon et al. 1991; Langlois et al. 1995). The object of this chapter is to address the regional variability of Hg levels in fish with respect to water quality.

5.1.1
Study Area

Fish Hg concentrations have been measured in three distinct regions of northern Québec: 1) the region of the La Grande hydroelectric complex; 2) the Grande Baleine region, including the drainage basins of the Grande Baleine river, the Petite Baleine river, the À l'Eau Claire and the Nastapoka rivers and 3) the watersheds of the Nottaway, Broadback and Rupert rivers (NBR region) (Figure 5.1).

The La Grande and NBR regions are inhabited by approximately 10 400 Cree occupying four villages on the coast of James Bay and four inland villages located in the southern part of the area. The Grande Baleine region is inhabited by approximately 600 Cree and 700 Inuit located in two villages along the coast of Hudson Bay. These three regions are subdivided into trapping lots where Cree populations exploit fish and wildlife year round.

The La Grande river watershed, located between latitudes 52°N and 54°N, occupies the northern section of the James Bay Territory and covers 175 000 km². The whole region is part of the Canadian Shield, one of the oldest known geological formations composed of igneous and metamorphic rock.

As shown in Table 5.1, waters of the La Grande complex are highly transparent, of low turbidity, well oxygenated, slightly acidic, poor in mineral content, relatively rich in organic matter and poor in nutrients. Water quality over the whole region is relatively homogeneous except for slightly increasing values from east to west for the parameters describing the organic and mineral content, as well as for turbidity and nutrients, because of the increasing presence of organic deposits over marine clay.

The most common fish species of the region include longnose sucker, white sucker, lake whitefish, cisco (*Coregonus artedii*), northern pike, walleye and lake trout (DesLandes et al. 1995). Walleye and cisco are found only in the western part of the territory, while dwarf populations of lake whitefish have been found in lakes of the eastern part (Doyon et al. 1998b).

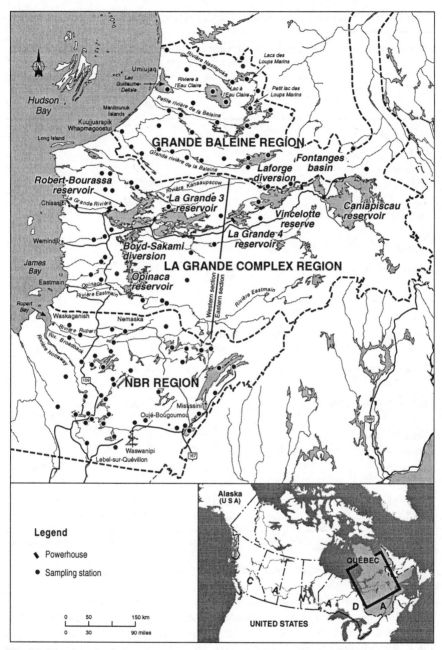

Fig. 5.1. Location of sampling stations in natural lakes and rivers of James Bay area

Table 5.1. Water quality measured in the La Grande and Grande Baleine regions

Parameters	La Grande river region (Western section)	La Grande river region (Eastern section)	Grande Baleine region (Grande and Petite Baleine basins)	Grande Baleine region (Nastapoka and À l'Eau Claire river basins)
Number of lakes (or rivers)	7	5	10	3
Turbidity (NTU)	0.4 - 3.7 [1] $\bar{X} = 1.2$ [2]	0.4 - 1.2 $\bar{X} = 0.7$	0.5 - 2.3 $\bar{X} = 1.2$	0.6 - 1.7 $\bar{X} = 1.1$
Total suspended solids (mg L^{-1})			0.4 - 2.9 $\bar{X} = 1.1$	0.4 - 1.7 $\bar{X} = 0.9$
Color (Cobalt - Pt)	21 - 55 $\bar{X} = 30$	16 - 25 $\bar{X} = 20$	10 - 32 $\bar{X} = 21$	6 - 10 $\bar{X} = 8$
Dissolved oxygen (% saturation)	91 - 105 $\bar{X} = 97$	92 - 98 $\bar{X} = 95$	96 - 110 $\bar{X} = 100$	97 - 109 $\bar{X} = 101$
Conductivity (µS cm^{-1})	12 - 36 $\bar{X} = 21$	8 - 11 $\bar{X} = 9$	12 - 34 $\bar{X} = 17$	15 – 19 $\bar{X} = 17$
pH (units)	6.4 - 7.1 $\bar{X} = 6.7$	5.9 - 6.5 $\bar{X} = 6.4$	6.5 - 6.9 $\bar{X} = 6.7$	6.7 - 6.8 $\bar{X} = 6.8$
Total organic carbon (mg L^{-1} of C)	5.2 - 11.4 $\bar{X} = 7.4$	3.9 - 6.1 $\bar{X} = 4.9$	3.5 - 5.1 $\bar{X} = 4.5$	2.5 - 3.0 $\bar{X} = 2.8$
Total dissolved carbon (mg L^{-1} of C)			2.7 - 4.3 $\bar{X} = 3.4$	1.9 - 2.3 $\bar{X} = 2.1$
Total phosphorus (µg L^{-1} of P)	7 - 14 $\bar{X} = 9$	4 - 6 $\bar{X} = 5$	5 - 9 $\bar{X} = 6$	5 - 7 $\bar{X} = 6$
Tannins (mg L^{-1})	0.9 - 2.1 $\bar{X} = 1.2$	0.6 - 0.9 $\bar{X} = 0.7$	0.4 - 1.0 $\bar{X} = 0.7$	0.3 - 0.4 $\bar{X} = 0.4$
Chlorophyll *a* (µg L^{-1})	1.1 - 2.4 $\bar{X} = 1.6$	1.0 - 1.7 $\bar{X} = 1.3$	0.5 - 1.5 $\bar{X} = 1.1$	0.3 - 1.1 $\bar{X} = 0.6$

[1] Range of mean values obtained during the ice-free period (generally 6 to 12 samples per period).

[2] $\bar{X}$ = Overall mean values per region, during the ice-free period.

The Grande Baleine region extends between latitudes 54°N and 56°N and is located on the eastern side of Hudson Bay. Its watershed covers a total area of approximately 70 000 km². Water quality of this region is very similar to that of the La Grande region (Table 5.1), as both regions overlie the same geologic formations and have been subjected to the same glacial activity. The same main fish species abound except for the absence of walleye and the scarcity of cisco.

The NBR region is located between latitudes 49°N and 52°N and its watershed covers an area of approximately 130 000 km² in the southern section of the James Bay Territory. In this region, characterized by the Canadian Shield to the east and marine and post-glacial lacustrine clays to the west, five different water quality types have been identified by multivariate analytical methods (Table 5.2, adapted

Table 5.2. Water quality measured in the NBR region according to water types

Water quality types	Type 1	Type 2	Type 3	Type 4	Type 5
Number of lakes (or rivers)	8	10	12	21	4
Turbidity (NTU)	0.4 - 2.8 [1] $\bar{X} = 1.0$ [2]	0.5 - 5.7 $\bar{X} = 0.9$	1.1 - 9.2 $\bar{X} = 6.2$	2.1 - 42 $\bar{X} = 9.6$	25.8 - 65 $\bar{X} = 42.0$
Total suspended sediments (mg L⁻¹)	< 0.5 - 5.5 $\bar{X} = 1.6$	0.8 - 7.0 $\bar{X} = 1.3$	1.1 - 8.9 $\bar{X} = 4.4$	2.6 - 15.2 $\bar{X} = 6.2$	13.5 - 38 $\bar{X} = 22.8$
Color (Pt-cobalt)	24 - 39 $\bar{X} = 31$	21 - 71 $\bar{X} = 39$	33 - 93 $\bar{X} = 62$	63 - 227 $\bar{X} = 108$	78 - 153 $\bar{X} = 121$
Conductivity (μS cm⁻¹)	19 - 26 $\bar{X} = 23$	10 - 36 $\bar{X} = 13$	17 - 28 $\bar{X} = 22$	23 - 76 $\bar{X} = 37$	38 - 187 $\bar{X} = 100$
pH (units)	7.0 - 7.2 $\bar{X} = 7.1$	5.9 - 6.6 $\bar{X} = 6.3$	6.2 - 7.0 $\bar{X} = 6.8$	6.3 - 7.2 $\bar{X} = 6.9$	6.8 - 7.4 $\bar{X} = 7.2$
Total organic carbon (mg L⁻¹ of C)	4.1 - 5.2 $\bar{X} = 4.5$	3.9 - 8.0 $\bar{X} = 5.3$	6.3 - 9.9 $\bar{X} = 8.0$	8.6 - 18.2 $\bar{X} = 11.6$	10.2 - 14.9 $\bar{X} = 12.5$
Total dissolved carbon (mg L⁻¹ of C)	3.3 - 4.4 $\bar{X} = 3.8$	3.2 - 7.0 $\bar{X} = 4.5$	5.6 - 8.7 $\bar{X} = 7.2$	7.8 - 16.2 $\bar{X} = 10.6$	9.3 - 15.4 $\bar{X} = 11.9$
Total phophorus (μg L⁻¹ of P)	3 - 8 $\bar{X} = 4$	4 - 11 $\bar{X} = 5$	10 - 18 $\bar{X} = 14$	10 - 40 $\bar{X} = 19$	24 - 55 $\bar{X} = 45$
Chlorophyll a (μg L⁻¹)	1.1 - 1.8 $\bar{X} = 1.3$	1.4 - 2.3 $\bar{X} = 1.8$	1.4 - 3.2 $\bar{X} = 2.2$	0.5 - 3.0 $\bar{X} = 1.7$	1.1 - 2.6 $\bar{X} = 1.9$
Tannins (mg L⁻¹)	0.7 - 1.1 $\bar{X} = 0.9$	0.6 - 2.0 $\bar{X} = 1.2$	1.1 - 2.5 $\bar{X} = 1.7$	1.7 - 4.9 $\bar{X} = 2.6$	2.5 - 3.7 $\bar{X} = 3.1$

[1] Range of mean values obtained during the ice-free period (generally 6 to 12 samples per period).
[2] $\bar{X}$ = Overall mean values per region (during the ice-free period).

from SOMER Inc. 1994). From type 1 to type 5, waters become richer in major ions, suspended sediments and nutrients, as well as in organic material (as shown by increasing total and dissolved organic carbon, tannins and water color). Local sources of pollution are restricted to a few pulp and paper mills in the Nottaway watershed in the southern part of the NBR region and to a chlor-alkali plant that, until the late 1970s, has discharged significant quantities of Hg into a tributary river. In addition to the fish species found in the La Grande region, the following are also common in the southern part of the NBR region: lake sturgeon (*Acipenser fulvescens*), sauger (*Stizostedion canadense*), goldeye (*Hiodon alosoides*) and mooneye (*Hiodon tergisus*).

5.2
Materials and Methods

Fish from over 180 stations located in lakes and rivers were sampled, for the most part, between 1987 and 1994 using experimental and single mesh size gill nets (mesh size ranging from 2.5 to 10 cm). A few lakes in the NBR region were sampled in 1976 with larger mesh sizes. For every fish species, the aim was to obtain 30 specimens evenly distributed in pre-selected length classes. For example, six specimens for each of the five 100-mm classes from 100 to 600 mm of length were selected for lake whitefish. A total of more than 8 000 fish were analyzed for Hg using samples of approximately 10 g of flesh. These samples were preserved frozen until analysis by cold vapor atomic absorption spectrophotometry (as described in Chapter 2.0). Concentrations are expressed in mg kg^{-1}, wet weight. The standard quality control procedures used by the analytical laboratory were enhanced by the addition of blind triplicates taken on ten percent of all fish sampled. The mean annual variation coefficients for the blind triplicates ranged from 5.2 to 9.2% over the 1986 to 1994 period.

The statistical approach used to follow the spatio-temporal evolution of fish muscle Hg concentrations is based on polynomial regression analysis with indicator variables, as described in Tremblay et al. (1998c). This method allows for the calculation of linear or curvilinear models to express muscle total Hg concentration (mg kg^{-1}, wet weight) as a function of a polynomial of total length (usually of the second order) such as:

$$[Hg] = a + b \times length + c \times length^2, \text{ where a, b and c are constants.}$$

It also allows for the rigorous statistical comparison of a number of sets of data pertaining to the shape and the position of the curve relating Hg to fish length, as well as pertaining to the estimated mean concentration for a standardized fish length.

The following standardized lengths were established for the main species, typically representing the mean length captured (the average weight for the corresponding standardized length is given in parentheses):

longnose sucker	: 400 mm (0.73 kg)	northern pike	: 700 mm (2.13 kg)
lake whitefish	: 400 mm (0.64 kg)	walleye	: 400 mm (0.63 kg)
		lake trout	: 600 mm (1.71 kg)

Water quality parameters were determined on integrated samples of the photic zone, taken from the surface to a depth corresponding to 2% of the surface light. Samples were taken once a month during the ice-free period, from June to October of each sampling year. Analytical and conservation methods complied with APHA's standard methods (APHA-AWWA-WPCF 1989). Principal component and clustering analyses, described in Legendre and Legendre (1984), were performed on data to classify bodies of water into 5 water quality types (SOMER Inc. 1994).

5.3
Results

For more than 250 available data sets, fish Hg concentrations nearly always correlated with total length. The typical relationship between Hg and length is slightly curvilinear with higher concentrations found in larger fish. As shown in Figure 5.2, concentrations in individual fish are often quite variable for a given size (in this case, for lake trout of Lac Hazeur, in 1987).

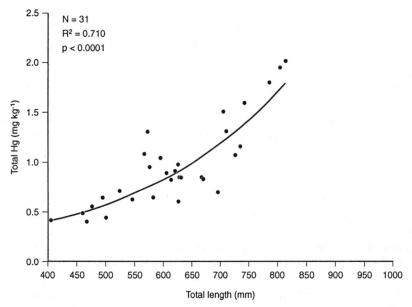

Fig. 5.2. Typical relationship between mercury and fish length observed in natural lakes (lake trout, Lac Hazeur, 1987)

For 5 lakes for which data are available for a number of years, no temporal trend was observed for all species monitored during periods of up to 12 years. For example, Figure 5.3 shows no significant differences ($p < 0.05$) in estimated mean concentrations obtained between 1984 and 1996, for standardized-length lake whitefish of Rond-de-Poêle lake (as 95% confidence intervals all overlap).

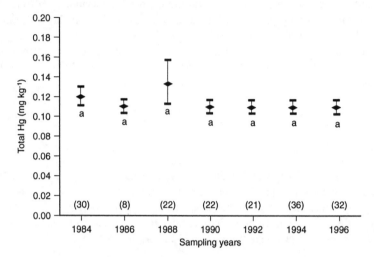

Note : Bars indicate 95% confidence intervals.
Sampling years with identical letters indicate that mean mercury concentrations are not significantly different as 95% confidence intervals overlap.
Number of fish per sample is given in parenthesis.

Fig. 5.3. Estimated mean mercury concentrations for standardized-length (400 mm) lake white-fish in a natural lake (Lac Rond-de-Poêle) over a 12 year period

Inter-lake variability in fish Hg concentrations was important for the lakes and rivers studied. Figure 5.4 shows mean concentrations and 95% confidence intervals estimated for lake whitefish for different lakes sampled in the western section of the La Grande region.

Estimated mean concentrations often varied significantly from one lake to another (in Figure 5.4, when confidence intervals do not overlap, p < 0.05). In this case, seven lakes yield the exact same values because their data sets could be described by the same polynomial equation, with the shape and position of each of their Hg-to-length relationships not being significantly different (p < 0.05).

Table 5.3 shows the range of estimated mean Hg concentrations for standardized length obtained for the most common species in the lakes and rivers sampled in the different regions of the James Bay Territory. As in the example of Figure 5.4, Table 5.3 also shows that Hg concentrations in fish of pristine bodies of water of the area vary greatly from one lake (or river) to another within the same region. Estimated mean concentrations for standardized length often vary by factors of 3 to 4 for neighboring bodies of water, as in the case of walleye in the western part of the La Grande region for which they vary from 0.30 to 1.02 mg kg^{-1}.

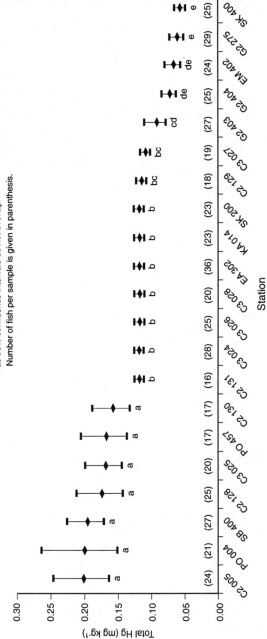

Fig. 5.4. Spatial distribution of estimated mean mercury concentrations for standardized-length (400 mm) lake whitefish in natural lakes (La Grande region, Western section)

Table 5.3. Range of mean mercury concentrations (mg kg^{-1} ww) measured in fish of standardized length in the study area

Fish species	Region	Nottaway Broadback Rupert	La Grande river (Western section)	La Grande river (Eastern section)	Grande Baleine region
	Latitude	49 - 52°N	52 - 54°N	52 - 54°N	54 - 56°N
Longnose sucker (400 mm)	Number of fish	81	182	282	284
	(Number of lakes)	(4)	(7)	(10)	(11)
	Range of [Hg] [1]	0.09 - 0.34	0.12 - 0.22	0.07 - 0.20	0.13 - 0.30
	Overall mean [Hg] [2]	0.19	0.12	0.12	0.16
	(confidence interval)	(0.17 - 0.21)	(0.11 - 0.15)	(0.10 - 0.13)	(0.14 - 0.17)
Lake whitefish (400 mm)	Number of fish	1 025	503	254	309
	(Number of lakes)	(38)	(21)	(9)	(18)
	Range of [Hg]	0.07 - 0.36	0.05 - 0.20	0.10 - 0.30	0.08 - 0.27
	Overall mean [Hg]	0.16	0.13	0.17	0.16
	(confidence interval)	(0.15 - 0.16)	(0.11 - 0.15)	(0.16 - 0.19)	(0.14 - 0.17)
Northern pike (700 mm)	Number of fish	873	373	145	23
	(Number of lakes)	(32)	(20)	(5)	(2)
	Range of [Hg]	0.33 - 1.81	0.30 - 0.93	0.36 - 0.92	0.45
	Overall mean [Hg]	0.88	0.64	0.58	0.45
	(confidence interval)	(0.86 - 0.91)	(0.60 - 0.68)	(0.55 - 0.61)	(0.33 - 0.59)
Walleye (400 mm)	Number of fish	1 356	353		
	(Number of lakes)	(46)	(13)		
	Range of [Hg]	0.34 - 1.41	0.30 - 1.02		
	Overall mean [Hg]	0.75	0.62		
	(confidence interval)	(0.73 - 0.77)	(0.59 - 0.66)		
Lake trout (600 mm)	Number of fish		140	285	335
	(Number of lakes)		(7)	(11)	(18)
	Range of [Hg]		0. 23 - 0.89	0.52 - 0.91	0.18 - 1.02
	Overall mean [Hg]		0.57	0.69	0.68
	(confidence interval)		(0.51 - 0.63)	(0.64 - 0.75)	(0.66 - 0.73)

[1] Range of mean Hg estimated for standardized length in the different lakes (or rivers) of each region.

[2] Overall estimated mean [Hg] and 95% confidence interval for standardized length in each region with fish of all water bodies pooled as one data set.

Hg concentrations measured for non-piscivorous species, such as longnose sucker and lake whitefish, were always well below the Canadian marketing standard of 0.5 mg kg^{-1}. In piscivorous species, concentrations for standardized lengths often exceeded this standard, as shown in Table 5.3.

For comparison purposes, the overall mean values given in Table 5.3 correspond to the estimated mean concentration, at the standardized length, for all the fish of a given species sampled in a region and treated as a single data set. Although Table 5.3 indicates some significant ($p < 0.05$) differences in the mean concentrations for each region calculated in this manner, with the highest values often corresponding to the NBR region, there is no consistent trend in fish Hg concentrations with latitude for the different species: concentrations in longnose sucker and lake whitefish show no south to north trend; concentrations in northern pike and walleye seem to decrease from south to north; and concentrations in lake trout seem to increase from south to north.

5.4
Discussion

As previously mentioned, there are no direct industrial sources of Hg to the James Bay Territory. A chlor-alkali plant, however, was in operation from 1967 until 1978 at Lebel-sur-Quévillon, in the southernmost part of the NBR region. During this period, Hg was discharged into the Bell river which flows into the Nottaway river approximately 150 km downstream. With this exception, the principal source of Hg to these pristine rivers and lakes is long range atmospheric transport (Lucotte et al. 1995a). Since the main atmospheric sources of Hg to the study area come from the south or southwest (Delisle 1978; Ouellet and Jones 1982), one would suspect that Hg concentrations in fish would decrease from the southern part (NBR region) to the northern part of the territory (Grande Baleine region). A number of authors have found such a trend for Hg concentrations in lake sediments (e.g., Nater and Grigal 1992 for the central U.S.; Lathrop et al. 1991 for southern Ontario; Iverfeldt 1991 and Johansson et al. 1991 for southern Scandinavia). As shown in Table 5.3, however, this is not evident for fish Hg levels in northern Québec. This absence of a clear latitudinal gradient in fish Hg concentrations is consistent with the results of Lucotte et al. (1995a) who found that the distribution of Hg concentrations in lake sediments, normalized for their organic carbon content, was clearly independent of latitude, along a 1 200 km transect (between latitudes 45°N and 55°N), across the NBR, La Grande and Grande Baleine regions (Chapter 3.0).

As shown in Table 5.3, the ranges of concentrations obtained in the different regions are extensive and overlap considerably, demonstrating important inter-lake variability within a region and for all fish species. Such inter-lake variability is a common feature, as shown by the wide range of concentrations reported by Strange et al. (1991) and Strange (1993) for the Southern Indian Lake region of Manitoba (from 0.22 to 0.45 mg kg^{-1} for 400-mm walleye). Furthermore, Driscoll et al. (1995) reported that average concentrations for 2+ year old yellow perch from 13 Adirondack lakes varied from 0.11 to 0.78 mg kg^{-1}.

A number of factors, other than distance from pollution sources, have been suggested by several authors to explain Hg concentrations in fish of pristine remote lakes. For our study area, water quality seems to be an important factor regulating Hg concentrations in fish. As indicated in the section relating to the study area, five different water quality types have been identified in the NBR region. Table 5.4 presents the range of estimated mean Hg concentrations for standardized-length fish obtained for each lake (or river), as well as the average of these values calculated for each water quality type. For all species, average Hg concentrations increase from water quality type 1 to type 5. As shown in Table 5.2, a similar trend is observed for turbidity, suspended sediments, color, conductivity, as well as concentrations of total and dissolved organic carbon, tannins and total phosphorous.

Table 5.4. Range of mean mercury concentrations (mg kg^{-1} ww) measured in fish of standardized length in the different water types identified in the NBR region

Species of fish		Water quality types				
		Type 1	Type 2	Type 3	Type 4	Type 5
Longnose sucker (400 mm)	Number of fish	30		36		
	(Number of lakes)	(2)		(2)		
	Range of [Hg] [1]	0.09 - 0.12		0.13 - 0.34		
	Average [Hg] [2]	0.11		0.24		
White sucker (400 mm)	Number of fish	288	47	152	206	11
	(Number of lakes)	(8)	(4)	(7)	(14)	(1)
	Range of [Hg]	0.07 - 0.16	0.07 - 0.11	0.12 - 0.19	0.13 - 0.29	0.18
	Average [Hg]	0.11	0.09	0.16	0.18	
Lake whitefish (400 mm)	Number of fish	200	91	136	318	13
	(Number of lakes)	(8)	(5)	(7)	(13)	(1)
	Range of [Hg]	0.08 - 0.12	0.07 - 0.16	0.11 - 0.20	0.12 - 0.36	0.21
	Average [Hg]	0.10	0.12	0.16	0.20	
Northern pike (700 mm)	Number of fish	127	53	220	291	28
	(Number of lakes)	(5)	(3)	(9)	(10)	(1)
	Range of [Hg]	0.33 - 0.73	0.55 - 0.73	0.51 - 1.42	0.73 - 1.81	1.56
	Average [Hg]	0.50	0.64	0.94	1.12	
Walleye (400 mm)	Number of fish	202	108	281	331	30
	(Number of lakes)	(7)	(4)	(9)	(13)	(1)
	Range of [Hg]	0.34 - 0.73	0.55 - 0.84	0.65 - 1.15	0.62 - 1.41	0.84
	Average [Hg]	0.53	0.67	0.84	0.95	

[1] Range of estimated mean concentrations for fish of standardized length for each lake (or river).

[2] Average of estimated mean concentrations for fish of standardized length for each water quality type.

Color, total and dissolved organic carbon (DOC) and tannins characterize the organic content of water which, by playing a role in the transport and availability of Hg (due to the stimulation of bacterial activity), is generally recognized as a major factor for elevated MeHg levels in fish (Verta 1984; Suns et al. 1987; Bowlby et al. 1988; McMurtry et al. 1989; Lee and Hultberg 1990; Haines et al. 1994).

The positive relationships shown in Table 5.4 between Hg concentrations in fish and the values of total phosphorus and conductivity seem surprising considering that a number of studies have shown higher fish Hg concentrations in oligotrophic lakes than in eutrophic ones (Huckabee et al. 1979; Håkanson et al. 1990a; Lange et al. 1993; Driscoll et al. 1995). Table 5.2 however shows that in the NBR region total phosphorus is not a good indicator of primary productivity, as chlorophyll a concentrations do not follow phosphorus concentrations. In this area characterized by the presence of post-glacial lacustrine clays, turbidity and suspended sediments have been positively correlated with total phosphorus (SOMER Inc. 1994).

The positive relationship between Hg levels in fish and the parameters of turbidity and suspended sediments is also surprising considering that Rudd and Turner (1983) suggested that the addition of clay-rich suspended sediments would render the Hg less bioavailable as a result of adsorption.

Although an apparent correlation between fish Hg levels and total phosphorus, conductivity, turbidity or suspended sediments is suggested by Tables 5.2 and 5.4, there is probably no cause and effect relationship between Hg in fish and these water quality parameters in the NBR region. Because of the nature of the surficial material of the area, where post-glacial lacustrine clays are covered with numerous peat bogs, these 4 parameters tend to be higher in lakes with high organic content. This particularity explains both the high organic content (due to peat bogs) and the high values of phosphorus, conductivity, turbidity and suspended sediments of surface waters (because of the clay). Thus, for this region, high Hg levels in fish would only be truly related to high organic content.

Hg content can also be linked to pH and Hg contamination of water systems (e.g., Håkanson 1980; Driscoll et al. 1995). The latter, among many others, found a negative correlation between pH and Hg concentrations in 1-kg pike (r = -0.61). Acidic waters would have a greater capacity to sustain monomethylmercury compounds. As shown in Tables 5.2 and 5.4, fish Hg levels do not seem to be related to water pH in the NBR region. This may again be due to the nature of the surficial material, as humic acids related to high organic content coming from peat bogs are compensated by the underlying clays which induce neutral or near neutral pH.

Many other factors have been identified as regulating fish Hg levels. Although they have not been investigated in our study, they may contribute to the great inter-lake variability observed. Driscoll et al. (1995) found that inputs of other metals, such as aluminum, which complex with dissolved organic carbon, appear to decrease the binding of MeHg with organic ligands, increasing its bioavailability. Thus, aluminum content appeared to be a better indicator of fish Hg concentrations than either pH or dissolved organic carbon alone for Adirondack lakes. They also found that hypolimnetic anoxia is another factor that may influence Hg bioavailability.

Bodaly et al. (1993), using data from six Ontario lakes, concluded that Hg concentrations in fish is inversely related to lake size. They suggested that higher water temperature in smaller lakes during the open-water season influenced methylation/demethylation ratios and were the cause of higher fish Hg levels. Håkanson et al. (1990b) also concluded that lake area is negatively correlated with fish Hg concentrations for Swedish lakes. Finally, Kelly et al. (1997) found that the Hg methylation process is stimulated by increased temperature to a greater degree than is the general microbial activity. These findings suggest that lakes in milder environments or small lakes, with higher summer temperatures, may lead to more elevated Hg levels in fish.

Björnberg et al. (1988) reviewed the possible mechanisms regulating the bioavailability of Hg in natural waters. They reported that the following factors are generally recognized as important: pH, lake trophic level, humic content, conductivity, calcium, oxygen conditions, Hg input, as well as zinc and selenium content. No single factor alone can explain inter-lake variability, but taken together, they may be good indicators of Hg in fish flesh.

Estimated mean Hg concentrations for fish of standardized length obtained for the study area are similar to those measured for other Canadian Shield lakes of the province of Québec (Schetagne et al. 1996), but are higher than those found in the Southern Indian Lake area of northern Manitoba. For example, estimated mean concentrations for 400-mm lake whitefish vary from 0.05 to 0.30 mg kg^{-1} in northern Québec, compared to 0.02-0.07 mg kg^{-1} for 350-mm specimens in northern Manitoba. For 400-mm walleye, corresponding concentrations range from 0.30 to 1.41 mg kg^{-1} in northern Québec, compared to 0.22 to 0.45 mg kg^{-1} in northern Manitoba (Strange et al. 1991; Strange 1993; Strange 1995). These differences may likely be explained by factors such as differences in regional geomorphology, in fish growth rates or in Hg inputs.

In addition to all of the preceding factors, which are mostly related to physical or chemical properties, several authors have identified biological factors which may also play a major role in regulating fish Hg levels. Since the primary source of Hg to fish is their food (e.g., Kelly et al. 1997), diet must play a key role

(Mathers and Johansen 1985; Allard and Stokes 1989; Lindqvist 1991). Furthermore, for a given lake, both the structure of the fish community and the length of the food chain must be important factors as Hg concentrations generally increase by a factor of about 3 from one trophic level to the next (Meili 1991b; Cabana et al. 1994; Lucotte et al. 1995b; Kidd et al. 1995; Chapter 4.0).

According to a number of authors, fish with high growth rates generally have less Hg than those with low growth rates, as a result of the biodilution phenomenon (Håkanson et al. 1990a; Verta 1990b; Wiener et al. 1990; Lathrop et al. 1991; Lange et al. 1993).

Our data also show wide concentration distributions in individual fish of the same size from a given lake. In Hazeur lake (Figure 5.2), lake trout of about 575 mm (567-580 mm), captured the same year, show concentrations ranging from 0.67 to 1.28 mg kg^{-1}. Individual physiology, growth rate or feeding habit, may explain such high indivivual variability. For instance, data from Chapter 10.0 show that pike from the western part of the La Grande region prey on a wide variety of fish (cisco, suckers, lake whitefish, walleye, pike, burbot, lake chub, sticklebacks, trout-perch, yellow perch and sculpins).

5.5
Conclusions

Hg concentrations in fish of natural lakes of northern Québec are relatively high compared to other regions of Canada. Concentrations measured for non-piscivorous species, such as longnose sucker and lake whitefish, are always well below the Canadian marketing standard of 0.5 mg kg^{-1}. In piscivorous species, concentrations for standardized lengths often exceed this standard.

Although long range atmospheric transport is an important source of Hg to the study area, where there is no direct industrial pollution, fish Hg concentrations show no clear latitudinal gradient between latitudes 49°N and 54°N, despite the fact that most atmospheric sources originate from the south or southwest. Furthermore, no temporal trend in fish Hg levels was observed for all species monitored in 5 natural lakes, over periods of up to twelve years.

Inter-lake variability within a region is important for all fish species, as estimated mean concentrations for standardized length often vary by factors of 3 to 4 for neighboring bodies of water. In the NBR region, where lakes and rivers show a wide range of physical and chemical properties, higher fish Hg concentrations are found in lakes with high organic content, as described by color and concentrations of tannins, as well as of total and dissolved organic carbon. In the James Bay Territory, inter-lake variability must be taken into account when determining consumption advisories for both native Cree practising subsistence fishing and for sport fishermen.

Acknowledgements

Funding for this research was provided by Hydro-Québec, the Société d'énergie de la Baie James and the James Bay Mercury Committee. We are grateful to Judit Vass of Corporation des services analytiques Philip for the quality control of the laboratory analyses, Jean-Louis Fréchette for the water quality sampling and several employees of Groupe-conseil Génivar who carried out the field work and the statistical analyses.

6 Mercury in Birds and Mammals

René Langis, Claude Langlois and François Morneau

Abstract

As part of the environmental impact studies of the Grande Baleine and the Notta-way–Broadback–Rupert (NBR) hydroelectric projects, Hydro-Québec has col-lected numerous data on the levels of mercury in wildlife from both regions be-tween 1989 and 1991. The collection of specimens was conducted with the col-laboration of the local Cree trappers and Inuit hunters. Species sampled included birds (waterfowl and Gulls), terrestrial mammals (marten, mink, ermine, fox, hare, and caribou) and aquatic mammals (freshwater and marine seals, belugas). Most laboratory analyses were done both on flesh and liver, but some were conducted on other tissues, such as feathers, eggs, blubber, kidney or brain. The results indi-cate that total mercury in flesh may reach high levels (> 0.5 mg kg^{-1}) in piscivorous birds and mammals (terrestrial and aquatic) from these pristine areas. The levels may be very high (> 20 mg kg^{-1}) in some tissues, such as the beluga and freshwater seal livers, but important food resources such as Canada geese, caribou and hare were determined to have mercury levels lower than the Canadian marketing stan-dard of 0.5 mg kg^{-1}.

6.1
Introduction

The transfer of mercury from aquatic systems to predatory birds and mammals, including humans, can be almost entirely attributed to the consumption of methyl-mercury-contaminated fish (Rodgers 1994). In northern Québec, mercury present in undeveloped areas can reach high concentrations in piscivorous fish species (Chapters 5.0 and 11.0) as well as in birds and mammals (Langlois et al. 1995). These concentrations have often been shown to exceed the Canadian marketing standard of 0.5 mg kg^{-1} established for fish and fish products by Health and Wel-fare Canada (1985).

Relatively little information is available on mercury levels in birds and land mammals in northern ecosystems, nor in marine mammals of northwestern Québec. The purpose of this study was to assess patterns of mercury contamination among several waterfowl and other aquatic bird species taking into consideration the types of tissues, their different feeding habits and breeding habitats. Waterfowl is an important part of the local native diet, especially during the spring and fall bird migration, of Cree and Inuit people living along the eastern shores of James Bay and Hudson Bay (Native Harvesting Research Committee 1982). In addition, mercury burdens in several terrestrial and marine mammals are considered in relation to their use as part of the native diet of the Cree and Inuit communities.

This chapter presents total mercury concentrations measured in twenty species of piscivorous and non-piscivorous birds collected in the Grande Baleine and NBR regions, as well as in six terrestrial and four marine mammal species from the Grande Baleine watershed. This study is part of a comprehensive investigation initiated by Hydro-Québec in 1991 to establish baseline levels of organic and inorganic contaminants in natural and pristine areas, prior to hydroelectric development, thus permitting subsequent monitoring of changes.

6.2
Study Area

The study area includes the Grande Baleine region, located between latitudes 54°N and 56°N on the Hudson Bay coast, and the NBR region, located between latitudes 49°N and 52°N in the James Bay Territory. The two zones cover areas of 70 000 km² and 130 000 km², respectively, and are inhabited by approximately 7 600 Native people (6 550 Cree and 750 Inuit) (Figure 5.1). The non-native population of the region is estimated to be 25 000, as indicated by information concerning the number of hunters, anglers and tourists who visited the area in 1990 (Nobert et al. 1992).

The two regions present important differences in physiography. For instance, till forms most of the superficial deposits of the region of Grande Baleine. Peat bogs cover less than 5% of the area of Grande Baleine, except for localized areas in the east and west ends of the region, where the highest density of peat bogs (> 10% of the area) is found. The NBR region is defined by three large zones according to the erodibility of superficial deposits: a zone of clayey plains, a zone of till situated between 80 and 290 m in altitude and a zone of hills at over 315 m in altitude. Peat bogs are prevalent in the lower reaches of the three main watersheds (Nottaway, Broadback, Rupert). Clay deposits, superficial drainage and the slightly inclined relief favors their development. Bog coverage varies from approximately 15% in the more elevated areas to 75% in the lower Rupert Bay.

Despite the relatively low densities of breeding waterfowl (Bordage 1988; Savard and Lamothe 1991) in the boreal and subarctic regions of Québec, several of these species breed in these regions (Gauthier-Guillemette–GREBE 1990, 1992a). The most abundant species include the Canada Goose (*Branta canadensis*), the Green-winged Teal (*Anas crecca*), the American Black Duck (*A. rubripes*), the Surf Scoter (*Melanitta perspicillata*), the Ring-necked Duck (*Aythya collaris*), the Greater Scaup (*A. marila*), the Lesser Scaup (*A. affinis*) and the Common Goldeneye (*Bucephala clangula*). A few other species, such as the Hooded Merganser (*Lophodytes cucullatus*), use the area for molting. Other aquatic bird species that live in the region include the Common Loon (*Gavia immer*), the Herring Gull (*Larus argentatus*), the Arctic Tern (*Sterna paradisaea*) and the Common Tern (*Sterna hirundo*) (Gauthier-Guillemette–GREBE 1990, 1992bc).

The Cree and Inuit diet includes an important portion of wild birds and mammals. Favorite bird species include the Canada Goose, the Ptarmigan (*Lagopus lagopus, L. mutus*), and several species of ducks, while the preferred terrestrial mammals are caribou (*Rangifer tarandus*), hares (*Lepus americanus, L. arcticus*), marten (*Martes americana*), and mink (*Mustela vison*), with the Inuit of Grande Baleine hunting sea mammals such as seals and beluga (*Delphinapterus leucas*).

6.3
Materials and Methods

Most bird and mammal specimens from the Grande Baleine region were collected with the collaboration of local Cree and Inuit hunters. All animals were collected between June and August 1989 and 1990. With the exception of marine mammals, the samples were preserved whole. Birds and terrestrial mammals were from several stations spread over the Grande Baleine study area (SOMER Inc. 1993).

Eggs from Herring Gulls, Great Black-backed Gulls (*Larus marinus*), Common Eider (*Somateria mollissima*), Black Guillemots (*Cepphus grylle*) and Arctic Tern were collected along the coast between the Manitounuk Islands and Long Island, a marine environment on the Hudson Bay coast.

Marine mammals were captured in Guillaume-Delisle lake and along the coast between the Manitounuk Islands and Long Island. In addition, three freshwater seals were captured accidentally, one in a gill net at Loups-Marins lake and two were beached on the shore of the Nastapoka river.

Birds of the NBR area were collected in 1990 and 1991. Between June and August 1990, birds were collected throughout the NBR area in collaboration with local Cree hunters. In the spring and fall of 1991, the collection was limited to the area of Rupert Bay and the specimens were collected by Cree hunters. Several specimens were grouped into composite samples (by species, sex and developmental stage when possible) to minimize cost and to secure critical amounts for other analyses and quality control. All bird specimens were shot.

Breast muscle, liver and feathers were removed and prepared for analysis in the laboratory. All appropriate standard precautions to avoid cross-contamination were followed (SOMER Inc. 1992). Total mercury analyses were performed using cold vapour atomic absorption spectrometry, as described in Chapter 2.0. The detection limit was 0.05 mg kg^{-1} for all muscle and liver tissues but depending on apparatus adjustments, reached 0.50 mg kg^{-1} in feathers (these prove, however, to be adequate for most samples since levels were well above detection limits). The quality control program included triplicate analyses, spiked and certified reference samples and inter-laboratory control. In this paper, all results are presented in mg of total mercury per kg of tissue, wet weight (ww).

Arithmetic means are used to present the results of total mercury concentrations. Values below the detection limit were assigned half that value (i.e., 0.025 mg kg^{-1} for muscle and liver samples) for all statistical calculations. Correlation coefficients were calculated using the non-parametric Kendall method while populations were compared with the Wilcoxon 2-sample test, since normality of the data distributions could not be assumed. All statistical tests were computed with SAS computer software.

6.4
Results and Discussion

6.4.1
Mercury in Birds

Positive total mercury levels (> detection limit) were found in at least some tissue samples from all bird species collected in both study areas (Table 6.1). In several species, total mercury levels of edible tissue samples (muscle and liver), for both areas, exceeded the Canadian marketing standard of 0.5 mg kg^{-1} (Health and Welfare Canada 1985). In either one or the other or both study areas, all bird species, except for the Canada Goose, had liver mercury levels which exceeded the marketing standard. Muscle tissue samples in all piscivorous species and some non-piscivorous species (i.e., Common Goldeneye and Black Scoter) also contained total mercury levels in excess of the marketing standard. All muscle and liver samples of the Common Merganser exceeded the marketing standard.

Total mercury levels found in feathers or livers of piscivorous birds indicate that these animals seem capable of tolerating relatively high total mercury burdens.

Table 6.1. Total mercury (mg kg^{-1} wet weight) in bird tissues collected from the Grande Baleine and NBR study areas (S.D. = Standard deviation, n = number of samples analyzed)

Species	Grande Baleine		NBR	
	Mean	S.D./n	Mean	S.D./n
Canada Goose				
Muscle	0.053	0.069/33	< 0.05	- /29
Liver	0.083	0.061/35	0.086	0.068/24
Feathers	< 0.5	- /35	< 0.5	- /24
American Black Duck				
Muscle	0.193	0.091/22	0.204	0.128/40
Liver	0.621	0.396/22	0.566	0.390/39
Feathers	1.441	0.895/23	1.751	1.174/39
Green-winged Teal[1]				
Muscle	0.138	0.095/41	0.330[**]	0.148/8
Liver	0.331	0.240/40	0.824[**]	0.339/8
Feathers	1.174	0.785/41	1.294	1.357/8
Mallard[2]				
Muscle	0.167	0.057/3	0.129	0.078/14
Liver	0.530	0.215/3	0.359	0.281/11
Feathers	1.500	0.361/3	0.855	0.484/11
Greater Scaup				
Muscle	0.239	0.235/30	0.224	0.128/7
Liver	0.495	0.201/30	0.751	0.366/10
Feathers	1.773	0.816/30	1.915	1.401/10
Ring-necked Duck				
Muscle	0.151	0.079/11	0.184	0.108/33
Liver	0.392	0.309/12	0.584[*]	0.325/35
Feathers	1.273	0.418/12	1.509	0.788/34
Common Goldeneye				
Muscle	0.645	0.105/4	0.366	0.215/31
Liver	3.130	1.817/4	1.544	1.791/32
Feathers	3.050	1.139/4	2.803	1.271/32
Black Scoter[3]				
Muscle	0.244	0.142/29	0.710	- /1
Liver	0.787	0.652/30	0.850	- /1
Feathers	1.670	0.735/30	1.200	- /1
Hooded Merganser[4]				
Muscle	0.674	0.446/19	1.008[*]	0.332/19
Liver	3.433	2.157/18	4.683	2.979/17
Feathers	7.329	7.329/19	10.365[*]	3.791/17
Common Merganser[5]				
Muscle	1.268	0.481/13	1.496	1.306/17
Liver	17.529	12.065/13	10.970	7.486/15
Feathers	10.179*	2.855/13	7.313	2.915/15

Table 6.1. Total mercury (mg kg^{-1} wet weight) in bird tissues collected from the Grande Baleine and NBR study areas (S.D. = Standard deviation, n = number of samples analyzed) (continued)

Species	Grande Baleine		NBR	
	Mean	S.D./n	Mean	S.D./n
Common Loon				
Muscle	0.618	0.357/4	0.968	0.474/17
Liver	3.132	2.712/5	7.746	7.841/15
Feathers	7.836	4.333/5	8.633	3.685/15
Herring Gull[6]				
Muscle	1.034	0.802/28	1.592	1.317/13
Liver	2.934	2.309/28	3.633	2.517/13
Feathers	12.132	8.958/28	19.062	15.317/13

[1] *Anas crecca.*
[2] *Anas platyrhynchos.*
[3] *Melanitta nigra.*
[4] *Lophodytes cucullatus.*
[5] *Mergus merganser.*
[6] *Larus argentatus.*
* Significantly higher (p < 0.05).
** Significantly higher (p < 0.01).

In a recent review, Thompson (1996) stated that it seems reasonable to conclude that mercury has been responsible for the death and/or poisoning of many individual birds from different species in the wild but that it is difficult to be unambiguous with respect to what concentrations are required to cause these effects.

Bird populations sampled for this study are probably not directly at risk since laboratory studies have indicated that lethal effects in birds of prey can occur at levels exceeding 30 mg kg^{-1} ww in the liver and kidney (Thompson 1996). However, effects on reproductive success have been suggested to occur at levels measured in some prey species of the Grande Baleine and NBR regions. Barr (1986 cited in Thompson 1996) found a negative correlation between the breeding success of Common Loons and the degree of environmental mercury contamination. The correlation suggested that total mercury concentrations as low as 0.3-0.4 mg kg^{-1}ww in prey could impair territorial fidelity and egg laying. A laboratory study using Mallards showed that dietary concentrations of total mercury of 3 mg kg^{-1} dry weight (or 0.6 mg kg^{-1} wet weight assuming 80% water content) can impair reproduction with little direct effect on breeding adults.

Among the different tissue types, feathers were generally more contaminated than liver or muscle tissues. On average, 65% (n = 597, S.D. = 19%) of the combined total mercury burden for the three tissue types was present in plumage (range of 27% to 87%), which is close to the average value of 70% of the total body burden reported by Honda et al. (1985, 1986, cited by Monteiro and Furness 1995). The percentage of the total mercury burden in plumage has been shown to reach 93% following completion of molt in Bonaparte's Gull (*Larus philadelphia*) (Braune and Gaskin 1987) and accounts for the total mercury accumulated between molts (Monteiro and Furness 1995). Because the birds in this study were collected between early June and late August, a period corresponding to the molting of most species (this phenomenon lasts for more than 30 days (Bellrose 1980)), one can assume that the high variability in total mercury levels in feathers may result, in part, from the presence of both pre- and post-molt birds.

The majority of species considered showed a significant positive correlation between mercury levels in feathers and levels in muscle (7/15) or in liver (9/15) samples (Table 6.2). The lack of correlation in other species and the negative correlation in two species (Canada Goose and American Black Duck) can be attributed to the large number of samples with mercury levels below the limit of detection. This is especially the case for the Canada Goose for which more than 50% of total mercury levels in feather samples were below the 0.5 mg kg^{-1} detection limit.

Table 6.2. Correlation coefficients for total mercury levels in feathers vs other bird tissue types

Bird species	Feather/Muscle	Feather/Liver
Canada Goose	-0.215	-0.407[***]
American Black Duck	-0.285[**]	0.286[***]
Green-winged Teal	0.179	0.253[*]
Northern Pintail[1]	0.508[*]	0.437[*]
Mallard	0.056	0.132
Lesser Scaup	0.089	0.082
Ring-necked Duck	0.086	0.086
Common Goldeneye	0.368[**]	0.302[*]
Black Scoter	0.323[*]	0.558[***]
Surf Scoter	0.055	0.011
Hooded Merganser	0.260[*]	0.251[*]
Common Merganser	0.029	0.149
Red-breasted Merganser[2]	0.867[***]	0.532[**]
Common Loon	0.446[**]	0.533[**]
Herring Gull	0.679[***]	0.712[***]

[1] *Anas acuter.*
[2] *Mergus serratos.*
[*] Significant at p < 0.05.
[**] Significant at 0.001 < p < 0.05.
[***] Significant at p < 0.001.

6.4.2
Relationship with Geographical and Habitat Distribution

For most species and tissue types, no significant difference in concentrations were found between the NBR and Grande Baleine regions (Table 6.1). This absence of relationship with latitude is consistent with fish contamination data since total mercury levels in NBR pike are higher than those measured in the same species collected from sites closer to industrialized areas of Montréal (Langlois et al. 1995; Chapter 5.0). In addition, data collected by Lucotte et al. (1995a) clearly indicate that contrary to distribution patterns observed with lead, mercury contamination is independent of latitude over the boreal forest domain. The presence of local sources, such as active or abandoned mines and a closed chlor-alkali plant on the Nottaway river watershed (SOMER Inc. 1994), is a possible factor in the observed differences.

However, a few species of waterfowl showed significantly higher mean total mercury concentrations in NBR than in Grande Baleine (i.e., Green-winged Teal, Hooded Merganser, Ring-necked Duck and Tern spp.) for at least one of the three tissue types (muscle, liver or feathers; Table 6.1). Only the Grande Baleine Common Merganser seems to have higher levels of total mercury (feathers and liver) than their NBR counterparts.

These differences could result from various types of water quality. The southern part of NBR is characterized by highly mineralized turbid waters while in the northern part waters are transparent and poorly mineralized, as are the waters of the Grande Baleine watershed. Such differences were used to explain the higher levels of total mercury in predatory fish from the north versus the south NBR area (SEEEQ–Environnement Illimité 1993).

Another explanation could be that species such as the Hooded Merganser and the Ring-necked Duck nest in the NBR area and use the Grande Baleine region mostly for molting (Gauthier-Guillemette–GREBE 1992a), thereby experiencing a shorter exposure time. What is more puzzling however is that no pattern is consistent for all species, suggesting that natural variability and different food sources may account for most of the differences observed between the two regions.

6.4.3
Diet Types

When waterfowl species are grouped according to diet types, three distinct groups emerge: (*i*) strictly herbivorous species, (*ii*) strictly piscivorous and omnivorous species and, (*iii*) benthivorous and other less specialized birds (Table 6.3). Bird feeding habits were obtained from Gauthier and Aubry (1995). For all tissue types, except for feathers from Herring Gulls, results show a general progression of total mercury levels with proportion of fish in the diet, as was shown by Jensen et al. (1997) for aquatic and terrestrial birds of northern Canada.

Table 6.3. Mean total mercury concentrations (mg kg^{-1} wet weight) in birds according to their feeding habits (n = number of samples analyzed)

Feeding group	Grande Baleine				NBR			
	Muscle	Liver	Feather	n	Muscle	Liver	Feather	n
Strictly herbivorous[1]	0.053	0.082	0.283	35	0.032	0.086	0.298	26
Herbivorous/benthivorous[2]	0.165	0.453	1.382	82	0.206	0.562	1.518	58
Benthivorous[3]	0.215	0.589	1.811	69	0.17	0.85	1.20	1
Mixed[4]	0.215	0.465	1.630	42	0.191	0.620	1.601	45
Piscivorous-invertivorous[5]	0.800	2.280	2.011	20	1.097	2.947	4.810	12
Piscivorous-omnivorous[6]	1.033	2.934	12.311	28	1.592	3.633	19.062	13
Strictly piscivorous[7]	0.900	8.314	7.402	76	0.976	6.492	6.683	50

[1] Canada Goose.
[2] American Black Duck, Northern Pintail, Mallard, Green-winged Teal.
[3] Black Scoter, Surf Scoter.
[4] Greater Scaup, Lesser Scaup, Ring-necked Duck, Common Goldeneye.
[5] Arctic and other terns.
[6] Herring Gull.
[7] Common Merganser, Red-breasted Merganser and Common Loon.

Of all species considered in this study, the herbivorous Canada Goose showed the lowest total mercury burden in any of its tissues with almost half of the samples lying below the detection limit. The Canada Goose, a confirmed breeder in both study areas, feeds almost entirely on vegetation such as *Scirpus* spp.

Another herbivorous species of northwestern Québec, the Willow Ptarmigan (*Lagopus lagopus*), was also found to contain no detectable levels (< 0.05 mg kg^{-1} ww) of total mercury in its muscle tissue, this finding being based on specimens of the Grande Baleine area (9 specimens analyzed) and 31 (out of 33) specimens captured in the La Grande area (located between the Grande Baleine and NBR regions). One individual sample and 2 composite samples (composed of 3 and 4 specimens, respectively) of Ptarmigan livers from the Grande Baleine region had levels of 0.22 mg kg^{-1} ww, 0.18 mg kg^{-1} and 0.13 mg kg^{-1} of total mercury. Ptarmigan samples from La Grande were also low, at 0.11 mg kg^{-1} and 0.05 mg kg^{-1} for composite samples of 11 individuals each (SOMER Inc. 1993).

The highest total mercury levels in feathers came from the piscivorous-omnivorous group; the Herring Gull being its only representative. Total mercury levels in feathers were also high in the strict piscivorous mergansers and the Common Loon, as well as in Terns from both the Grande Baleine and NBR areas.

If comparisons are made on the basis of levels in the liver and muscle tissues, Common Mergansers show the highest levels. The elevated total mercury burden of the Herring Gull can be explained by its dietary preference of carrion over live fish, since large dead fish are more likely to have high total mercury levels.

Intermediate total mercury concentrations are found in birds which feed on small fish (Terns) or a diet of mollusks, shellfish and/or insect larvae (Scoters, Scaups, Goldeneye). Many of these birds can be considered opportunists, since they choose their food according to availability.

6.4.4
Eggs

Total mercury levels in egg homogenates collected along the shore of the Grande Baleine area are shown in Table 6.4. Levels are below the Canadian marketing standard except for the Great Black-backed Gull. In the absence of replicate samples, significant differences in total mercury concentrations could not be established; however, the relatively high levels found in the eggs of Great Black-backed Gulls may be associated with their feeding habits which include eggs and chicks in addition to fish and invertebrates (Gauthier and Aubry 1995).

Table 6.4. Concentrations of total mercury in composite egg homogenate samples (mg kg^{-1}ww) for bird species from the Grande Baleine study area

Species	n[1]	Total Hg concentration
Herring Gull	7	0.36
Great Black-backed Gull	7	0.67
Common Eider	5	0.22
Black Guillemot	8	0.47
Arctic Tern	7	0.24

[1]   Number of eggs in each composite sample.

In gulls and Arctic Tern of the Grande Baleine region, the lowest total mercury concentrations of all tissue types analyzed were found in eggs, regardless of sex. Nonetheless, total mercury concentrations in eggs are roughly one third of those measured in muscle tissues (Tables 6.1 and 6.4). Egg laying has been postulated to be a major elimination path for total mercury in female sea birds (Lewis et al. 1993) especially if multiple egg clutches are produced. Since, according to Gauthier and Aubry (1995), sea birds collected in the area typically produce 2 to 3 egg clutches (3 to 5 eggs/clutch for the Common Eider), female birds could be eliminating nearly 20% of their total mercury body burden through egg laying, as has been previously demonstrated for female Herring Gulls (Becker 1992).

6.4.5
Terrestrial Mammals

Terrestrial mammals of northwestern Québec are exposed to elevated levels of mercury through their food. Here again, levels vary according to their diet. Herbivorous species, such as hare and caribou, have low levels, while fish eating mustilids, such as mink, have higher values (Table 6.5). Other carnivorous animals, such as ermine, marten and red fox have intermediate values.

Table 6.5. Total mercury burden in tissues (mg kg^{-1} ww) of terrestrial mammals from the Grande Baleine study area (S.D. = Standard deviation, n = number of samples analyzed)

Species	Mean	S.D./n	Range
Hare			
Muscle	0.07	0.01/22	< 0.05 - 0.09
Liver	0.18	0.22/9	< 0.05 - 0.62
Mink			
Muscle	2.40	2.24/6	0.41 - 6.2
Liver	8.34	7.86/5	2.21 - 20
Ermine			
Muscle	0.21	0.13/2	0.11 - 0.30
Liver	0.26	0.09/2	0.19 - 0.32
Marten			
Muscle	0.28	0.13/8	0.12 - 0.55
Liver	0.39	0.11/6	0.21 - 0.54
Red fox			
Muscle	0.15	0.07/5	0.09 - 0.24
Liver	0.30	0.10/5	0.20 - 0.47
Caribou			
Muscle	0.02	0.02/36	< 0.01 - 0.03

The concentrations determined in this study are within the range of values reported in AMAP (1998) and in a comprehensive review of mercury accumulation and toxicity in wild terrestrial mammals for non-industrialized areas (Wren 1986). Animal species having mercury tissue concentrations similar to this study included martens from James Bay (Québec) and Muskoka (Ontario), as well as mink collected in Québec (exact location not specified). Mink from other areas of North America, however, had much lower levels (ranging from 0.052 mg kg^{-1} to 9.78 mg kg^{-1} in liver and 0.135 mg kg^{-1} to 1.26 mg kg^{-1} in muscle). Red fox from Muskoka (Ontario) also had lower muscle concentrations (0.05 mg kg^{-1}) than those found in this study, while liver concentrations were similar. A much higher value was, however, reported by Wobeser (1976 cited in Wren 1986) near the mercury contaminated South Saskatchewan river with one specimen having 58.2 mg kg^{-1} in the liver and 15.2 mg kg^{-1} in muscle. These values are known to be lethal in experimentally poisoned mink (Wren 1986).

Total mercury levels in caribou, an important native food source, are very low (0.02 mg kg^{-1}) despite the fact that their main food source, lichen (Scotter 1967; Gauthier et al. 1989), has been shown to accumulate atmospheric pollutants. Lichen of the Grande Baleine zone contained 0.04 mg kg^{-1} dry weight to 0.15 mg kg^{-1} of total mercury (n = 9) (SOMER Inc. 1993), which is in the same range as the levels reported by AMAP (1998) for Canada, Greenland, Finland and Russia (0.01-0.15 mg kg^{-1}) and in Chapter 3.0 for *Cladina* sp. (0.022-0.12 mg kg^{-1}). Levels of total mercury in Grande Baleine caribou muscle are similar to levels reported by Wren (1986) for Québec and the Canadian Northwest Territories caribou, but AMAP (1998) reported much higher values in kidney for northern Canada (max. 6.76 mg kg^{-1}) and in muscle and liver for Sweden (respective maxima of 3.65 and 80.7 mg kg^{-1}).

Although Wren (1986) recognized the difficulty in establishing useful background levels in view of the prominent variability in total mercury levels in wild mammals, he suggested that average concentrations of total mercury in mink from pristine areas should be less than 2.5 mg kg^{-1} in livers and kidneys and less than 2.0 mg kg^{-1} in muscle.

6.4.6
Marine Mammals

Table 6.6 shows that levels of total mercury measured in muscle and liver tissues of ringed seals (*Phoca hispida*) and bearded seals (*Erignathus barbatus*) collected off the coast of the Grande Baleine region are relatively low. Only one individual had a muscle concentration that exceeded the Canadian marketing standard of 0.5 mg kg^{-1}. Total mercury levels in ringed seals are comparable to values obtained by Smith and Armstrong (1978) and Wagemann (1989) in the Canadian Arctic and to values reported by AMAP (1998) for Canada, Greenland and Norway.

In belugas (*Delphinapterus leucas*), mercury levels are higher than the marketing standard for all muscle and liver samples. These levels are within the range of total mercury concentrations in tissues of all western Arctic belugas collected between 1981 and 1994 (Wagemann 1994) and values obtained by Beak Consultants (1978) in the Mackenzie delta. They are, however, generally higher than concentrations reported by Bligh and Armstrong (1971) for Hudson Bay, by Wagemann et al. (1990) for other Canadian Arctic areas and by AMAP (1998) for most locations reported.

As with seals, levels of total mercury in blubber of belugas appear to be relatively low compared to levels reported by Gaskin et al. (1973) for harbor seals (*Phoca vitulina*) captured of the coast of Maine. These findings support the observation that total mercury accumulates more in the liver, and to a lesser degree in muscle tissue, than in blubber.

Table 6.6. Total mercury levels (mg kg^{-1} ww) in marine mammal tissues of the Grande Baleine study zone (S.D. = Standard deviation, n = number of samples analyzed)

Species	Mean	S.D./n	Range
Ringed seal			
Muscle	0.32	0.19/8	0.08 - 0.74
Liver	5.12	4.64/8	0.93 - 15.03
Brain	0.19	0.09/5	< 0.05 - 0.30
Blubber	0.08	0.02/8	< 0.05 - 0.08
Kidney	0.49	0.07/2	0.44 - 0.54
Bearded seal			
Muscle	0.21	- /1	2.13 - 2.7
Liver	2.42	0.440/2	<0.05
Brain	0.05	- /1	
Blubber	<0.05	- /2	
Kidney	0.34	- /1	
Beluga			
Muscle	2.6	2.06/6	0.92 - 6.2
Liver	20.34	19.5/5	5.2 - 54
Brain	2.63	2.87/3	0.36 - 5.85
Blubber	< 0.10	- /2	< 0.10 - 0.12
Freshwater seals			
Muscle	1.10	0.37/2	0.83 - 1.36
Liver	28.80	37.06/2	2.59 - 55
Brain	0.37	0.21/2	0.22 - 0.51
Blubber	0.08	0.02/2	0.06 - 0.09
Kidney	6.13	1.65/2	4.97 - 7.3

6.4.7
Freshwater Seals

Total mercury concentrations in muscle and liver tissues of the two freshwater seal specimens (*Phoca* sp.) are comparable to levels measured in beluga tissues (Table 6.6) and levels reported by Kari and Kauranen (1978) for freshwater seals captured in Finland. As with belugas, blubber and brain tissues have low concentrations, while kidney levels were higher than the 0.5 mg kg^{-1} marketing standard. The higher total mercury concentrations observed (compared to their marine counterparts) likely result from their specific diet (i.e., fish vs fish and benthic organisms) and from the generally higher levels of total mercury observed in predatory fish of the freshwater food chain in the Grande Baleine region (Chapter 5.0).

6.5
Conclusions

This study confirms the occurrence of relatively high concentrations of total mercury in piscivorous wildlife of northern Québec. These pre-development values are however generally comparable with values collected in other pristine areas such as the Canadian Arctic and will prove useful in future monitoring of mercury levels in the NBR and Grande Baleine ecosystems.

Levels of mercury potentially hazardous for consumers were found in flesh and livers of piscivorous ducks, mink, beluga and freshwater seals. Some duck species and the marine seals had potentially toxic levels in livers only. Since several of these animals are part of the regular diet of Native people, wildlife monitoring programs should be conducted in coordination with surveys of mercury burdens in humans. However, important food resources such as Canada Geese, caribou and hare proved to have mercury levels much lower (an order of magnitude) than the Canadian marketing standard of 0.5 mg kg^{-1}.

Acknowledgements

This study was funded by Hydro-Québec. The participation of Hydro-Québec staff and consultants in the planning and realization of the study is also gratefully acknowledged. We are thankful to the Inuit hunters from Kuujjuarapik and the Cree trappers from Whapmagoostui and Waskaganish who contributed to the collection of bird and mammal specimens.

Mercury Dynamics at the Flooded Soil-Water Interface in the Reservoirs

7   *In Vitro* Release of Mercury and Methylmercury from Flooded Organic Matter

Normand Thérien and Ken Morrison

Abstract

Dominant vegetation and soil types found within the La Grande complex of northern Québec were subjected to flooding under controlled laboratory conditions of temperature, pH and levels of dissolved oxygen. For all substrates, the cumulative quantities of mercury (Hg) released showed an asymptotic behaviour as a function of time. Under the conditions set for the experiments, Hg releases to the water column occurred rapidly but were of relatively short duration (< 1 year). Of the different vegetation types, sphagnum moss (*Sphagnum* sp.) released the greatest amount per unit dry weight of total Hg, ~ 40 ng g^{-1} dry weight. Alder (*Alnus* sp.), lichen (*Cladonia* sp.) and spruce (*Picea mariana*) were all within the range of 4-10 ng g^{-1} regardless of the treatment applied. Alder released the greatest amount of methylmercury (MeHg), 5 ng g^{-1}, representing ~ 60% of total Hg. For sphagnum moss, MeHg released 3 ng g^{-1}, ~ 8% of total Hg. Alder and spruce were within the range of 1-2 ng g^{-1}. Globally, there were no clear effects of temperature, pH and levels of dissolved oxygen on releases of total Hg or MeHg, although for individual substrates there were some differences. For the humus samples, feather moss (*Pleurozium* sp.) humus yielded greater levels of both total Hg and MeHg per unit area, ~ 45 ng m^{-2} and ~ 1.2 ng m^{-2} respectively, as compared to lichen (*Cladonia* sp.) humus, ~ 4.3 ng m^{-2} and ~ 0.35 ng m^{-2} respectively. For both types of humus, ambient temperature treatment always produced the highest quantities of total Hg relative to colder temperature. This was also observed in the case of feather moss humus for MeHg. The results permit estimation of the contributions of total and MeHg from the dominant vegetation and soil types of proposed reservoirs.

7.1
Introduction

In hydroelectric reservoirs created in Québec within the last two decades, marked increases have been noted in fish Hg concentrations as compared to background levels (Doyon et al. 1996; Chapter 11.0). The Hg cannot have come from a sudden increase in atmospheric fallout since adjacent natural lakes have not had parallel increases in fish Hg (Chapters 10.0 and 11.0). These reservoirs are also far from any direct pollution sources. The Hg must, therefore, have come from *in situ* sources, the most likely being the flooded vegetation and soils. Indeed, it has already been established that the presence of flooded vegetation is associated with elevated fish Hg (Hecky et al. 1986, 1987), that flooded vegetation and soils release Hg directly to the water column immediately after flooding (Verta et al. 1986b; Thérien and Morrison 1995), and that the materials released from flooded vegetation favor methylation of Hg (Morrison and Thérien 1991ab; Porvari and Verta 1995).

Due to the large surface areas of the reservoirs created in northern Québec (on the order of 10^5 ha), it was impractical to remove the vegetation or the organic layer of soil before flooding, so this material was left in place to decompose. The terrestrial material flooded released organic carbon and nutrients to the water column, as well as Hg. It thus appeared important to better assess the temporal aspects of the release of Hg to the water column following reservoir creation because it represents a direct pathway to many components of the aquatic system, including suspended particulate matter, part of which is involved in fish diets (Doyon et al. 1996) and other parts in the cycling of Hg in the aquatic system (Hudson et al. 1994).

Unfortunately, not until the nineties have analytical methods become sufficiently sensitive (Chapter 2.0) and ultra-clean methods been effectively applied to the sampling and transport of water samples from northern reservoirs to reliably measure changes in dissolved Hg as a result of flooding. Consequently, Hg levels in water following the creation of two of the largest of the reservoirs, namely the Robert-Bourassa and Opinaca reservoirs (for which Hg data on accumulation in fish are reported in Chapter 11.0), were not known in the early years following flooding. Total and MeHg concentrations in samples of water originating from different reservoirs located in the James Bay territory have been reported in the past (Thérien 1990,1994) but none of these results represented a systematic follow-up of recently created reservoirs. Only recently have Lucotte et al. (Chapter 8.0) reported dissolved Hg levels six weeks and a year after initial flooding of peatbogs and podzols that were flooded during construction of the LA-40 containment basin.

In general, the quantities of dissolved materials released to the water column, including Hg, are dependent upon the nature of the vegetation substrates and soil types being flooded (Maystrenko and Denisova 1972; Thérien and Morrison 1995; Morrison and Thérien 1996). The resulting changes observed in Hg concentrations are also a function of time since flooding, of the flushing rate and dilution process occurring in the aquatic system being investigated, and of the history of the environmental conditions in the system since flooding (Thérien and Morrison 1995; Morrison and Thérien 1996). Since the levels of Hg obtained from *in situ* measurements reflect a combination of environmental conditions (temperature, pH, dissolved oxygen, etc) and hydrologic processes which have acted since flooding occurred, positive identification of controlling variables and quantification of the direct effect of specific factors is not possible. Furthermore, specific contributions from distinct vegetation substrates and types of soil are not discernable. These observations are the main reasons why, in order to better understand the changes of Hg levels occurring in the water column following flooding, two sets of separate *in vitro* experiments were carried out to quantify Hg release from selected vegetation and humus types.

7.2
Materials and Methods

7.2.1
Sample Collection

Vegetation samples were collected in unperturbed areas adjacent to the La Grande 4 reservoir (54°N 73°W). Species were selected on the basis of dominance, and are generally representative of ground cover throughout the region. Tree samples consisted of small branches with attached foliage, the species being black spruce (*Picea mariana*) and alder (*Alnus* sp.). Ground cover samples were sphagnum moss (*Sphagnum* sp.) and lichen (*Cladonia* sp.). Tree and lichen samples were kept frozen until used, while sphagnum samples were maintained at 4°C.

For humus, the organic horizon of the soil immediately beneath the vegetation and on top of the mineral soil, samples were collected in unperturbed areas adjacent to the Robert-Bourassa reservoir (54°N 77°W) and consisted of 2 humus types: lichen (*Cladonia* sp.) and feather moss (*Pleurozium* sp.). Lichen was by far the dominant soil cover type with feather moss being found only in well-drained upland spruce stands. Samples 30 cm in diameter x 15 cm thick were carefully collected to maintain the soil structure. They were kept frozen until used.

7.2.2
Experimental Set up

Both series of experiments were conducted in 26 L Pyrex® containers (Figure 7.1). Humus samples were placed at the bottom of the containers and cleaned Ottawa sand was used to fill any remaining space around the sample. The containers were sealed with glass covers having Teflon® TFE plugs through which passed Teflon® tubes for water addition-withdrawal and for headspace gas analyses, as well as a permeable Teflon® FEP tube for oxygenation. Water was then added to the containers.

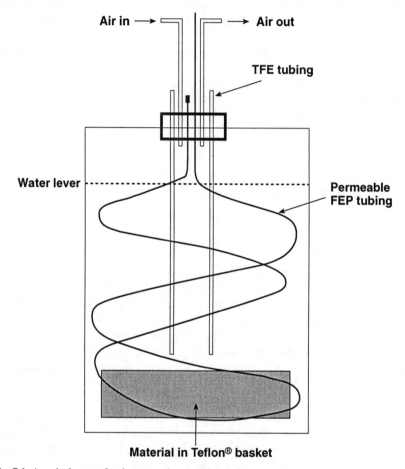

Fig. 7.1. A typical set-up for the vegetation experiments

The main differences for the vegetation experiments were that the material was in a Teflon® TFE mesh basket and no Ottawa sand was used. All vegetation samples were 201.5 ± 1 g wet weight. Dry weights were determined on separate subsamples. Three parameters, each of which was controlled for two different conditions, were examined at two levels: temperature (T1 & T2), dissolved oxygen (A1 & A2) and pH (P1 & P2). The treatments for each parameter were:

T1 :	4-5°C	A1 :	anaerobic	P1 :	pH 4.5-5.0
T2 :	18-20°C	A2 :	oxygenated	P2 :	pH 6.0-6.5

Except for the fact that surface water attains 0°C during winter, the experimental conditions correspond fairly well to the extremes found in reservoirs of northern Québec, both on a temporal basis (e.g., winter versus summer) and on a spatial basis (e.g., bays versus open water). Only incomplete factorial design experiments were used since the studies were exploratory and the design of experiments differed somewhat between the two series. Cold temperatures were maintained by keeping the containers in a cooled bath for the vegetation series and in refrigeration units for the humus series. Oxygenation was carried out by keeping O_2 (zero grade) under pressure in the permeable FEP tubing. Deoxygenated water was used for all water additions to the anaerobic tanks and since these were sealed they stayed anaerobic. The water in the containers was recirculated periodically by pumping it through a closed loop system (all wetted parts in Teflon®) for dissolved oxygen (D.O.) and pH measurements. At these times, pH was adjusted by the addition of small amounts of dilute ultra-pure LiOH or NaOH using a syringe.

For the vegetation series, alder and sphagnum were kept only at T2A2P1 conditions, while spruce and lichen (the dominant vegetation types) were kept at the following conditions T1A2P1, T2A1P1, T2A2P1 and T2A2P2. For the humus series, both lichen humus and feather moss humus were kept at the following conditions T1A1P1, T1A2P2, T1A1P2, T2A1P2, and T2A2P2. These combinations of conditions were chosen to reflect some environmental limits known to exist in the newly created reservoirs on the James Bay territory. Unfortunately, the limited number of experimental set-ups available precluded a more extensive and complete set of experiments.

On each sampling date water samples were withdrawn and replaced with equal volumes of ultra-pure water. The samples were then filtered on 5 μm filters. Finer filters could not be used because the membranes plugged quickly. Therefore the reported quantities of Hg include that which was sorbed to particles up to 5 μm in size. Experiments performed with the vegetation substrates lasted 170 days while those for the humus series were 320 to 340 days.

7.2.3
Analytical Methods

Different analytical methods were used for the two experimental series for various reasons. Details of the methods can be found in Chapter 2.0.

7.2.3.1
Total Hg

For the vegetation series, BrCl oxidation (Robertson et al. 1987) was inadequate due to high concentrations of organic material therefore an acid-permanganate-persulfate digestion (APHA-AWWA-WPCF 1989) was used. For the humus series, BrCl oxidation was adequate. In both series acid digestion was followed by reduction, amalgamation on gold traps, thermal desorption and CVAAS detection.

7.2.3.2
MeHg

For the vegetation series, MeHg was measured using SCF pre-concentration followed by desorption into HCl-NaCl, extraction into C_6H_6, and analysis by capillary GC with an ECD detector (Lee 1987; Lee and Mowrer 1989). This was the most sensitive technique available at the time (1989-1990). For the humus series, the method used was CH_2Cl_2 extraction followed by evaporation in water, aqueous-phase ethylation, Carbotrap pre-concentration, GC separation, pyrolysis and CVAFS detection (Bloom 1989; Brooks-Rand 1990).

7.3
Results

Results for the vegetation series are shown in Figures 7.2 to 7.4 and for the humus series in Figures 7.5 and 7.6. For both series, the results are presented as cumulative quantities of Hg released as a function of time since the beginning of the experiments. Such representation is easily adaptable to a model (Morrison and Thérien 1996) and can be related to the dynamics of the physical process acting since the slope of the curve represents the rate of release at any given time. Moreover, the quantity released from time t_1 to time t_2 is simply the quantity read from the curve at time t_2 minus the one at time t_1. For the vegetation series, results are presented as quantities of Hg per unit of biomass dry weight, while for the humus series, they are presented as quantities of Hg per unit surface area.

Of the different vegetation types, sphagnum released the greatest amount per unit dry weight of total Hg, ~ 40 ng g^{-1}. Alder, lichen and spruce were all within the range of 4-10 ng g^{-1} regardless of the treatment applied. Alder had the greatest MeHg release, 5 ng g^{-1}, representing ~ 60% of total Hg released. For sphagnum, MeHg release was 3 ng g^{-1}, ~ 8% of total Hg released. For lichen the different treatments affected the timing of the release of total Hg, but the effects on asymptotic final quantities were less clear. Lower temperature greatly reduced the quantity of MeHg released from lichen, and anaerobic conditions reduced it somewhat as compared to oxygenated and ambient temperature treatments. In all cases, MeHg released never exceeded 1 ng g^{-1}. For spruce, total Hg was highest for the ambient temperature-oxygenated-low pH combination, the others being similar. For MeHg from spruce, the two ambient temperature-oxygenated treatments (T2A2P1 and T2A2P2) gave similar quantities (~ 2 ng g^{-1}), while the other two treatments gave lower results (~ 1 ng g^{-1}).

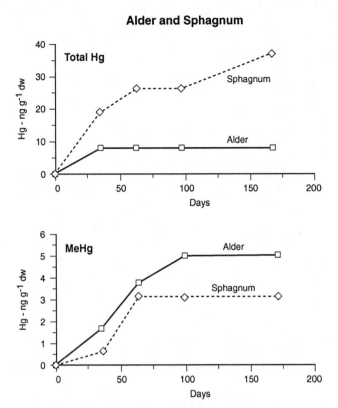

Fig. 7.2. Cumulative quantities of total Hg and MeHg released from alder and sphagnum

For the humus samples, feather moss humus gave greater quantities of both total Hg and MeHg than did lichen humus. For the former, total Hg ranged from 8 ng m^{-2} to 45 ng m^{-2} and MeHg from 0.30 ng m^{-2} to 1.2 ng m^{-2} while for the lichen humus, total Hg ranged from 1.7 ng m^{-2} to 4.3 ng m^{-2} and MeHg from 0.20 ng m^{-2} to 0.35 ng m^{-2}. For both types of humus, ambient temperature treatment always produced the highest quantities of total Hg relative to colder temperature. This was also observed in the case of feather moss humus for MeHg. With the exception of lichen humus, for which there were no clear effects of any one factor on MeHg because of the small range of variation observed, lower temperature reduced somewhat both total Hg and MeHg.

For both vegetation substrates and types of humus, the rates of maximum release of Hg always occurred during the first 100 days of flooding under the environmental conditions applied. Except for the oxygenated treatments of lichen, for which the release of total Hg did not level off after more than 170 days, the cumulative quantities of Hg released by all other vegetation substrates and types of humus showed an asymptotic behaviour as a function of time. MeHg release

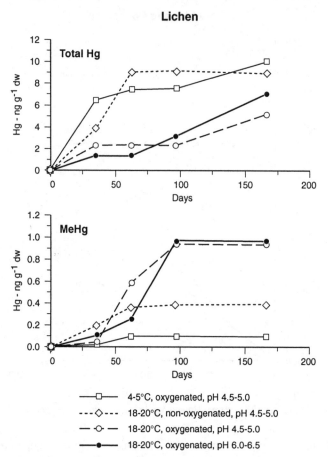

Fig. 7.3. Cumulative quantities of total Hg and MeHg released from lichen

generally slowed after 50-100 days but was still observed for both humus types even after 250 days. When considering total Hg, release from sphagnum did not level off and both humus types generally kept giving up Hg.

7.4
Discussion

A point of interest is to compare the cumulative quantities of Hg released to the water column at the end of the experiments with respect to the quantities of Hg initially contained in the vegetation and humus samples. Under the most favorable environmental conditions used in the experiments, the maximum cumulative quantities of total Hg released to the water column after 320-340 days, as shown

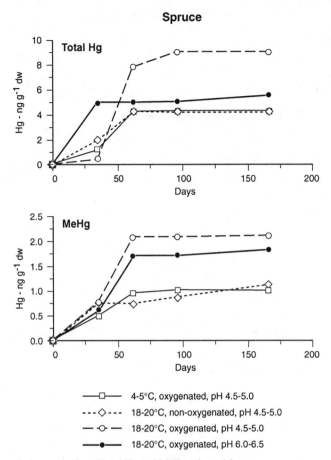

Fig. 7.4. Cumulative quantities of total Hg and MeHg released from spruce

in Figures 7.5 and 7.6 for feather humus moss (*Pleurozium* sp.) and lichen humus (*Cladonia* sp.), were 45 µg m^{-2} and 4.3 µg m^{-2} respectively. Levels of total Hg in the organic horizon of these types of podzols, sampled near the Robert-Bourassa reservoir, averaged 133 ng g^{-1} and 98 ng g^{-1} respectively (Grondin et al. 1995). Considering the distinct volumes and densities of the samples used in the two series of experiments and volumetric Hg burdens reported by Caron (1997), the Hg burdens for these samples were calculated as 1 200 g m^{-2} and 710 g m^{-2}, respectively.

Therefore, the percentages of the initial content of total Hg released in the water column from both types of humus are small and represent approximately 4% for the feather moss humus and less than 1% for the lichen humus. Similarly, for MeHg, using values of 0.1-0.2 ng g^{-1} reported for the organic horizons of podzols

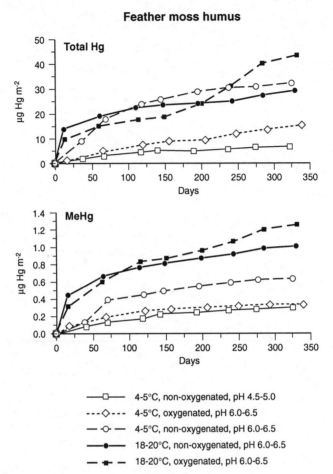

Fig. 7.5. Cumulative quantities of total Hg and MeHg released from feather moss humus

(Caron 1997), MeHg burdens for the samples used were calculated as 0.23-0.45 $\mu g\ m^{-2}$ for the feather moss humus and 0.18-0.34 $\mu g\ m^{-2}$ for the lichen humus. These quantities are less than the cumulative quantities of MeHg found in the water column over time, 1.2 $\mu g\ m^{-2}$ and 0.35 $\mu g\ m^{-2}$ for the two types of humus considered (Figures 7.5 and 7.6). This indicates that MeHg has been produced from the methylation of the inorganic Hg present in the system (Morrison and Thérien 1991b; Porvari and Verta 1995) which would have been promoted by the parallel release of organic carbon and nutrients to the water column (Thérien et Morrison 1995).

For sphagnum moss, levels of total Hg ranged from 20 to 150 ng g^{-1} (Moore et al. 1995; Caron 1997; Moore et al. 1994) with concentrations being closer to

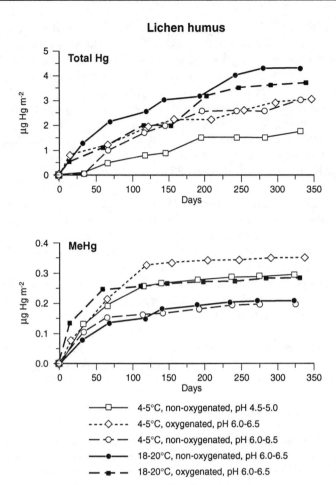

Fig. 7.6. Cumulative quantities of total Hg and MeHg released from lichen humus

80-100 ng g^{-1} for samples originating from terrestrial stations near the Robert-Bourassa reservoir (Grondin et al. 1995). For MeHg, values reported for various types of sphagnum range from 0.2 to 40 ng g^{-1} (Moore et al. 1995) with values for peatlands before flooding being in the 0.2-1.5 ng g^{-1} range (Bégin 1997). For lichen (*Cladonia* sp.), values of 45 ng g^{-1} for total Hg and 10 ng g^{-1} for MeHg have been reported (Moore et al. 1994). For spruce needles, literature values range from 20 to 34 ng g^{-1} for total Hg and from 0.20 to 0.25 ng g^{-1} for MeHg (Zhang et al. 1995; Moore et al. 1994). For alder leaves, the values reported range from 8 to 15 ng g^{-1} for total Hg and from 0.09 to 0.20 ng g^{-1} for MeHg (Moore et al. 1994). The maximum cumulative quantities of total Hg released to the water column after 170 days, were 37.5 ng g^{-1} for sphagnum, 10 ng g^{-1} for lichen, 9 ng g^{-1} for spruce and 8 ng g^{-1} for alder. The percentages of the initial total Hg burdens released from the vegetation substrates during the experiments are 37-47% for sphagnum,

22% for lichen, 26-45% for spruce, and 53-100% for alder. It is important to note that for sphagnum and lichen, these percentages could be higher since release of total Hg continued even after 170 days (see Figures 7.2 and 7.3). Cumulative quantities of MeHg found in the water column after 170 days were 5 ng g^{-1} for sphagnum, 1 ng g^{-1} for lichen, 2 ng g^{-1} for spruce, and 5 ng g^{-1} for alder. With the exception of lichen, all of these quantities are considerably larger than the MeHg burdens of the vegetation. Again, this clearly indicated that the quantities of MeHg in the water column are not the result of a simple physical leaching from the vegetation substrates but that they have been produced from the methylation of inorganic Hg in the system (Chapter 8.0).

With respect to the dynamics of Hg release, there were marked differences between vegetation and humus. For the former, release happened more quickly and more erratically while for the latter it was steadier and continuous. This can be understood when one takes into account the nature of the materials. The original organic debris transformed into humus have already been subjected to some degradation and leaching with the most labile material having already been removed, so the remaining material is more refractory and decomposes slowly and steadily. In contrast, the vegetation still has a high proportion of labile material and initial decomposition can be higher. Tables 7.1 and 7.2 show the maximum rates of release of Hg observed at some time during the first 100 days of the experiment for the vegetation substrates and soil types, respectively. Average rates of release of Hg during the second period of observation (> 100 days) are also reported in these tables.

Table 7.1. Maximum rate of release of total Hg and MeHg from substrates of the flooded vegetation occurring early in the experiment (< 100 days) and average rate of release thereafter

Flooded Vegetation	Total Hg		MeHg	
	Maximum rate < 100 days ng g^{-1} per day	Average rate 100-168 days ng g^{-1} per day	Maximum rate < 100 days ng g^{-1} per day	Average rate 100-168 days ng g^{-1} per day
Spagnum moss (*Spagnum* sp.)	0.537	0.153	0.089	0
Alder (*Alnus* sp.)	0.250	0	0.078	0
Spruce (*Picea mariana*)	0.261	0.002	0.046	0.001
Lichen (*Cladonia* sp.)	0.188	0.035	0.021	0

Generally, there were no clear effects of the different experimental conditions on quantities of total Hg or MeHg, although for individual substrates there were some differences. For treatments where the effects of two distinct temperatures were used, low temperatures seemed to reduce somewhat the quantities of MeHg in all cases except possibly lichen humus. For treatments where the effects of two distinct D.O. levels were used, anaerobic conditions did not favor quantities of MeHg for any of the substrates and may actually have had an attenuating effect.

Table 7.2. Maximum rate of release of total Hg and MeHg from flooded soil humus occurring early in the experiment (< 100 days) and average rate of release thereafter

Flooded Soil Humus	Total Hg		MeHg	
	Maximum Rate < 100 days g m^{-2} per day	Average Rate 100-335 days g m^{-2} per day	Maximum Rate < 100 days g m^{-2} per day	Average Rate 100-335 days g m^{-2} per day
Feather moss (*Pleurozium* sp.)	0.956	0.055	0.030	< 0.001
Lichen (*Cladonia* sp.)	0.054	0.004	0.009	< 0.001

There were marked differences among the substrates within the two series (vegetation and soil types). For the humus series, feather moss humus released higher levels of Total Hg and MeHg as compared to lichen moss humus, but is far less prevalent as a soil cover type in northern regions. Sphagnum released much more total Hg than the other vegetation types per unit dry weight. However, this observation must be tempered by the fact that sphagnum was ~ 90% water by wet weight as compared to ~ 50% for the others. Although peatbogs in northern Québec may cover less than 10% of the land surface, the overall contribution from sphagnum could be more important since it is representative of ground covers associated with other types of wetlands which may globally represent a larger percentage of the land that has been or might be flooded (Poulin-Thériault–Gauthier-Guillemette 1993).

Except for sphagnum, the units used to express results (ng Hg g^{-1} dry weight or μg Hg m^{-2}) are easily related to field measurements of quantities of vegetation or surface areas of soil. In retrospect, it would have been preferable to treat sphagnum in the same way as the humus samples, except for the fact that sphagnum does not always remain in place after flooding but may, at times, rise to the surface of the reservoir to form floating islands. Furthermore, since in reality, the cover thickness is highly variable, it becomes difficult to quantify the amount of sphagnum decomposing per unit surface area. However, once the average thickness of the cover is defined for a given territory and its average density calculated, the results expressed as mass of Hg released per unit dry weight can easily be related to the surface areas flooded.

Filtration for these studies was only carried out at a 5 μm pore size, finer pore sizes being impractical due to rapid plugging. Therefore the quantities of Hg reported here include fractions sorbed onto particles smaller than 5 μm. Given the propensity of Hg to be associated with particulate organic matter, the proportion of Hg that was actually sorbed rather than truly dissolved may have been very high (Hurley et al. 1994; Mucci et al. 1995).

For extrapolation of these results to release occurring in reservoirs it may be prudent to use either the average results for all treatments or the highest results per substrate. A better approach would be to take into account the variability observed (Ruiz and Thérien 1997) as is done when models are used for environmental and ecological risk assessments (Goodwin 1989; Suter et al. 1993).

Vast surfaces of land are subjected to flooding with large quantities of phytomass in the James Bay territory. Therefore, because of the relatively high percentage of the initial Hg content of the vegetation substrates that would be released to the water column over time, a significant quantity of Hg would be released to the aquatic system as dissolved Hg, Hg forming complexes with organic substances in solution or Hg sorbed on suspended particulate matter (Chapter 8.0). To illustrate this point, Table 7.3 gives the relative contribution of flooded vegetation and soil to the quantity of Hg released to the water column for three distinct vegetation groups (Poulin-Thériault–Gauthier-Guillemette 1993) over a period of 168 days taking into account the characteristic vegetation substrates and densities associated with the given soil types for a flooded area of 1 m^2. These calculations are somewhat hypothetical since they assume uniform and constant conditions favoring the maximum release of Hg (the highest quantities after 168 days from Figures 7.3 to 7.6) during the whole period of time. Also, only the top 100 mm of peat moss is considered active for the release of Hg to the water column. Nonetheless, the main observation for the vegetation groups associated with the forest soils is that the release of Hg from the flooded vegetation (leaves from trees, shrubs and groundcover) over that period of time is a significant, if not dominant, contributor to the release of Hg to the water column. For the vegetation group composed of peatlands (including wetlands), the results show that the flooded soil would contribute more than the flooded vegetation (leaves from shrubs and ground cover). However, this is debatable since sphagnum moss represents the «living or green» part of peat and may also be considered part of the vegetation substrates (Poulin-Thériault–Gauthier-Guillemette 1993). Another point of interest is that, as time passes, the relative contribution of the flooded vegetation to the release of MeHg to the water column would fall. This is also true for total Hg, except for sphagnum and lichen, since the rates of release of Hg from the vegetation substrates are zero after the 100th day while, for humus, the rates are still positive at that time and even after 335 days (although the rates are only a small fraction of the maximum rate observed).

Table 7.3. Relative amounts of Hg ultimately released from the flooding of combinations of labile vegetation substrates and flooded organic soil types corresponding to selected vegetation groups

	Flooded Vegetation Group		
	Peatland	Lichen Dominated Forest Soil	Feather Moss Dominated Forest Soil
Labile component of the flooded soil considered	Sphagnum	Lichen humus	Feather moss humus
Basis for comparison	2800 g m^{-2} [1]	1 m^2	1 m^2
Labile component of the flooded vegetation considered	Leaves from shrubs (> 5 cm)	Leaves from trees and shrubs (> 5 cm)	Leaves from trees and shrubs (> 5 cm)
Representative vegetation type	Alder	Spruce	Spruce
Density of the vegetation component used [2]	120 g m^{-2}	475 g m^{-2}	525 g m^{-2}
Ground cover components	Leaves from shrubs (< 5 cm)	Lichen moss and leaves from shrubs (< 5 cm)	Feather moss and leaves from shrubs (< 5 cm)
Density of ground cover [2]	200 g m^{-2}	1200 g m^{-2}	400 g m^{-2}
Ultimate amount of total Hg released per m^2 from [3]:			
- vegetation + ground cover	2.6 µg	18.7 µg	9.5 µg
- soil	103.6 µg	3.1 µg	27 µg
Ultimate amount of MeHg released per m^2 from [3]:			
- vegetation + ground cover	1.6 µg	2.2 µg	1.5 µg
- soil	8.7 µg	0.34 µg	0.9 µg

[1]   Average mass (dry weight) / cross section area ratio of 100 mm thick samples used in the experiment.
[2]   Typical values reported for the Robert-Bourassa reservoir (Poulin-Thériault–Gauthier-Guillemette 1993).
[3]   Maximum cumulative quantity released after a period of 168 days in all cases (from Figures 7.2 to 7.6).

Due to the large volume of water contained in reservoirs such as the Robert-Bourassa reservoir, the significant quantities of Hg released to the water column from flooded vegetation and soils would not greatly increase Hg concentrations in the water (Ruiz and Thérien 1997). Nonetheless, the contribution to the water column could last for a number of years. In effect, the favorable environmental factors (e.g., temperature) which promote the release of Hg to the water column would occur only during a few months of the year and would act in conjunction with the dynamics of the drawdown that would expose new surfaces and subject new material to flooding.

Thus, the ultimate quantities of Hg that would have been continuously released under constant and favorable conditions over a period of a year (as observed in this study) would then be released as cyclic pulses corresponding to the periodic occurrence of favorable conditions in the reservoir but over a period of many years (Ruiz and Thérien 1997). In shallow reservoirs, the increases in Hg would generally be higher and more easily observable from *in situ* measurements, but possibly for even shorter periods due to higher flushing rates. In recent years, such cyclic release of Hg in the water column has been observed for the yearly flooding of a wetland (Kelly et al. 1997). Nonetheless, because of the low concentration resulting from dilution in large reservoirs and because of the natural variability of the concentrations of Hg in surface water systems (Table 7.1; Chapter 8.0), the change in Hg level in the water column may remain undetected. This would even be more so if Hg in solution was adsorbed by suspended particulate matter and if determinations of Hg were made on filtered water samples. This appears to have been the case when Montgomery et al. (1995) found that the average concentration, the range of dissolved total Hg concentrations and the ratios of maximal to minimal concentrations of water samples did not differ significantly from concentrations measured in the Robert-Bourassa reservoir (14 years after impoundment) and six weeks and a year after initial flooding of peatbogs and podzols during construction of the containment basin for the Laforge 1 reservoir (see also Chapter 8.0). The data reported showed high variability with ratios of maximal to minimal concentrations ranging from 2 to 5 in the case of total Hg and up to 28 for MeHg. This variability may have hidden any transient increase in Hg effectively present at that time. In effect, flooding of a wetland done in 1993 as part of the Experimental Lakes Area Reservoir Project (Rudd et al. 1994) revealed a transient increase in Hg over a few weeks with ratios of post-flood maximum concentration to pre-flood average concentration of 3.6 for total Hg and of 17 for MeHg. Peaks of similar amplitude would not have been detected considering the variability in concentration reported by Montgomery et al. (1995). Also, measurements of Hg by Montgomery et al. (1995) were done on water samples, filtered on a 0.45 µm pore size, when measurements at ELARP were done on unfiltered water samples thus including Hg sorbed on suspended particulate matter.

For the reasons indicated above, the mere fact that Hg may not be detected in the water column of a recently flooded reservoir should not be interpreted as indicating no release of Hg to the water column when it is known that vegetation substrates would release a significant fraction of their initial Hg contents over time. Systems dynamics (Forrester 1968) teach us that the variation of the level or the state variable (e.g., Hg) of any given system (e.g., the water column) with time is the integral of the sum of incoming rates from a source (ex. flooded vegetation and soils) minus the sum of the outgoing rates towards a sink (e.g., sorbtion by particulates, accumulation by biota, etc). The magnitude of the level or of the state variable of a system with time does not reveal anything specifically about the incoming or the outgoing rates. If the average level or magnitude of a state variable is observed to not vary significantly, it is not an indication that the incoming rates are negligible but rather that the magnitude of the outgoing rates is as important as that of the incoming rates. When it is known that important quantities of Hg would be released from flooded vegetation substrates and soils humus, as is the case reported here, the water column then becomes a vector of Hg transfer towards other components of the aquatic system, specifically including the fine particulate material, and would contribute to the cycling of Hg in this system.

Finally, because the work is based on laboratory experiments, it is always difficult to know how realistic or applicable the results are when applied to actual reservoirs. For example, calculation of Hg and MeHg releases assume no biological activity other than microbial, and thus no activity by invertebrates in directly consuming organic matter and taking up MeHg in the process. Thus, it may be argued that the direct release of Hg to the water column may be much less than these calculations indicate. On the other hand, it may also be argued that bioturbation and biodiffusion resulting from such activity may effectively increase the rate of release of Hg to the water column (Toms et al. 1995). Furthermore, the releases of Hg and MeHg in the water column under the somewhat quiescent conditions of the experiment may in fact represent minimal releases that would be present under conditions of flooded soil erosion caused by the drawdown and the wind induced wave action in an actual reservoir.

7.5
Conclusions

All substrates showed some asymptotic leveling of Hg release in these experiments, particularly for MeHg. This indicates that the contributions of Hg to the water column during the experiments occurred rapidly and were finite. *In vitro* results show that the release of Hg would be of short duration (< 1 year) under the constant conditions of the experiments performed. However, the release of Hg under field conditions is more likely to be cyclic in nature and of longer duration when considering the yearly occurrence of favorable factors in the reservoir and the dynamics of the drawdown on exposing new surfaces and submitting new materials to flooding as a function of time. The results obtained permit estimation of the contributions of total and MeHg from the dominant vegetation and soil types for proposed reservoirs in the same region, and provide at least a starting point for estimates for reservoirs in other regions.

Acknowledgements

Major funding for this work was provided by Hydro-Québec, and some funding was provided by the Canadian Electrical Association. From Hydro-Québec, Richard Verdon and Marcel Laperle were involved in all stages of the vegetation study while Roger Schetagne and André Chamberland were involved in various stages of both studies. Greg Pope of Ontario Hydro helped in the planning stages of the vegetation study as the CEA technical advisor. Financial support was also provided by a Fonds pour la formation de chercheurs et l'Aide à la Recherche doctoral bursary from the Québec government.

8 Mercury Dynamics at the Flooded Soil-Water Interface in Reservoirs of Northern Québec: in Situ Observations

Marc Lucotte, Shelagh Montgomery and Maxime Bégin

Abstract

A partial and gradual transfer of natural and anthropogenic mercury (Hg) accumulated over time in boreal forest soils is the principal origin of Hg contamination of the aquatic organisms living in northern Québec reservoirs. After a decade or more of flooding, little structural change is evident in flooded soils, with the exception of those eroded on the banks of the reservoirs. In non-eroded flooded soils, both the total carbon (C) and Hg burdens in the humic horizon of the podzolic soils and in the entire organic layer of the peatland appear to remain stable through time. The major change in the Hg biogeochemistry in northern soils after their impoundment is represented by the progressive methylation of their initial inorganic Hg content, which reaches approximately 10% and 30% after 13 years for peatlands and podzolic soils, respectively. Once Hg is methylated in the soils, no net demethylation can be clearly observed during winter periods or from one year to the next over the first 13 years of flooding. Low concentrations of dissolved methylmercury (MeHg) (in the few tenths of ng L^{-1} range) in the reservoirs are indicative of the important dilution of the fraction of this compound which is readily transferred from surface layers of flooded soils to the water column of the reservoirs. In order to explain the rapid Hg contamination of the entire food chain of the reservoirs, one should first take into account that Hg-rich organic particles of the flooded soils represent a source of food for organisms at the base of the aquatic food chain. Insects ingest Hg while burrowing as larvae in the surface of the flooded soils, and zooplankton do so while feeding on particles at the flooded soil-water interface. Moreover, the rapid erosion of the organic horizon of the soils situated in shallow and exposed zones of the reservoirs leads to a higher bioavailability of Hg-rich particles to filter feeding invertebrates. Although phytoplankton production as such does not appear to clearly contribute to the high bioaccumulation of Hg in higher organisms of the reservoirs, the sedimentation of autochthonous particles at the flooded soil-water interface promotes bacterial activity in the biofilm which in turn entrains an additional methylation of the Hg present in the surface layer in which invertebrates often feed. In addition to the initial release of Hg from the flooded soils, processes of active Hg methylation in the biofilm and efficient mechanisms of bioaccumulation of the pollutant by invertebrates must be invoked to explain the Hg contamination of the aquatic organisms of northern reservoirs.

8.1
Introduction

The extensive impoundment of forested regions during the creation of reservoirs results in a rapid increase in Hg concentrations in all aquatic organisms thriving in these artificial water bodies (Potter et al. 1975; Cox et al. 1979; Meister et al. 1979; Bodaly et al. 1997; Tremblay and Lucotte 1997). For instance, the level of Hg bioaccumulation in adult fish of commercial value living in newly impounded hydroelectric reservoirs in temperate climates largely exceeds the recommended maximum threshold for regular human consumption without risk to human health (Bodaly et al. 1984; Hecky et al. 1991; Jackson 1991; Verdon et al. 1991). The problem may persist for two or more decades for predatory species (Chapter 11.0). It is commonly accepted that upon impoundment, the natural and anthropogenic Hg accumulated in the vegetation and soils prior to flooding (Chapter 3.0) repre-sents the main source of Hg being partially transferred to the aquatic organisms living in the reservoirs (Lodenius et al. 1983; Jackson 1991; Grondin et al. 1995; Bodaly et al. 1997). This is clearly supported by the positive relationship between the rate of Hg accumulation in fish and the proportion of flooded land in reservoirs (Bodaly et al. 1984; Hecky et al. 1991; Johnston et al. 1991).

Upon flooding, the terrigenous organic matter of the vegetation and the soils is submitted to bacterial degradation. This microbial activity is responsible for nutri-ent release to the water column, enhancing planktonic development, and thus pro-moting higher productivity in the entire aquatic food chain (Pinel-Alloul 1991; Schetagne 1992; Tremblay et al. 1998a). After the death of these organisms, a fraction of the locally produced organic matter is deposited on the bottom of the reservoirs and contributes to the high bacterial activity at the surface of the flooded soils (Louchouarn et al. 1993; Lebeau 1996; Duchemin et al. 1996). This bacterial activity has been noted as being responsible for the transformation of inorganic Hg into methylmercury, the Hg form most easily bioassimilated by the living organ-isms (Jackson 1988a, 1991; Hecky et al. 1991; Mucci et al. 1995; Lebeau 1996; Bodaly et al. 1997).

Particulate organic matter offers numerous and strong binding sites for total Hg in general and MeHg in particular (Aastrup et al. 1991; Johansson et al. 1991; Lindqvist 1991; Schuster 1991; Steinnes 1995; EPRI 1996). This appears particu-larly valid for the partially degraded terrigenous matter found in the soils (Dmy-triw et al. 1995; Chapter 3.0). Thus, with the exception of the initial pulse of Hg released to the water column upon flooding, associated with the rapid degradation of the most labile organic matter of the vegetation and soils (Chapter 7.0), most of the MeHg then produced over the years through the bacterial activity at the bottom of the reservoirs seems to remain fairly stable and not readily liberated to the water column (Montgomery et al. 1995, 1996). Alternative processes other than diffusion must be found in order to explain the long term transfer of MeHg from the flooded soils to the biota of the reservoirs. In this chapter, we present a synthesis of our knowledge of the processes of Hg methylation over several years of impoundment in several flooded soils representative of the various environments of northern Québec reservoirs. We follow the onset of this Hg transformation and its consequences for the presence of Hg in the water column over the first four years of impoundment of a new reservoir and we compare these results to the levels attained after more than a decade in an older reservoir of the same region. We then present the most plausible spectrum of biogeochemical processes responsible for the release of Hg from the flooded soils to the water column and ultimately to the aquatic organisms.

8.2
Materials and Methods

8.2.1
Study Sites

In order to follow the temporal evolution of flooded environments in terms of early biogeochemical transformations they undergo, we regularly visited, during the first four years of impoundment, the new Laforge 1 reservoir (54°N, 73°W) and a small nearby flooded diversion (LA-40). In parallel, we also sampled the Robert-Bourassa reservoir (53°N, 77°W) during its 13th to 17th years of impoundment. With the ambient climate and general vegetation types being similar at both sites we can compare, with some precautions, the results obtained for the two reservoirs.

8.2.2
Sampling

The pedology of northern Québec is dominated by two major features, i.e., well drained podzolic soils and ombrotrophic peatlands. These soil types cover about 60% and 10%, respectively, of the boreal forest/taïga domain in the region of the two hydroelectric reservoirs studied (Poulin-Thériault–Gauthier-Guillemette 1993). The remaining 30% of soil cover is represented by organic forest soils, with poor drainage conditions intermediate between those of podzolic soils and peatlands. In both reservoirs, we sampled flooded soils of both types, taking care in choosing the least perturbed sites, not affected by wave or ice action on the bottom. We collected samples of podzolic soils and ombrotrophic peatlands at various depths below the water surface, the shallow samples (< 3 m depth) receiving sunlight and the deep samples (> 15 m depth) lying below the photic zone. The deep samples retrieved were nevertheless always lying above the seasonal thermocline which could only be found at 20 m depth or more in the regions of both reservoirs. Flooded soil cores were collected with a 15 cm in diameter core tube manually inserted by divers, the roots of the vascular plants being progressively cut with a knife on the external side of the tube. Samples of periphyton mats at the surface of shallow but non eroded flooded soils were collected with a syringe equipped with a brush, according to the technique described by Loeb et al. (1981). The periphyton samples used for Hg and chlorophyll *a* determinations were placed in plastic bags and kept frozen in the dark until analyses were performed, while those used for bacterial activity were kept at 4°C until the bacterial activity was determined. Redox potential (Eh) profiles in the flooded soils were obtained by inserting a platinum-Ag/AgCl combination electrode at each sampled depth prior to subsampling. The soil samples were then extracted from the tube, sliced at one centimeter intervals and the subsamples placed in plastic bags. Dry densities measured for each subsample allowed us to make elemental burden calculations per unit of surface.

The collection of water samples for the dissolved fraction was conducted following an ultra-clean procedure (described in detail in Montgomery et al. 1995). Briefly, water was collected into Teflon® bottles using a manually operated peristaltic pump and a short length of Masterflex® silicone tubing to which was attached Teflon® tubing and, at the outlet, a 47 mm Millipore® in-line filter holder assembly. For each sample, two filters, a glass fiber (GF) filter and a 0.45 μm GN-6 mixed cellulose ester Gelman® filter, were placed one atop the other in the filter unit. Four stations in the Robert-Bourassa reservoir and two in the Laforge 1 reservoir were sampled using an *in situ* multi-port sampler designed for close-interval sampling in shallow areas (approximately 2 m). The main body of the sampler is a length of plastic pipe into which lengths of Tygon® tubing (ID = 3 mm) were fixed at 25-cm intervals. Each sampler was installed vertically in the water, with its base resting at the flooded soil-water interface and the Tygon® sampling ports accessible from the water surface. Using the peristaltic pump, water was drawn through the Tygon® tubing from the desired depth, and following extensive rinsing, the samples were collected. The samplers remained in place for the duration of the sampling season.

For the particulate fraction, 200 L of water collected with the aid of an electrical pump and prefiltered on-line with 64 and 210 μm Nytex® filters, was stored in 50-L carboys for the subsequent concentration of the fine particulate fraction. The 64 and 210 μm fractions were rinsed from the filters and stored in Sardstedt® tubes while the fine particulate matter (FPM) was concentrated the same day by continuous flow centrifugation at a speed of 10 000 rpm or 8 500 G. Previous laboratory tests have indicated that about 72 to 85% of particles retained by 0.45 μm GN6 filters are recovered by centrifugation (Mucci et al. 1995). Following centrifugation, the FPM was stored in Teflon® tubes and all the samples were stored frozen.

8.2.3
Analyses

Total Hg concentrations in the periphyton samples, as well as lacustrine sediment or flooded soil samples, were determined using cold vapor atomic fluorescence spectrometry (CVAFS), following the methodology described in Louchouarn et al. (1993) and in Chapter 2.0. In brief, 0.5 to 1 g (dw) aliquots of sediment or flooded soils and 1 to 10 mg (dw) periphyton aliquots were transferred to glass tubes and digested in 10:1 HNO_3:HCl for 6 hours at 120°C. As fully described in Chapter 2.0 and in Bégin (1997), MeHg concentrations on solid samples were obtained following a digestion with $CuSO_4$ 25% and HCl 6N. This digestion allows the liberation of MeHg as MeHgCl, which is then recuperated by three consecutive extractions with toluene. After concentration, the toluene extract was injected in a chromatographic column and eluted at 140°C. MeHg was then decomposed as Hg(0) in a quartz tube heated at 500°C and finally quantitatively measured by atomic fluorescence.

Total dissolved Hg concentrations ($[Hg_T]_D$) were measured following a modification of the technique perfected by Bloom and Fitzgerald (1988) where dissolved Hg(II) is reduced to elemental mercury Hg(0) vapor by a tin chloride ($SnCl_2$) solution in a reaction vessel, and Hg(0) analyzed by CVAFS (Chapter 2.0). Our system, however, involved direct volatilization and therefore does not include a preconcentration step with a gold trap. The detection limit for the method is 0.3 ng L^{-1}, with procedural blanks of 0.5 ng L^{-1}.

The analysis of dissolved MeHg is described in detail in Chapter 2.0. Briefly, subsamples of filtered water were concentrated via distillation in the presence of nitrogen. The MeHg in the distillate was trapped on a Tenax® column following a stirring and bubbling step and then, while being desorbed by heating, was thermally decomposed in a GC column before finally being detected as Hg by atomic fluorescence.

Total C and N determinations were conducted using a Carlo-Erba® type analyzer with a precision on the order of 1%. Given the absence of carbonates in the samples, total C represents total organic C. For stable C isotope (^{13}C) analysis, the organic matter of samples was converted to CO_2 by combustion with copper oxide and copper in vacuum-sealed quartz tubes at 550°C for one hour. The $^{13}C/^{12}C$ ratios were determined using a VG-Prism mass spectrometer with a precision of ± 0.1‰ and are expressed relative to the Pee Dee Belemnite (PDB) standard.

8.3
Results and Discussion

8.3.1
Temporal Evolution of Mercury in Flooded Soils

Within the first few years following the initial impoundment of northern reservoirs, intense erosion of shallow flooded soils is observed along the perimeter of the reservoirs, mostly in the drawdown zone. We will consider later the role these zones, exposed to the wave and ice action, play on the Hg cycle in the reservoirs. On the other hand, large flooded areas in reservoirs remain in place and are not subjected to active erosion. Whether put in suspension in the water column or remaining in place, the most labile compounds of the vegetation and soils flooded at the bottom of the reservoirs are partially biodegraded within the first 200 days after impoundment. This reaction is responsible for an initial pulse of Hg released to the water column (Chapter 7.0).

The rapid biodegradation of organic matter only affects a small fraction (less than 5%) of the total C burden initially present in the soils (Duchemin et al. 1996; Bégin 1997). Indeed, a first examination of the cores of flooded podzolic soils which have not been eroded shows little structural change after flooding. After a decade or more of flooding, forest soils still exhibit their thin humic horizon overlying the mineral horizon as described in Chapter 3.0 (Figure 8.1a). The fragile ground cover, composed for the most part of lichen (*Cladina* sp.), is still recognizable in most samples, even in the ones flooded for more than 15 years, as in the Robert-Bourassa reservoir. Similarly, aside from the death of the former living sphagnum cover, the general structure of peatlands flooded for several years remains identical to the pristine ones, with a gradual transition with depth of little degraded sphagnum to homogeneous humus (Figure 8.1b).

The only observable physical difference between pristine and flooded soils is the sedimentary and/or periphyton layer developing at the surface of the former soils. Even though the pedologic features of the former soils are apparently preserved, active bacterial degradation of the flooded terrigenous matter is taking place as sharp drops in redox potential in the flooded soils indicate that oxygen demand has increased (Figures 8.2a and 8.2b) and as CO_2 and methane (CH_4) are steadily produced at the soil-water interface (Duchemin et al. 1995). In spite of the active bacterial decomposition of organic matter, no significant decrease of the C content in the humic horizon of the podzolic soils or in the entire peat layer of the peatlands could be detected through time after impoundment (Figures 8.3a and 8.4a) (Bégin 1997). This confirms that most of the flooded terrigenous organic matter is quite refractory to remineralisation under flooded conditions. One must actually consider the importance of biodegradation of the autochthonous matter

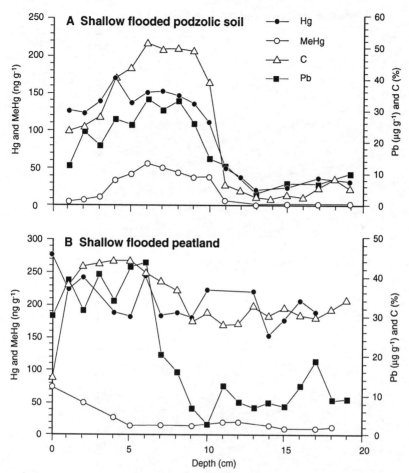

Fig. 8.1. Typical profiles of total mercury, methylmercury, lead and carbon in flooded soils of the Robert-Bourassa reservoir 13 years after its impoundment

produced within the reservoirs and deposited on top of the flooded soils in order to account for the benthic production rates of CO_2 and CH_4 over several decades (Duchemin et al. 1996).

Throughout the first fifteen years of flooding, the total Hg burdens in both the humic horizon of the flooded podzolic soils and the entire organic layer of the flooded peatlands do not appear to significantly evolve (Figures 8.3b and 8.4b) and remain comparable with those in equivalent non flooded soils (Chapter 3.0; Grondin et al. 1995; Bégin 1997). This observation underlines that the initial release of Hg associated with the rapid degradation of the labile organic matter in vegetation and soils (Chapter 7.0) only concerns a negligible fraction of the Hg

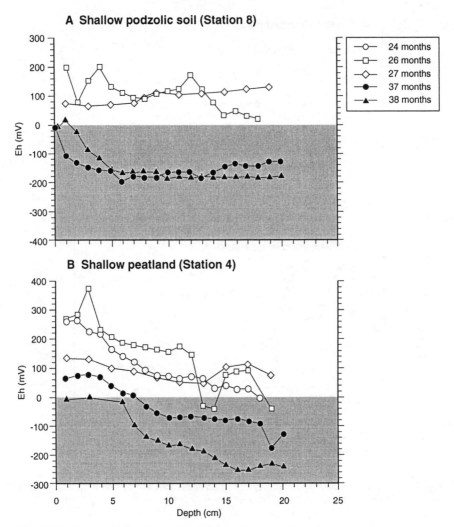

Fig. 8.2. Redox potential profiles in flooded soils of the LA-40 impoundment

reservoir originally present in the soils. It also means that the evolutionary trend of the Hg burden in non eroded flooded soils would be masked by the inherent variability of that burden observed in nearby soils (Chapter 3.0; Grondin et al. 1995; Bégin 1997). After the flooding of podzolic soils, the only slight remobilization of Hg was found associated with the reduction of the iron oxides in the Bf horizon (Dmytriw et al. 1995). Unlike for flooded tropical soils (Roulet and Lucotte 1995), little Hg seems to be directly released to the water column after its desorption from iron oxides as the heavy metal has a strong affinity for the organic matter of the humic layer of the podzolic soil (Johansson et al. 1991; Aastrup et al. 1991; Steinnes 1995; Dmytriw et al. 1995; Grondin et al. 1995, Bégin 1997).

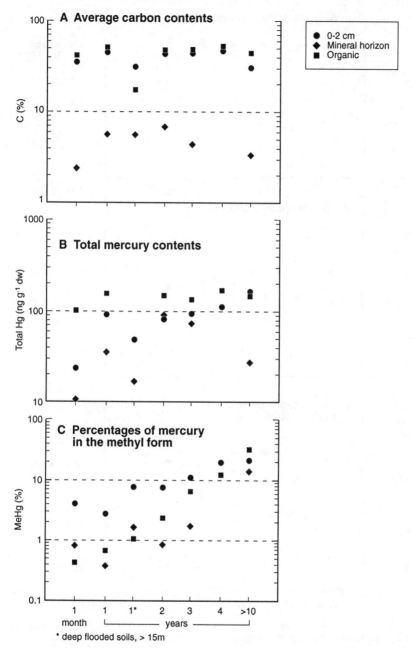

Fig. 8.3. Evolution through time of C, total Hg and MeHg in the various horizons of flooded podzolic soils from the La Grande hydroelectric complex

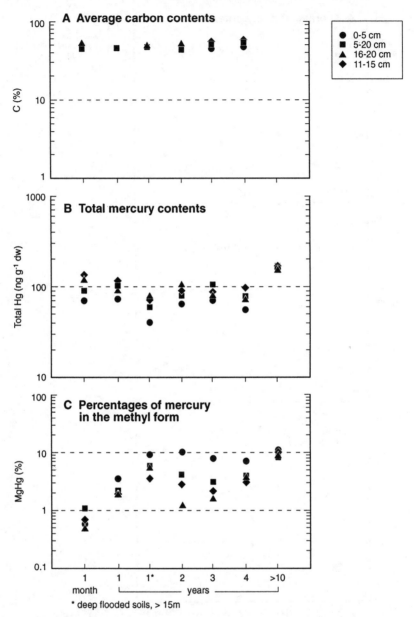

Fig. 8.4. Evolution through time of C, total Hg and MeHg in various layers of flooded peatlands from the La Grande hydroelectric complex

The major change in the Hg biogeochemistry of northern soils after their impoundment is the progressive methylation of their inorganic Hg content over time (Bégin 1997; Figures 8.3c and 8.4c). The transformation of inorganic Hg into

MeHg appears to be fairly intense within the first year following impoundment, both in the humic horizon of the podzolic soils and the surface layers of the peatlands. Whereas less than 1% of the total Hg in natural soils is MeHg before impoundment (Chapter 3.0), this amount increases up to about 10% over the first year of flooding in both types of northern soils. After this initial pulse, the MeHg proportion stabilises around this value in the surface layer of the flooded peatlands, while it still progressively increases in the deeper peat layers to also reach about 10% after 10 years of flooding (Figure 8.4c). In contrast to the peatlands, the methylation of inorganic Hg progresses through time in the humic horizon of the podzolic soils as MeHg represents up to 30% of the total Hg after 10 years of flooding.

In the literature, the Hg methylation rate (M) has been compared to the Hg demethylation rate (D) in various sedimentary environments (Ramlal et al. 1993). A net increase in MeHg concentrations in a given environment is thus described with a M/D ratio > 1. In the flooded soils of northern reservoirs, it seems that the M/D ratio never goes below 1 after more than a decade of impoundment, since once Hg has been methylated, we do not observe a decrease of the MeHg concentrations through time. At most, the MeHg concentrations stabilize in the surface layers of flooded peatlands after one year or so of impoundment. Furthermore, based on sampling conducted just after the spring melt, our results suggest that no winter demethylation of the MeHg formed during the warm summer months occurs. These findings suggest that once formed, MeHg seems to progressively accumulate in the flooded soils. This behavior is most probably attributable to the strong affinity of MeHg for the terrigenous, partially degraded organic matter commonly found in the flooded soils (Mucci et al. 1995; Dmytriw et al. 1995; Bégin 1997).

While 10 and more years of flooding do not seem to significantly affect the organic C burdens initially present in northern soils (Figures 8.3a and 8.3b), this perturbation promotes the biodegradation of the organic matter present, as decreasing redox potentials suggest. Without entirely leading to full remineralisation, the original organic compounds of the soils are progressively denatured as evidenced by the marked decreases of the C/N atomic ratios (Bégin 1997) and loss of cinnamyl and syringyl compounds with respect to vanillyl ones (Farella 1998). The progressive methylation of the Hg initially present in its inorganic form in the pristine soils is then explained as a direct consequence of the maturation of the organic matter in the flooded soils. Strains of anaerobic bacteria distinct from the methanogenic strains but capable of methylating Hg under the environmental conditions of flooded soils have recently been identified.

The flooding of peatlands only leads to a moderate degradation of organic matter as this matter, with the exception of the living sphagnum, is already water saturated under pre-flood conditions. In contrast, the organic matter in the humic horizon of the podzolic soils is more sensitive to degradation upon flooding, as indicated by the strong shifts in the C/N ratios and in the partition of molecular biomarkers in the top few centimetres of these soils after a few months of flooding. Consequently, the methylation of Hg in flooded podzolic soils is more intense and extends over a longer period of time than in peatlands (Figures 8.3c and 8.4c).

Moreover, for either type of flooded soil (podzol or peatland), Bégin (1997) showed that there is no marked difference in Hg methylation between shallow (< 3 m) and deep sampling sites (> 15 m). Considering the often reported influence of temperature on the methylation rate of Hg in sedimentary environments (Parks et al. 1989; Bodaly et al. 1993; Plourde et al. 1997), the absence of differences in MeHg concentrations between shallow and deep sites is probably due to the fact that the deep sites sampled in northern reservoirs were not exposed to much cooler temperatures than the shallow ones, as they were all still lying above the depth of the seasonal thermocline (Bégin 1997). As large surfaces of northern reservoirs are underlain by non eroded flooded soils situated above the seasonal thermocline, one may conclude that the important methylation of Hg described here in the top layers of flooded soils is a major feature of the Hg biogeochemistry in these man made lakes. Finally, Lebeau (1996) demonstrated that protected shallow zones of northern reservoirs, within the photic layer, support the development of periphyton mats. Dense communities of heterotrophic bacteria and microalgae are thriving in these mats, and coincide with MeHg concentrations even higher than those found in the humic horizon of flooded podzols.

8.3.2
Mercury in the Water Column of Flooded Systems

8.3.2.1
Dissolved Fraction

Concentrations of dissolved total Hg (Hg_{T-D}) and dissolved methylmercury ($MeHg_D$) were monitored over a four year period, from 1993 to 1996. The general results are presented in Table 8.1. As noted in Chapter 3.0, investigations concerned with the cycling of Hg in the water column have been hampered by difficulties associated with the sampling and analysis of subnanogram levels of Hg. While in the discussion concerned with natural settings (Chapter 3.0) the time frame is limited to 1994 to 1996 due to better detection limits, the total Hg data presented for the flooded sites span 1993 to 1996. Even with improvements made to the measurement technique no significant difference is observed between 1993 and 1994 to 1996 (Table 8.1). Overall, the concentrations of both Hg_{T-D} and $MeHg_D$ in flooded environments are significantly greater (t-test, $p = 0.05$) than those of natural systems. Furthermore, the proportion of Hg_{T-D} which is in the methylated form is, on average, 4 times greater in the reservoir sites as compared to those of the lakes (i.e., 12% vs 3%). With respect to variability within reservoirs

Figures 8.5 and 8.6 show the results for Hg_{T-D} and $MeHg_D$, respectively. As observed in Figure 8.5 there is no significant difference in inorganic Hg levels despite a nearly two-fold increase in the water column temperature. For the case of MeHg, however, there is a marked increase in concentrations with increased water column temperature (Figure 8.6). This is particularly noted for the well protected shallow sites (Station 4 of LA-40 and Station 184 of Laforge 1) and may be attributed to enhanced bacterial activity as well as a reduced influence of dilution.

Table 8.1. General results of Hg_{T-D} and $MeHg_D$ analyses for the flooded systems (Robert-Bourassa and Laforge 1 reservoirs) sampled in 1993 and between 1994-1996. The average values for neighboring lakes (as presented in Chapter 3.0) are also shown. The units for the average, standard deviation and max./min. are ng Hg L^{-1}

	Hg_{T-D}	$MeHg_D$
1993:		
Average	2.70	n.a.
Standard deviation	0.11	n.a.
Max./min.	5.30/1.10	n.a.
Number of analyses	75	n.a.
1994-1996:		
Average	2.35	0.28
Standard deviation	0.12	0.02
Max./min.	5.34/0.96	0.85/0.03
Number of analyses	79	87
Neighbouring lakes:		
1993:		
Average	2.39	n.a.
Standard deviation	0.15	n.a.
Max./min.	3.67/0.77	n.a.
Number of analyses	53	n.a.
1994-1996:		
Average	1.51	0.05
Standard deviation	0.06	0.004
Max./min.	2.6/0.4	0.11/0.02
Number of analyses	70	30

Another important observation made from this data set is that flooding has a direct influence on MeHg levels. In Figure 8.6b, showing the results for the LA-40 flow-through system, we observe a significant increase in MeHg concentrations in the 95A samples from the flooded zone as compared to those of the upstream

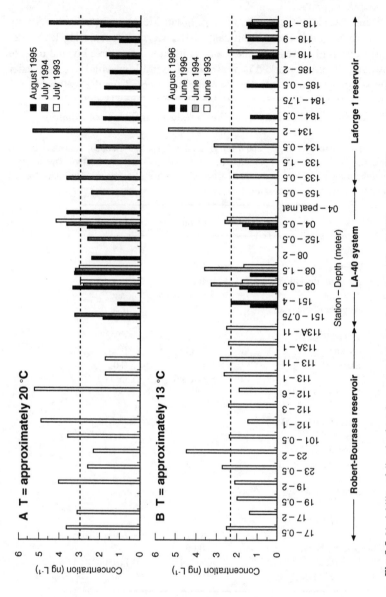

Fig. 8.5. Variability of dissolved total mercury concentrations among flooded sites at two different water column temperatures. Sampling was conducted in the Robert Bourassa reservoir, the LA-40 flooded system and the Laforge-1 reservoir. The sampling depth is indicated along the x-axis. The dashed line represents the average concentration

natural lake. Likewise, lower concentrations are recorded in the downstream lake, however the return to background levels is gradual. That is, at the intermediate sites, Stations 122 and 154F (physically affected during impoundment of LA-40), the concentrations remain elevated. Further downstream (Stations 154B and 188) the MeHg levels decrease but a return to background is not observed until the

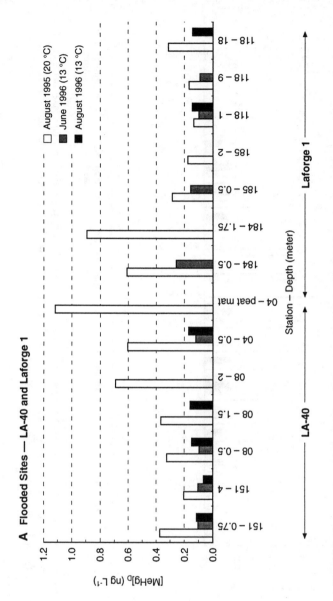

Fig. 8.6A. Concentrations of dissolved methylmercury in the LA-40 and Laforge 1 system. The average water column temperature at the time of sampling is shown in parentheses. The sampling depth is indicated along the x-axis and, in plot B, the station numbers (as referred to in the text) are also shown

station furthest from the flooded zone is reached (i.e., Station 154D). This trend suggests that upstream flooding can have an adverse effect on Hg levels in downstream systems (Chapter 11.0).

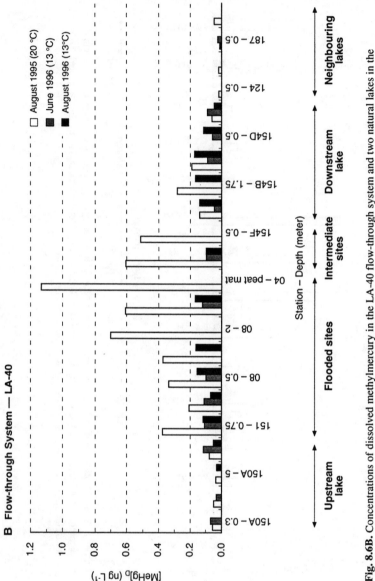

Fig. 8.6B. Concentrations of dissolved methylmercury in the LA-40 flow-through system and two natural lakes in the region (neighboring lakes). The sampling depth is indicated along the x-axis and, in plot B, the station numbers (as referred to in the text) are also shown (continued)

8.3.2.2
Particulate Fraction

During August 1995 large-volume water samples collected from the LA-40 flow-through system were divided into three main fractions: (1) X < 64 µm, (2) 64 µm < X < 210 µm and (3) X > 210 µm, representing, for the most part, unidentified fine suspended particulate matter (FPM), phyto- and microzooplankton, and zoo-plankton, respectively. In the case of the fractions, $X < 64$ μm and $X > 210$ μm, the concentrations of total mercury (Hg_T) were significantly higher in the samples collected from the flooded zone than those from the lakes (mean = 1 167 vs 626 ppb and 882 vs 363 ppb, respectively) (Figure 8.7a and 8.7c). No significant difference was found for the phyto-microzooplankton fraction of the different sampling stations. Similarly, with respect to MeHg, the concentrations determined for the smallest and largest fractions were consistently higher in the samples of the flooded environment (mean = 102 vs 21 ppb and 410 vs 97, respectively), while for the intermediate fraction no significant difference was detected (Figures 8.7a, 8.7b and 8.7c). As with the dissolved fraction, comparing the results for the up-stream and downstream lakes we found that for the $X < 64$ μm and $X > 210$ μm fractions the Hg levels were greater downstream from the flooded region, sug-gesting that the transport (or mobility, in the case of zooplankton) of MeHg ad-sorbed to, or incorporated by, suspended material from flooded environments may be important in the contamination of otherwise pristine systems (Chapter 11.0).

As a first attempt to evaluate the relative importance of the dissolved and par-ticulate fractions in the bioamplification of MeHg in the food chain, the two were compared by converting the concentrations in the FPM fraction for several stations in the LA-40 system to ng L^{-1} units. With the results for Hg_T and MeHg being calculated, we find that in both instances the dissolved concentrations are greater than those of the particulate, with the latter making up 12 - 38% of the former for Hg_T and 14 - 25% in the case of MeHg. While these results seem to point to the potential importance of the dissolved fraction in the transfer of MeHg to higher trophic levels, the findings of Plourde et al. (1997), which show an insignificant difference between lakes and reservoirs with respect to Hg concentrations in a 20-to 150-μm sized fraction (composed principally of phytoplankton, organisms which derive their nutrition in the dissolved form) suggest the contrary. As a con-sequence of this difference, C uptake and trophic interactions in the aquatic food chain were evaluated by determining $\delta^{13}C$ in the dissolved and particulate frac-tions. Measurements of $\delta^{13}C$ for water samples and the $X < 64$ μm fraction show a significant difference between the upstream lake and the flooded site, with a de-crease from about -9‰ to -14‰ for the water column and -29‰ to -36‰ for the particles (the particulate results are shown in Figure 8.7a). Despite the water col-umn difference of 5‰ from the pristine to flooded sites no such variation was observed for the 64 μm $< X <$ 210 μm samples, where $\delta^{13}C$ values show this frac-tion to be little affected by flooding ($\Delta\delta^{13}C = 1$‰). With respect to the $X > 210$ μm fraction, a difference of 4‰ was calculated between the upstream lake and the flooded region suggesting, again, an influence of flooding. Furthermore, compara-ble $\delta^{13}C$ values of the $X < 64$ μm (-35‰) and $X > 210$ μm (-34.5‰) samples col-lected from flooded sites suggest that these fractions are trophically linked, i.e., that zooplankton organisms not only feed on phytoplankton but also heavily on

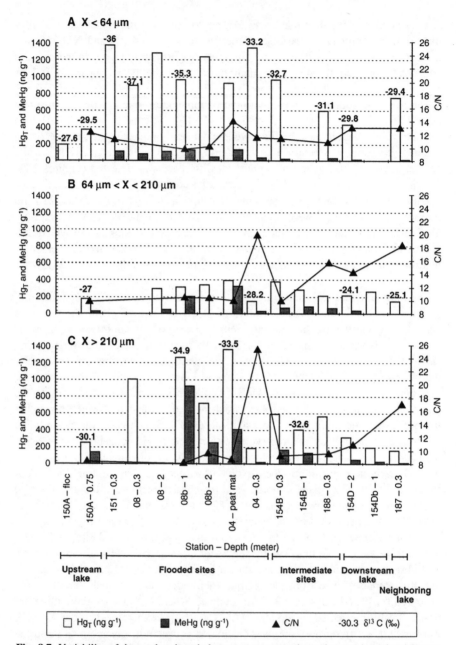

Fig. 8.7. Variability of the total and methylmercury concentrations, the atomic carbon/nitrogen ratios, and the carbon-13 stable isotope ratio in the three principal particulate fractions. Sampling was conducted in the LA-40 flow-through system during August 1995. The sampling depth is indicated along the x-axis. The intermediate sites are those influenced by upstream flooding without being physically part of the flooded zone

detrital organic matter in suspension in the water column or deposited at the bottom of the reservoirs (Chapter 9.0).

8.3.3
Influence of Flooded Soils on the Biogeochemical Cycle of Mercury in Reservoirs

8.3.3.1
Diffusion of Mercury from Flooded Soils

Analytical and contamination problems have hampered direct measurements of benthic fluxes of Hg at the flooded soil-water interface. To our knowledge, Montgomery et al. (1996) is the only study that has been conducted to determine total Hg concentrations in interstitial waters of flooded soils in a reservoir. In that study, the dialysis technique used revealed weak, but not accurately quantifiable fluxes of Hg from flooded soils to supernatant waters in the Robert-Bourassa and Laforge 1 reservoirs, 14 years and 1 year old, respectively, at the time of the study. The *in situ* measurements of the evolution of total Hg and MeHg concentrations in the water column of the Laforge 1 reservoir throughout its first three years of impoundment revealed small but significant increases in dissolved Hg under both total and organic forms in the protected zone of the LA-40 impoundment where the dilution factor is minimal (Figures 8.5 and 8.6). These observations suggest that a fraction of the heavy metal initially bound to the vegetation and soils is leached to the water column during the first few months following impoundment, corroborating the *in vitro* findings reported in Chapter 7.0. Nevertheless, the amounts of Hg released to the water column by diffusion are exceedingly small as compared to the total Hg burden found in the humic layer of podzolic soils or organic layer of peatlands, and consequently, no significant departure of Hg can be evidenced in the soil even after a decade flooding (Figures 8.3b and 8.4b).

The exact role played by dissolved inorganic Hg and dissolved MeHg directly released to the water column of the reservoirs on the contamination of the aquatic food chain still remains uncertain. As demonstrated above, differences in dissolved Hg concentrations do not seem to have a clear influence on the Hg content in phytoplankton (Chapter 9.0; Plourde et al. 1997). On the other hand, it is probable that dissolved Hg readily released from the flooded soils to the supernatant water column rapidly re-binds to fine organic particulate matter in suspension, thus rendering this Hg more bioavailable to filter feeding organisms at the base of the food chain.

8.3.3.2
Burrowing Organisms and Erosion

Forest soils represent the major depository of Hg in northern reservoirs. In spite of this, non-eroded flooded soils appear to retain most of their initial Hg burden as a result of the strong affinity of inorganic and organic Hg for the humic matter of the soil (Figures 8.3b and 8.4b). This observation holds true even after the partial biodegradation of the flooded organic matter in the surface layers of the soils rapidly leads to the methylation of a substantial fraction of the initial inorganic Hg burden of the soils (Figures 8.3c and 8.4c). In the absence of erosion, insect larvae located in the first few centimeters of the flooded soils are the main aquatic organisms directly exposed to the higher concentrations of total and organic Hg. In some cases, such as for dipterans, larvae rapidly and densely colonise the first centimeters of the shallow flooded podzolic soils (Tremblay and Lucotte 1997). As the larvae feed on partially degraded and Hg-rich organic matter, they rapidly bioaccumulate significant concentrations of the heavy metal and then transfer it to higher aquatic organisms as they emerge (Chapter 9.0; Tremblay et al. 1998b).

During the first years of creation of the northern reservoirs, most of the shallow and exposed flooded soils of the drawdown zone are progressively eroded (Louchouarn et al. 1993). In particular, the organic horizon of podzolic soils is readily subjected to loosening by wave or ice action. A rapid sorting of the eroded soil particles then occurs, resulting in the suspension of the fine, Hg-rich organic particles in the water column of the reservoirs (Figure 8.8; Mucci et al. 1995). Filter feeding organisms ingesting these organic particles may then constitute a primary entry route of Hg to the aquatic food chain of the reservoirs (Grimard 1996; Plourde et al. 1997; Tremblay et al. 1998a). The fine organic particles in suspension may then eventually settle out to the surface of deeper flooded soils, where they become MeHg-rich food available for benthic feeders, thus continuing to participate in the transfer of Hg from the former flooded soils to the organisms of the reservoirs.

8.3.3.3
Enhanced Biological Production and Methylation of Mercury

The gradual release of nutrients from the newly flooded vegetation and soils to the water column represents one of the major effects of impoundment on the biogeochemistry of northern reservoirs. Within the first few years of impoundment, this release promotes an increase of the phytoplankton production in the reservoirs when compared to nearby pristine lakes (Schetagne 1992). This increase is particularly strong in the shore zones of the reservoirs, where the effects of the nutrient release are maximal due to their minimal dilution in the water column. In turn, the enhanced primary production has positive repercussions on secondary production and fish stocks of the newly impounded reservoirs (Verdon et al. 1991;

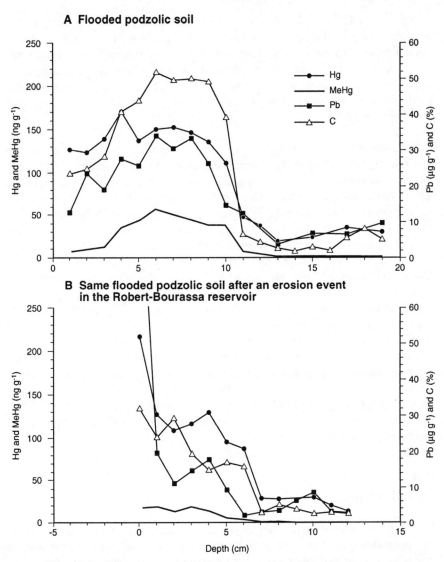

Fig. 8.8. Organic carbon, total mercury, methylmercury, and lead distribution (adapted from Mucci et al. 1995). Note that the organic horizon of the eroded soil has been reduced by approximately 5 centimetres with respect to the originally flooded soil

Chapters 9.0, 10.0 and 11.0). At the end of the biological cycle, the general increase in biological production in the water column translates into higher sedimentary fluxes of autochthonous organic matter at the bottom of the reservoirs (Louchouarn et al. 1993).

This new sedimentary layer, composed of fairly labile organic C as compared to the ligneous compounds of the flooded soils, thus promotes a fairly sustained bacterial activity (Duchemin et al. 1996). In addition, in the shallow and protected zones of the reservoirs, the combined effects of light penetration and mineralisation of nutrients leads to the development of benthic algal mats growing in symbiosis with the bacterial colonies degrading the autochthonous organic matter (Lebeau 1996).

After the initial pulse of Hg released to the water column upon the degradation of labile organic matter in the flooded vegetation and soils, little new Hg (either inorganic or organic) appears to escape the non-eroded flooded soils towards the water column and to ultimately accumulate in aquatic organisms. Rather, most of the Hg already bioconcentrated in aquatic organisms appears to be rapidly recycled and reenters the food chain. In this respect, zooplankton at the base of the food chain appear to heavily feed on detrital, bacteria-rich organic matter in suspension in the water column rather than simply on phytoplankton (Figure 8.7; Chapter 9.0; Grimard 1996; Plourde et al. 1997; Tremblay et al. 1998a). Similarly, bacterial mats at the flooded soil-water interface may also play an important role in certain shallow and protected areas. Not only do they rapidly degrade the labile organic matter, but they promote the methylation of the Hg content of the degraded organic matter, particularly when in symbiosis with algae (Figures 8.9a and 8.9b; Lebeau 1997). The periphyton layer appears to constitute a prime source of food for zooplankton grazers and insect larvae during their burrowing stage (Chapter 9.0, Tremblay and Lucotte 1998a). Finally, the periphyton layer is easily loosened up during storm or drawdown events, enabling the ingestion of these MeHg-rich fine particles by the filter feeding organisms living in the water column of the reservoirs.

8.4
Conclusions

Following an initial release of Hg to the water column during the first few weeks of impoundment and in association with the rapid degradation of labile organic compounds in the vegetation and soils, Hg contamination of the aquatic organisms living in northern Québec reservoirs is fuelled by a partial and gradual transfer of the Hg accumulated over time in the soils of the boreal forest. Natural and anthropogenic Hg originally contained in the soils prior to their flooding is transferred to the aquatic organisms. Although the long range atmospheric Hg pollution is far from being negligible in the region of the reservoirs, the anthropogenic Hg load remains small in the northern soils as compared to the natural Hg accumulated over millennia (Chapter 3.0). Thus, the Hg transferred from the flooded vegetation and soils to the aquatic organisms of the reservoirs is primarily natural Hg, being only slightly enhanced by the recent deposition of anthropogenic Hg.

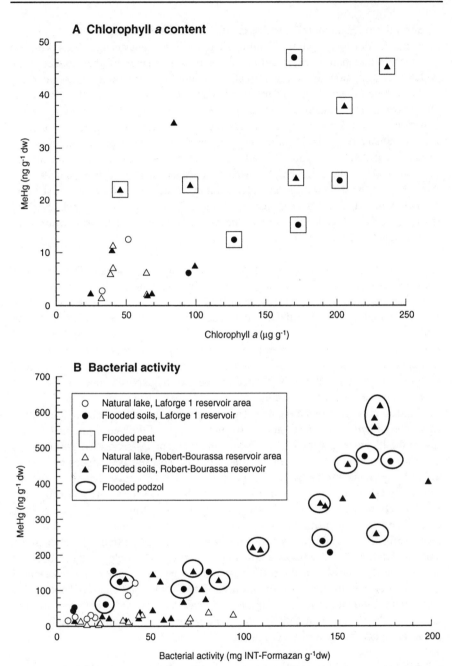

Fig. 8.9. Methylmercury concentrations at the end of the summer time as a function of chlorophyll-a content and bacterial activity in the periphyton layer developing at the surface of shallow flooded soils of the La Grande complex reservoirs (from Lebeau 1996)

Several biotic and abiotic processes affect the behavior of Hg in forest soils and in peatlands during long-term flooding. The most striking change deals with the progressive and partial degradation of the organic matter of the flooded soils which is accompanied by a transformation of significant fractions of inorganic Hg in the surface organic layers of the flooded soils into MeHg. Even though the organic form of Hg is much more bioavailable to living organisms than its inorganic counterpart, little Hg thus methylated in the surface layers of the flooded soils is readily transferred to the biota living at the bottom or in the water column of the reservoirs. Instead, newly methylated Hg progressively accumulates in the surface organic layers of the soils. Even if weak increases of dissolved inorganic and organic Hg can be detected in the water column of the reservoirs in the first years after flooding, these releases alone appear far too small to account for the rates of increase of the Hg concentrations in the higher organisms of the reservoirs over several years (Chapter 12.0).

In order to explain the general Hg contamination of the aquatic food chain of the reservoirs, one should take into account a combination of several indirect processes in addition to the free diffusion of Hg from the flooded soils and vegetation to the water column. One of these may be represented by the action of insect larvae burrowing in the flooded environments, particularly the podzolic soils, and emerging as prey in the water column after having bioaccumulated considerable amounts of Hg (Chapter 9.0). A second process appears to be the rapid erosion of the organic horizon of the soils situated in shallow and exposed zones of the reservoirs. Opportunistic filter feeding organisms benefit from this action, but also accumulate considerable amounts of Hg by ingesting detrital organic matter rather than solely phytoplankton (Chapter 9.0). Finally, one of the major processes leading to the bioaccumulation of MeHg in aquatic organisms seems triggered by the significant nutrient release from the flooded soils. This release induces an overall enhanced primary productivity in a reservoir, which of course readily translates into a higher production of zooplankton and higher organisms. The generally high bioproduction in a reservoir following its impoundment (including the burrowing insects and the organisms feeding on the eroded organic matter) contributes to high levels of bacterial activity associated with suspended particulate matter in the water column and at the flooded soil-water interface. Not only do these bacteria seem to be responsible for the methylation of the Hg in circulation, but they also serve as food for several invertebrates at the base of the food chain (Chapter 9.0). Thus, with minimal amounts of Hg newly released to the water column from the flooded soils, Hg in circulation appears to be progressively methylated and further bioaccumulated by the aquatic organisms.

In order to establish a mechanistic model of the Hg cycle in northern reservoirs, one must now quantify at the megascale of these man made lakes the relative importance of all of the above described processes in terms of Hg transfer from the flooded soils and vegetation to the aquatic organisms.

Acknowledgements

This research was supported by a grant from Hydro-Québec, the Conseil de recherches en sciences naturelles et en génie (CRSNG) du Canada and the Université du Québec à Montréal (UQAM) to the Chaire de recherche en environnement HQ/CRSNG/UQAM. Additional financial support, in the form of student scholarships, was supplied by the fund for the Formation de Chercheurs et l'Aide à la Recherche (Fonds FCAR) and the Programme de Formation scientifique dans le Nord (PFSN). The laboratory and field work for this project could not have been accomplished without the assistance of many motivated people, particularly, P. Pichet, P. Ferland and I. Rheault, as well as, R. Canuel, B. Caron, L. Cournoyer, É. Duchemin, B. Fortin, C. Guignard, M. Kainz, D. Lebeau, Y. Plourde, L.-F. Richard, S. Tran, and A. Tremblay.

Evolution of Mercury Concentrations in Aquatic Organisms from Hydroelectric Reservoirs

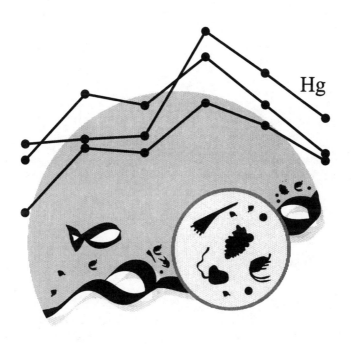

Hg

9 Bioaccumulation of Methylmercury in Invertebrates from Boreal Hydroelectric Reservoirs

Alain Tremblay

Abstract

Methylmercury (MeHg) concentrations in invertebrates (insects and zooplankton) increased rapidly in the first years of flooding and remained high in non-eroded littoral zones of reservoirs 16 years after flooding with values ranging from 45 ng g^{-1} dry weight (dw) to 650 ng g^{-1} dw. In general, both biomass and MeHg content in invertebrates from the reservoirs were 3-5 times (up to 10 times) higher than in their counterparts from natural lakes. Plankton collected from pelagic zones 8 years after impoundment showed MeHg concentrations similar to those of natural lakes with values ranging from 20 ng g^{-1} dw to 140 ng g^{-1} dw. The results indicate a biomagnification of MeHg along the invertebrate food chain, since MeHg concentrations increase when moving from suspended particulate matter (SPM) to > 150 µm-mesh plankton for the pelagic food chain and from the substrates to the predators insects for the benthic food chain. The biomagnification factors are independent of the type of environment (natural lake or reservoir). The results also suggest, in reservoirs, that SPM eroded from flooded soils by wave and ice action, and the biofilm at the soil-water interface may constitute the active route for the transfer of MeHg from flooded soils and vegetation to the invertebrates. In the long term, the fluctuation of reservoir water levels, which gradually erodes the flooded soils, results in a loss of habitat for insects, a decrease of the amount of SPM and a reduction of the overall invertebrate biomass and MeHg burden. These processes probably explain the decrease of Hg levels observed in non-piscivorous fish, as both the food and the MeHg sources provided by the invertebrates decrease over time.

9.1
Introduction

Numerous researchers working in remote temperate regions have documented elevated concentrations of mercury (Hg) in fish of hydroelectric reservoirs (Potter et al. 1975; Cox et al. 1979; Meister et al. 1979; Bodaly et al. 1984). Hg from both natural and anthropogenic sources is known to be widely dispersed via atmospheric transport and then accumulated in the humic horizon of soils following wet and dry deposition (e.g., Lucotte et al. 1995a; Grondin et al. 1995). It is generally recognized that the accumulation of Hg in fish of reservoirs is related to increases in bacterial activity and the release of MeHg from flooded soils to the water column where it is made available for aquatic organisms (Chapter 8.0; Verdon et al. 1991; Schetagne et al. 1996).

Since the ingestion of food is the dominant pathway of Hg accumulation in fish, and invertebrates constitute a major part of their diet, zooplankton and aquatic insects probably play a key role in the contamination of non-piscivorous fish (e.g., Spry and Wiener 1991; Schetagne et al. 1996, Tremblay et al. 1998a). This chapter will show that invertebrates play a key role in the contamination of fish of hydroelectric reservoirs of northern Québec, by the indirect and active MeHg transfer from flooded soils to non-piscivorous fish. To assess the accumulation of MeHg in invertebrates and thus in the fish community, extensive sampling was carried out during the first 3 years of a newly created reservoir as well as in reservoirs of various ages and in several natural lakes.

9.2
Materials and Methods

9.2.1
Study Area

Field work was carried out during the ice free periods of 1991-1995 in the La Grande region of northern Québec. The study focussed on five reservoirs and two river diversions of the La Grande hydroelectric complex created in 1978-1984 (Figure 9.1). Also included in the study was a small water body, LA-40, immediately outside the perimeter of the Laforge 1 reservoir, impounded in June 1992 (Table 9.1). Two sampling stations (4 and 8) were established in the LA-40 impoundment, fed by a pristine lake upstream (Lake 150) and drained through a dredged channel to Lake 154. The residence time of the flow-through systems was 7-15 days (Figure 9.2, Table 9.1).

9.2.2
Sampling

9.2.2.1
Insects

Insect larvae were collected from the littoral zones (< 2 m deep) using a 250 µm mesh hand-held net, which enables sampling over flooded soils in the presence of broken trees and root systems. Insect larvae were sorted by family or genus, counted and frozen.

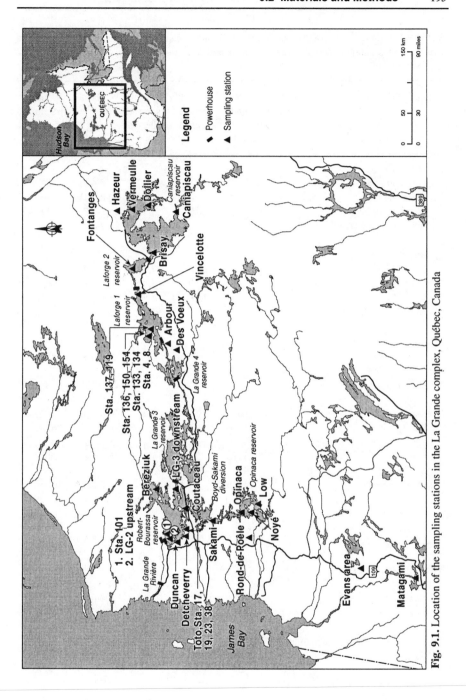

Fig. 9.1. Location of the sampling stations in the La Grande complex, Québec, Canada

Table 9.1. Characteristics of stations in the La Grande complex area. Values represent ranges over the entire sampling period (Type of zone sampled, L:littoral, P:pelagic; Type of organisms, I:Insects, P:plankton)

Name of the station	Year of flooding	Age of impound-ment (years)	Type of zone sampled	Type of organisms collected	pH	Cond. (μS cm^{-1})
NATURAL LAKES						
LA GRANDE 2 AREA						
Duncan lake			L, P	I, P	6.8-7.1	30
Detcheverry lake			L, P	I, P	6.8-7.3	37
Rond-de-Poêle lake				P	6.8-7.0	29
LAFORGE 1 AREA						
Lake 136			L, P	I, P	6.9-7.3	12
Lake 150			L, P	I, P	7.0-7.3	13
Lake 154			L, P	I, P	6.9-7.2	13
Des Voeux lake			L, P	I, P	6.9-7.1	12
CANIAPISCAU AREA						
Hazeur lake			P	P	5.9-6.8	11
ROBERT-BOURASSA RESERVOIR	1978					
Sta. 17		5, 16	L	I, P	6.6-6.9	20-21
Sta. 19		14, 15, 16	L	I, P	6.6-6.9	20-21
Sta. 23		14, 15, 16	L	I, P	5.7-6.2	19-21
Sta. 101		14, 15, 16	L	I, P	5.7-6.2	19-21
Sta. 38		14	L	I	6.0-6.2	18-20
LG2 upstream		12, 14, 16	P	P	4.7-5.1	10-30
Bereziuk		12, 14	P	P	4.8-5.2	13-30
Coutaceau		12, 14, 16	P	P	4.8-5.2	10-28
Toto		14, 16	P	P	4.8-5.1	5-15
LG-3 downstream		9, 11	P	P	5.3-5.5	9-28
OPINACA RESERVOIR	1980					
Opinaca		12	P	P	4.3-6.2	-18
Noyé		12, 14	P	P	4.4-6.6	10-40
Low		12	P	P	5.1-6.7	4-14
BOYD-SAKAMI DIVERSION	1980					
Sakami		13	P	P	4.9-6.5	10-21
LA GRANDE 4 RESERVOIR	1983					
Arbour lake		10	P	P	6.0-6.8	9-29

Table 9.1. Characteristics of stations in the La Grande complex area. Values represent ranges over the entire sampling period (Type of zone sampled, L:littoral, P:pelagic; Type of organisms, I:Insects, P:plankton) (continued)

Name of the station	Year of flooding	Age of impound-ment (years)	Type of zone sampled	Type of organisms collected	pH	Cond. (μS cm^{-1})
CANIAPISCAU RESERVOIR	1981					
Dollier		11, 14	P	P	5.1-5.7	4-14
Caniapiscau		11	P	P	5.1-6.1	4-14
Brisay		11	P	P	5.8-6.5	10-14
Vermeulle		11, 14	P	P	4.7-6.6	5-17
LAFORGE 1 AREA	1993					
Sta. 4		2, 3	L	I, P	5.5-6.1	11-12
Sta. 8		2, 3	L	I, P	5.9-6.4	11-13
Channel 4-8		3	L	I, P	5.9-6.4	11-14
Sta. 133		1,2	L	I, P	6.3-6.5	10-12
Sta. 134		1,2	L	I, P	6.3-6.6	10-12
Sta. 137		1, 2	P	P	6.2-6.5	9-11
Sta. 119		1, 2	P	P	6.1-6.4	9-11
LAFORGE DIVERSION	1982					
Fontanges		8	P	P	5.4-5.7	7-9
Vincelotte		8	P	P	6.2-6.7	9-11

Adult insects were collected from the littoral zone (1-2 m deep) of four flooded stations (4, 8, 133 and 134) and a natural lake (Lake 136), using 1 m^2 floating emergence traps with a 250 μm mesh size net. Emerging adults were retained in a flask containing a 1:1 solution of ethylene glycol and distilled water. Upon return to the field laboratory, the insect samples were rinsed with both tap and distilled water to remove the excess ethylene glycol, and kept in 85% ethanol until further analysis. Adult insects were sorted by species using a binocular microscope and tweezers with stainless steel tips.

For adult insects, annual relative abundance (number of individuals m^{-2} year^{-1}) was calculated using the mean total number of individuals collected from 2 emergence traps for the entire sampling season from June 20 to September 15, 1994 (Table 9.3). Annual insect biomass (mg m^{-2} year^{-1} dw) was calculated from the abundance and the mean dry weights of 10 individuals of the most common taxa: *Agrypnia* spp. (11 mg), *Polycentropus* spp. (1 mg), other trichopterans (5 mg), *Chironomus* spp. (0.5 mg), *Ablabesmyia* spp. (0.13 mg) and other dipterans (0.05 mg) (Table 9.3). The manipulation of insects is described in greater detail in Tremblay et al. (1998b).

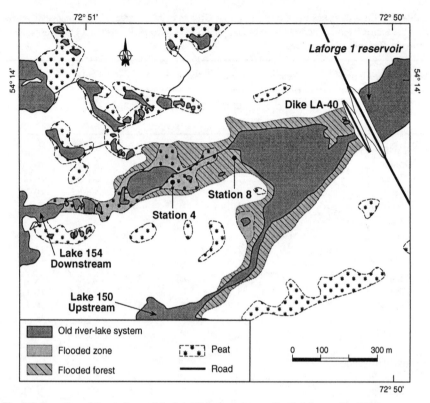

Fig. 9.2. Location of the stations of the LA-40 impoundment, flooded September 1993

9.2.2.2
Plankton

Plankton sampling was carried out in surface waters during the day (littoral zone, depth < 3 m, or pelagic zone, depth > 10 m) using conical nets of either 20, 75, 150 or 225 μm mesh. Suspended particulate matter (SPM < 40 μm) was collected by continuous flow centrifugation of about 50 L of water. Additional information describing sampling techniques is available in Chapter 4.0, Plourde et al. (1997) and Doyon and Tremblay (1997).

During the summer 1994 sampling trip, zooplankton abundance (mean number of individuals L^{-1} of 2 sampling dates) was determined using 50 L water samples collected from four flooded littoral zone stations (4, 8, 19 and 133) and two flooded pelagic zone stations (137 and 119), as well as for four natural lakes.

Ten individuals of the most common taxa were measured and their dry weight calculated from a length-weight regression (Malley et al. 1989). Zooplankton biomass ($mg\,L^{-1}$ dw) was calculated from the abundance and the mean dry weight.

9.2.2.3
Flooded Soils and Lake Sediments

Sublittoral sediment cores from lakes and flooded soils were collected by SCUBA divers using 15 cm-diameter PVC tubes. The cores were sectioned at 1 cm intervals, and the sections were kept frozen until their analyses. The manipulation of flooded soils and sediments cores is described in greater details in Tremblay (1996).

9.2.3
Analysis

Pooled insect samples consisting of at least 10 individuals and bulk plankton samples were used since this procedure enables all analyses (total Hg, MeHg, C, N) to be conducted on the same sample. The total Hg and MeHg concentrations, determined by cold vapor atomic fluorescence spectrophotometry (CVAFS), as well as the quality control of the samples are described in details in Chapter 2.0. The preservative solutions, ethylene glycol and 85% ethanol, were analyzed for traces of total Hg and MeHg and in both cases the concentrations were below the detection limits.

9.3
Results and Discussion

9.3.1
Methylmercury Concentrations in Invertebrates

Our studies show a range in MeHg concentrations in insect larvae collected from natural lakes that was comparable to those reported for different taxonomic groups of Swedish and Finnish lakes, with concentrations from 13 ng g^{-1} to 550 ng g^{-1} (dw) (Figure 9.3) (Parkman and Meili 1993; Rask et al. 1994). In reservoirs, insect MeHg concentrations increase rapidly within the first two years after flooding to reach levels ranging from 45 ng g^{-1} to 776 ng g^{-1} (Figures 9.3 and 9.4). Similarly, Surma-Aho et al. (1986) for Finnish reservoirs, as well as Hall et al. (1998) for the Experimental Lakes Area Reservoir Project (ELARP) in Ontario, reported increases in MeHg concentrations in insects, with levels ranging from 60 ng g^{-1} to 694 ng g^{-1}.

For plankton, our studies show a range in MeHg concentrations from 18 ng g^{-1} to 180 ng g^{-1} (dw) in natural lakes and from 25 ng g^{-1} to 625 ng g^{-1} in reservoirs (Figures 9.5 and 9.6). Back et al. (1995), Tsalkitzis (1995) and Westcott and Kalff (1996) reported values of MeHg ranging from 40 ng g^{-1} to 244 ng g^{-1} for different lakes of Canada and United States. Surma-Aho et al. (1986) for Finnish

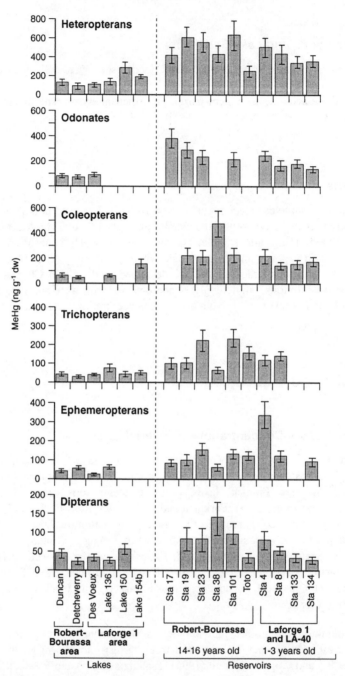

Fig. 9.3. Mean methylmercury concentrations (ng g^{-1} dw) in various insect orders from the shallow, non eroded and protected areas of different lakes and reservoirs in Québec. Error bars represent one standard deviation of the mean (modified from Tremblay and Lucotte 1997)

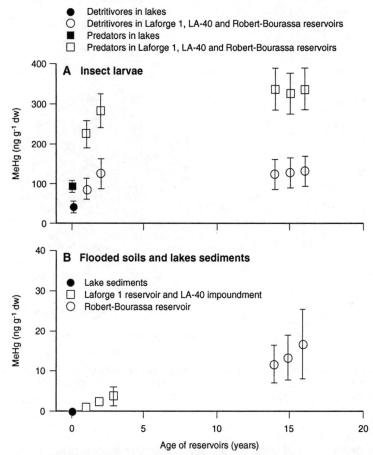

Fig. 9.4. Mean methylmercury concentrations (ng g^{-1} dw) in insect larvae collected from the littoral zone (depth < 2 m) (A) and in the substrate (B) as a function of the age of reservoirs (years, age 0 represents the mean value for natural lakes). Error bars represent one standard deviation of the mean (modified from Tremblay et al. 1996ab)

reservoirs, and Kelly et al. (1997) for the ELARP project, reported plankton MeHg concentrations ranging from 50 ng g^{-1} to 730 ng g^{-1}. As is the case for insects, the MeHg concentrations in zooplankton from reservoirs rose rapidly after flooding and stayed at high levels in protected littoral zones, for as long as 16 years (Figure 9.6B), but zooplankton collected in the pelagic zone of reservoir 8 to 10 years after flooding had similar MeHg concentrations to those of natural lakes (Figure 9.6A).

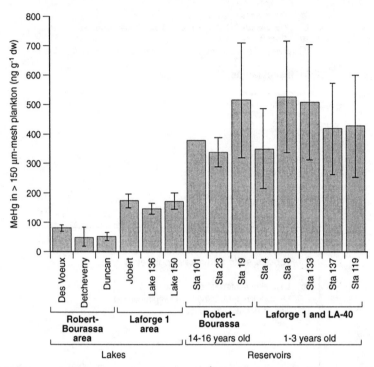

Fig. 9.5. Mean methylmercury concentrations (ng g^{-1} dw) in plankton (> 150 µm-mesh) from different lakes and reservoirs in Québec. Error bars represent one standard deviation of the mean (modified from Tremblay 1996)

9.3.2
Biomagnification of Methylmercury Along the Invertebrate Food Chain

Two distinct trophic levels for the benthic invertebrate food web structure were defined by Tremblay et al. (1996ab). Detritivore-grazer and herbivore-predator organisms represent, respectively, the first and the second trophic levels. Similarly, three trophic levels for the pelagic food web were defined by Tremblay et al. (1998a) using the mesh-sizes of the collecting plankton nets. The suspended particulate matter smaller than 40 µm corresponds to the first trophic level. The 20-75 µm mesh collected many groups of small organisms such as rotifers and large phytoplankton, but excluded the nanoplankton, and corresponds to the second trophic level. The fractions larger than 150 µm mesh net collected mainly large zooplankton but also large phytoplankton and possibly organic debris, and corresponds to the third trophic level.

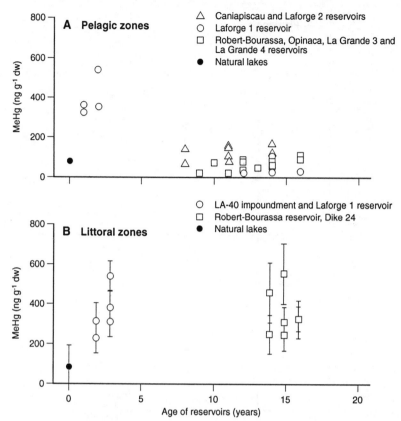

Fig. 9.6. Mean methylmercury concentrations (ng g^{-1} dw) in plankton collected from (A) pelagic zones (> 150 μm-mesh. depth < 10 m) (B) and littoral zones (> 150 μm mesh. depth < 3 m) of reservoirs of varying age (years, age 0 represents the mean value for natural lakes). Error bars represent one standard deviation of the mean

The results of our studies indicated a biomagnification of MeHg along the invertebrate food chain, since MeHg concentrations increase when moving from SPM to > 150 μm-mesh plankton for the pelagic food chain and from the substrate to the predator insects for the benthic food chain (Figure 9.7). Biomagnification factors (BMF, concentration of MeHg in one trophic level divided by the concentration in the level below) of about 3 between two adjacent trophic levels were obtained for both natural lakes and reservoirs (Table 9.2). We also observed a constant ratio of mean MeHg concentrations between given taxa in the insects (e.g., odonates/heteropterans, odonates/trichopterans, etc) and in the plankton communities (> 150 μm mesh/20 μm mesh, etc). These ratios give an idea of the relative position of a taxon with respect to another and indicate that the relationships between given taxa are the same for both natural lakes and reservoirs although the MeHg concentrations differ by a factor of 2-3 (Tremblay and Lucotte

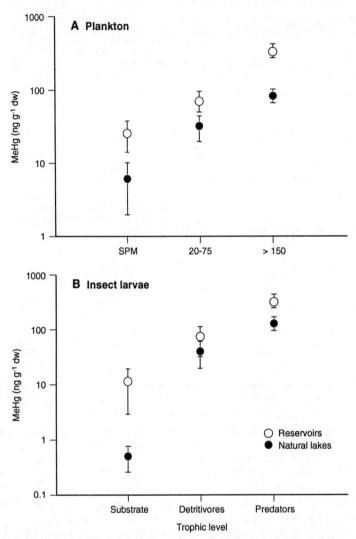

Fig. 9.7. Methylmercury concentrations (ng g^{-1} dw) in plankton (A) and insects (B) versus the trophic level of the organisms. For plankton; trophic levels refer to suspended particulate matter and various plankton mesh sizes (see text). For insects; trophic levels refer to substrates (flooded soils and lake sediments), detritivore (dipterans and trichopterans) and predator (heteropterans and odonates) organisms (see text). Error bars represent one standard deviation of the mean

1997; Tremblay et al. 1996b). These results suggested that MeHg biomagnification along an invertebrate food chain works similarly in both reservoirs and natural lakes and gives us insight into the differences between these two environments with respect to the bioavailability of MeHg or its transfer to the base of the food chain.

Table 9.2. Mean methylmercury concentrations in insect larvae (ng g^{-1} dw) (detritivores and predators) and predators/detritivores MeHg ratio in insect larvae

	Mean MeHg concentrations in detritivores (dipterans, trichopterans and ephemeropterans) (ng g^{-1} dw)	Mean MeHg concentrations in predators (odonates, heteropterans and coleopterans) (ng g^{-1} dw)	Ratio predators/ detritivores
NATURAL LAKES			
Duncan	41.67	90.33	2.17
Detcheverry	31.67	70.00	2.21
Des Voeux	26.67	96.00	3.60
Lake 136	50.33	98.00	1.95
Lake 150	49.50	280.00	5.66
Lake 154b	45.00	164.00	3.64
Lake 154a	110.00	305.00	2.77
Mean	**50.69**	**157.62**	**3.14**
RESERVOIRS			
Sta. 17	89.50	396.50	4.43
Sta. 19	96.00	370.67	3.86
Sta. 23	151.67	331.67	2.19
Sta. 38	86.67	447.50	5.16
Sta. 101	148.67	364.00	2.45
Sta. 112	99.33	242.00	2.44
Sta. 4	173.33	316.67	1.83
Sta. 8	99.67	243.67	2.44
Sta. 133	37.00	214.33	5.79
Mean	**109.09**	**325.22**	**3.40**

9.3.3
Bioavailability of Methylmercury to Insects

Upon flooding, forest soils are subjected to biochemical and physical modifications that affect the Hg cycle and lead to the release of nutrients, the methylation of Hg, its accumulation in the flooded soils over time and its transfer to the food chain (Chapter 8.0; Bégin 1996; Tremblay and Lucotte 1997; Plourde et al. 1997). These processes are most active in the first 2-3 years after flooding when MeHg concentration in flooded soils, insect larvae and plankton increase rapidly (Figures 9.4 and 9.6). The key role of flooded soils in the bioaccumulation of MeHg is further supported by Bodaly et al. (1984), Johnson et al. (1991) and Verdon et al. (1991) who all reported a positive relationship between the rate of Hg accumulation in fish and the proportion of flooded land in a reservoir.

The release of nutrients which enhance microbial productivity also stimulates the production of phytoplankton, periphyton and zooplankton (Pinel-Alloul 1991; Schetagne 1994; Lebeau 1996). Part of this autochthonous productivity eventually settles to the bottom and, along with the abundant degradable organic carbon (C) of the flooded soils and vegetation, may contribute to the long-term maintenance of high microbial productivity in certain parts of reservoirs (Chapter 8.0). This soil-water interface is composed of microbes, phytoplankton, periphyton and soil particles, and is characterized by lower C/N and $^{13}C/^{12}C$ ratios and by higher MeHg concentrations than in the underlying layers (Louchouarn et al. 1993; Grondin et al. 1996; Chapter 8.0). Insects may feed at this soil-water interface and within the soil itself and thereby constitute an active biological pathway for the transfer of MeHg from flooded soils to the food chain (Tremblay and Lucotte 1997).

Another important active process is the physical erosive action of waves and ice which suspends particles originating from the flooded soils (e.g., organic matter) (Mucci et al. 1995; Plourde et al. 1997). These suspended particles which are rich in MeHg can settle to the bottom where they are mixed with the algae and microbes present in the biofilm at the soil-water interface and serve as food for insect larvae (Mucci et al. 1995; Lebeau 1996; Plourde et al. 1997; Tremblay and Lucotte 1997; Chapter 8.0). These suspended particles represent another pathway, along with the soil-water interface and the soils themselves, that could contribute to the active transfer of MeHg from the flooded soils to the food chain. Dissolved Hg compounds released to the water column, following intense flooding and drawdown action, could also sorb to SPM and represent a transfer route of MeHg to invertebrates (Chapters 7.0 and 8.0).

The reservoir insect community was dominated by a few abundant taxa that were present in all stations (namely: Chironomini, *Chironomus* spp., *Ablabesmyia* spp., Coryneurinae, Orthocladiinae and *Psectrocladius* spp.). *Chironomus* spp. was the most abundant group in this study, representing 45-65% of the total number of individuals (Table 9.3). In natural lakes, chironomids represented around 40-45% of the total number of individuals, but lakes differed from the reservoirs mainly by the fact that trichopterans were very important making up about 42% of the total number of individuals. Chironomini is the most constant element of colonization in new reservoirs and their dominance is probably related to their ability to exploit flooded soils and vegetation, as well as to their resistance to low dissolved oxygen concentrations caused by decaying organic matter during the initial years following reservoir formation (Cloutier and Dufort 1979; Rosenberg et al. 1984; Boudreault and Roy 1985; Rosenberg and Wiens 1994; Tremblay 1996; Tremblay et al. 1998b).

Table 9.3. Abundance, biomass and methylmercury burden in insects collected from the littoral zone for a lake and the LA-40 and Laforge 1 impoundments during summer of 1994

Taxon	Abundance (Ind. m^{-2} $year^{-1}$)					Biomass (mg m^{-2} $year^{-1}$)					MeHg burden (ng m^{-2} $year^{-1}$)				
	Lake	LA-40		Laforge 1		Lake	LA-40		Laforge 1		Lake	LA-40		Laforge 1	
	136	4	8	133	134	136	4	8	133	134	136	4	8	133	134
TRICHOPTERA															
LIMNEPHILIDAE															
Anabolia bimaculata		4	3	-	1		63	45	-	9		9	6	-	1
Asynarchus curtus		8	16	-	2		135	279	-	27		19	39	-	4
Limnephilus argenteus		-	3	-	1		-	54	-	9		-	8	-	1
Limnephilus externus		11	11	-	1		198	198	-	9		28	28	-	1
POLYCENTROPODIDAE															
Agrypnia straminea	74	24	8	4	5	814	423	135	72	90	73	59	19	10	13
Polycentropus smithae		60	65	9	48		1 071	1 170	153	855		150	164	21	120
Others trichopterans	20	12	17	4	5	137	207	306	63	90	12.2	29	43	9	13
DIPTERA															
Chironomini		1 213	1 798	532	713		606	900	266	357		85	126	37	50
Chironomus decorus		728	1 532	4	22		364	766	2	11		51	107	-	2
Chironomus riparius		77	66	402	131		38	33	201	66		5	5	28	9
Dicrotendipes modestus		122	38	-	11		6	2	-	1		1	1	-	-
Einfeldia dorsalis		137	22	-	2		7	1	-	-		1	-	-	-
TANYPODINAE															
Ablabesmyia monilis		99	52	32	65		5	3	2	3		-	1	-	-
Ablabesmyia spp.		15	52	12	1		1	3	1	-		1	1	-	-
Procladius spp.	10	175	102	42	16		9	5	2	1		-	-	-	-
ORTHOCLADIINAE	30	510	961	662	153	0.1	25	48	33	8	<0.1	4	7	5	1
CORYNONEURINAE		310	50	60	14	1.5	15	2	3	-	<0.1	2	-	-	-
Psectrocladius spp.	17	860	427	224	161	0.9	43	21	11	8	<0.1	6	3	2	1
Others dipterans	12	38	73	20	20	0.1	2	4	1	-	<0.1	-	1	-	-
Total	179	4 406	5 299	2 038	1 377	954	3 218	3 975	812	1 546	85.3	451	557	114	216

Abundance is expressed as a mean integrated total emergence over the entire sampling season (Ind. m-2 year-1). The mean biomass (mg m-2 year-1 dw) was estimated from the mean dry weight of individuals (see text). We use a mean methylmercury concentration of 140 ng g-1 dw for the insects.

9.3.4
Bioavailability of Methylmercury to Plankton

Suspended particles can be directly ingested by zooplankton. In fact, our studies indicated that they play a significant role in zooplankton contamination since MeHg concentrations in phytoplankton do not seem to differ between natural lakes and reservoirs and yet both the quantity of suspended particles, as well as their MeHg concentrations, were 3 to 5 times higher in reservoirs as in natural lakes (Doyon et al. 1996; Plourde et al. 1997; Chapter 8.0).

Evidence for the effect of SPM on Hg contamination of invertebrates is provided by a gradient of increasing MeHg concentrations in both insect larvae and plankton of the LA-40 drainage system (Figure 9.8). Water flows from the upstream Lake 150 to stations 8 and 4 of the flooded zone and exits via a dredged channel to Lake 154a. Although invertebrates and sediments collected in the upstream Lake 150 showed MeHg concentrations similar to those in natural lakes, values for invertebrates in the downstream lake (154a) were higher, demonstrating export from the LA-40 impoundment. Johnston et al. (1991) and Schetagne et al. (1996) (see Chapter 11.0) also suggested that particulate Hg may be transported far downstream.

In protected littoral zones, MeHg concentrations in plankton collected from the 16 year old reservoir were similar to those in plankton collected in recently flooded reservoirs (Figure 9.6B). The non-eroded soils of these protected shallow areas are subjected to very little exchange with the main waterbody of the reservoir and hence, for a relatively long period of time, can maintain high microbial production and provide a sustained quantity of contaminated particles which can be repeatedly resuspended. On the other hand, in the pelagic zone, MeHg concentrations in plankton sampled from 8 to 16 year old reservoirs were similar to those in plankton collected from natural lakes (Figure 9.6A). Thus, once reservoir water level fluctuations have completely eroded away flooded soils in exposed littoral zones, and the resulting particles settle to deeper areas where they are removed from circulation, the source of contaminated solids and microbes to zooplankton decreases (Tremblay et al. 1998a). Another consequence of the erosion of flooded soils is a decrease in nutrients leading to a global decline in the primary (phytoplankton, bacteria, periphyton) and secondary production (insects, zooplankton).

It has been reported that dissolved humic substances or detritus might serve as food for microbes and influence the pelagic food chain by providing an alternative food base for the energetic and nutritional support of consumer organisms (in addition to autotrophic primary production) and, by increasing microbial demand for limiting nutrients at the expense of phytoplankton, play a role in depressing autotrophic primary production (e.g., Hessen 1992; Tranvik 1992; Jones 1992).

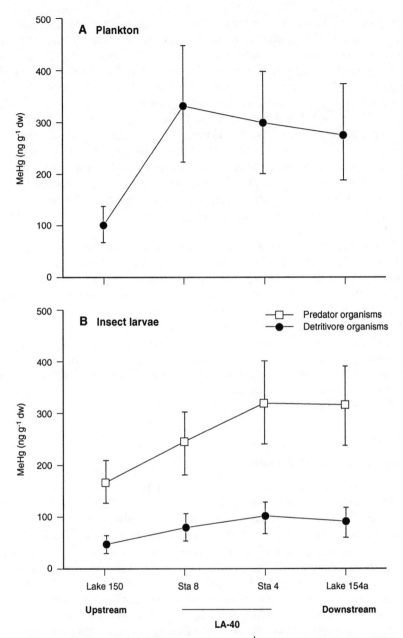

Fig. 9.8. Mean methylmercury concentrations (MeHg ng g^{-1} dw) in (A) zooplankton (> 150 μm mesh) and (B) insect larvae of the LA-40 drainage system. Water flows from the upstream Lake 150 to Stations 8 and 4 of the flooded zone and exits via a dredged channel to Lake 154a (first basin). Error bars represent one standard deviation of the mean (MeHg in insect larvae are modified from Tremblay and Lucotte 1997)

In reservoirs, the structure of the zooplankton community varied from littoral to pelagic zones and from recently flooded reservoirs (1-2 year old) to older ones (16 years old) (Table 9.4; Chapter 4.0; Tremblay et al. 1998a). The zooplankton community in the pelagic zone of reservoirs changed over time from a community dominated by *Daphnia* spp. and *Holopedium* spp. (Laforge 1 reservoir) to one dominated by *Senecella* spp. and *Bosmina* spp. (Robert-Bourassa reservoir) (Table 9.4; Chapter 4.0; Tremblay et al. 1998a). The pelagic zone of the 16 years old reservoir was similar, in terms of plankton abundance and composition, to that of natural lakes. As for the shallow zones of reservoirs protected from drawdown, there was a change in the plankton community structure: the proportions of *Senecella* spp., *Bosmina* spp. and *Daphnia* spp. were higher, while those of *Cyclops* spp. were lower in the 16 years old reservoir as compared to the 2 years old reservoir (Table 9.4; Chapter 4.0; Tremblay et al. 1998a).

The discrepancy in the plankton composition of reservoirs with respect to natural lakes, could be related to a switch from a planktonic food web based on phytoplankton (natural lakes or 16 years old reservoirs) to one based on bacteria-detritus (in newly flooded reservoirs) (Chapter 8.0). This transition from a phytoplankton to a microbe-detritus food base has also been reported by Pinel-Alloul and Méthot (1984) in the Robert-Bourassa reservoir and Paterson et al. (1997) in the ELARP experimental reservoir. In protected, shallow areas (e.g., Digue 24 of Robert-Bourassa reservoir or LA-40) the transition from phytoplankton to microbes-detritus and back to a phytoplankton food base could be longer because of the recycling of the nutrients and the poor water exchange with the main waterbody of the reservoir (Chapter 8.0).

9.3.5
Biomass and MeHg Burdens

The mean number of adult insects emerging from the Laforge 1 reservoir and the LA-40 impoundment varied from 1 300 Ind. $m^{-2} yr^{-1}$ to 5 200 Ind. $m^{-2} yr^{-1}$. These abundances were similar to those measured for the Tglwys Nunydd reservoir in South Wales (5 000 Ind. $m^{-2} yr^{-1}$, Potter and Cardiff 1974) and for the shallow regions of the Kempton Park East reservoir (500-7 000 Ind. $m^{-2} yr^{-1}$, Mundie 1957). These values were about 5 to 20 times higher than those determined for a natural lake where the mean abundance was about 180 Ind. $m^{-2} yr^{-1}$ and corresponds to the findings of Vallières and Gilbert (1992) for insect larvae collected in Bob-Grant lake in eastern Québec as well as to those of Magnin (1977) for 7 natural lakes of northern Québec. Based on the literature values, average insect larvae biomass lost by emergence is estimated at 30-35 % and the average mortality due to predation is about 50-60 % (Miller 1941; Hayne and Ball 1956; Jonasson 1965). Since there is no such information available for reservoirs, we used the values for natural lakes. Emerging insect biomass can therefore be multiplied by two to

Table 9.4. Abundance, biomass and mercury burden of zooplankton collected with a 225 µm mesh in an integrated water column (0-10 m) during the fall 1994

	Laforge 1			LA-40		R-B[1]	Lakes[2]
Stations	119	133	137	4	8	19	
Abundance **(Ind. 50 L^{-1})**							
Bosmina spp.	293	761	78	32	127	106	74
Cyclops spp.	80	163	51	62	221	79	44
Daphnia spp.	455	146	226	5	160	109	32
Diacyclops spp.	7	109	35	103	144	1	40
Epischura spp.					5		3
Holopedium spp.	446		8	2	18		10
Leptodora spp.	8	1				40	1
Senecella spp.	154	25	21	103	259	179	201
Others species	59	9	23	15	21	19	8
Total	1 502	1 214	442	217	955	533	419
Biomass **(ng L^{-1})**							
Bosmina spp.	12 076	31 365	3 214	1 318	5 234	4 368	530
Cyclops spp.	107	180	82	92	230	106	199
Daphnia spp.	7 468	2 452	3 751	164	2 680	1 755	2 190
Diacyclops spp.	18	276	89	261	365	3	73
Holopedium spp.	3 7805	85	678	169	1 525		460
Senecella spp.	1642	267	224	1 098	2 763	1 909	2 260
Total	59 116	34 625	8 038	3 102	12 797	8 138	5 712
Mean levels in the zooplankton (ng g^{-1})	421	504	413	348	523	196	147
Hg burden **(femto g Hg L^{-1})**							
Bosmina spp.	508.4	1 580.8	132.7	45.9	273.7	85.6	5.7
Cyclops spp.	4.5	9.1	3.4	3.2	12.1	2.1	2.5
Daphnia spp.	314.4	123.6	154.9	5.7	140.2	34.4	24.3
Diacyclops spp.	0.8	13.9	3.7	9.1	19.1		1.2
Holopedium spp.	1591.6	4.3	28.0	5.9	79.8		7.6
Senecella spp.	69.1	13.4	9.2	38.2	144.5	37.4	27.4
Total	2 488.8	1 745.1	331.9	108.0	668.7	159.5	68.6

Abundance represents the number of individuals in 50 L of water and the biomass was determined by measuring the length of the individuals while the dry weights were estimated by weight-length regression (see text).

Table 9.4. Abundance, biomass and mercury burden of zooplankton collected with a 225 μm mesh in an integrated water column (0-10 m) during the fall 1994

Stations	Laforge 1			LA-40		R-B[1]	Lakes[2]
	119	133	137	4	8	19	
Mean levels in the zooplankton (ng g^{-1})	358	389	363	293	417	69	63
MeHg burden (femto g Hg L^{-1})							
Bosmina spp.	432.3	1 220.1	116.7	38.6	218.3	30.1	3.2
Cyclops spp.	3.8	7.0	2.9	2.7	9.6	0.8	1.3
Daphnia spp.	267.3	95.4	136.1	4.8	111.7	12.11	13.4
Diacyclops spp.	0.6	10.7	3.2	7.7	15.2		0.7
Holopedium spp.	1 353.5	3.3	24.6	4.9	63.6		4.4
Senecella spp.	58.8	10.4	8.1	32.8	115.2	13.2	14.9
Total	2 116.3	1 346.9	291.6	91.5	533.6	56.4	37.9

[1] Robert-Bourassa reservoir.
[2] Mean of four natural lakes.
Abundance represents the number of individuals in 50 L of water and the biomass was determined by measuring the length of the individuals while the dry weights were estimated by weight-length regression (see text).

estimate an annual biomass of insect potentially ingested by the fish. Since zooplankton are organisms which have a short life cycle (2-4 weeks) we have considered five generations over the summer (Wetzel 1983) and multiplied the biomass by the same number.

Both biomass and MeHg burden of the invertebrate communities of recently flooded reservoirs (1-2 years old) were 3 to 10 times higher than their counterparts in natural lakes (Tables 9.3 and 9.4). Thus, assuming that the abundances are representative of reservoir littoral zones, despite the spatial heterogeneity of the latter, we estimated a conservative MeHg burden for the invertebrate community of a generic waterbody of 1 km^2 and 5 m of depth (lake or reservoir).

MeHg burdens of insects and zooplankton would represent respectively, 3 345 mg km^{-2} and 634 mg km^{-2} of MeHg in such a generic reservoir and 853 mg km^{-2} and 48 mg km^{-2} of MeHg in such a generic natural lake. The reservoir levels would therefore be 4 times higher in the case of insects and 12 times higher for the zooplankton. Of the total invertebrate MeHg burden, the proportion that is made up by zooplankton would be about 3 times higher in reservoirs (16%) than in natural lakes (5%), indicating that the zooplankton community may play a more important role in the MeHg mass balance in shallow areas of reservoirs. Moreover, the significance of the role of zooplankton in the cycling of MeHg in reservoir invertebrates increases when considering the pelagic zones, since the volume of water increases more rapidly than the surface area of the flooded soils. Although both zooplankton abundances and MeHg levels are lower in pelagic zones, the overall biomass and MeHg burden would be greater.

9.4
Conclusions

Podzolic soils as well as peat represent a large reservoir of Hg accumulated over decades by atmospheric deposition. Flooding causes physical and biochemical changes in the soils, enhancing microbial degradation of the organic matter and the release of nutrients. These changes increase the overall biological production of reservoirs by a factor of 5 to 10. On the other hand, they also contribute to the methylation of a small fraction of the Hg pool of flooded soils and its transfer to the food chain where it is bioaccumulated.

The parallelism of the invertebrate food chains as well as the similarity of the biomagnification factors between reservoirs and natural lakes, suggest a higher transfer of MeHg from flooded soils to organisms at the base of the reservoir invertebrates food chain. The suspended particulate matter and the biofilm at the soil-water interface, which have high MeHg levels, are probably routes to actively transfer MeHg from the flooded soils to invertebrates.

The consequence of a higher transfer of MeHg to invertebrates and the higher productivity of a reservoir, in comparison to natural lakes, is an overall higher MeHg burden which then results in higher Hg exposure for the fish community. Monitoring of Hg concentrations in fish of the La Grande complex since 1978 supports this as it revealed an increase of Hg concentrations in fish, reaching a maximum 5 to 13 years after flooding (depending on the species), followed by a significant and continuous decrease of Hg levels (Chapter 10.0 and 11.0; Doyon et al. 1996; Schetagne et al. 1996). The loss of habitat for insects and the gradual depletion of decomposable organic matter intensified by the gradual erosion of flooded soils in exposed littoral zones of reservoirs by repeated water level fluctuation would reduce the release of nutrients, the amount of SPM and the microbial production which, in return, would reduce both overall invertebrate production and MeHg concentrations. These processes probably explain the decrease of Hg levels observed in non-piscivorous fish, as both the food and the MeHg supply provided by the invertebrates decreases over time.

Acknowledgements

I am thankful to Maxime Bégin, Bernard Caron, Louise Cournoyer, Hugo Poirier, Isabelle Rheault and Louis-Filip Richard from Université du Québec à Montréal for their assistance in the field and laboratory. This research was supported by a grant from Hydro-Québec (HQ), the Conseil de recherches en sciences naturelles et en génie (CRSNG) du Canada, and the Université du Québec à Montréal (UQAM) to the Chaire de recherche en environnement (HQ/CRSNG/UQAM).

10 Mercury Accumulation in Fish from the La Grande Complex: Influence of Feeding Habits and Concentrations of Mercury in Ingested Prey

Richard Verdon and Alain Tremblay

Abstract

The total mercury (total Hg) concentrations in ingested prey, as well as the mercury levels in fish, are 2 to 5 times higher in reservoirs than those in natural lakes. Feeding habits of non-piscivorous and piscivorous fish species from the La Grande complex were studied 2 to 4 years after filling (1980 to 1982) and several years later (1992 to 1994). Results showed that non-piscivorous fish from both natural lakes and reservoirs have similar diets. In both environments, lake whitefish shift from a zooplankton dominated diet to one dominated by benthos with increasing fish size. Cisco and longnose sucker feed mainly on zooplankton and benthos, respectively, regardless of size. Thus, our results suggest that the differences in the total Hg concentrations in fish between natural lakes and reservoirs, and the variations over time in reservoirs, are more related to the temporal changes in the concentration of methylmercury (MeHg) in the organisms of the food web than to changes in feeding habits. There are, however, a few exceptions such as large lake whitefish captured below the Robert-Bourassa powerhouse that eat small fish stunned by their passage through the turbines. These small fish represent more than 80% of the stomach content in comparison to only 2% for whitefish captured above the powerhouse. The higher proportion of fish in their diet leads to higher Hg levels in whitefish captured below the powerhouse than those above it. Also, piscivores from reservoirs generally eat more of their congenere or other piscivorous fish, up to 50% of the stomach content volume, than those from natural lakes. This lengthens the food chain which tends to further increases in total Hg levels in piscivores from reservoirs.

10.1
Introduction

In the La Grande complex, fish populations developed similarly among most reservoirs after filling. After a temporarily decrease in biomass, associated with the dilution of natural populations in large water volumes, an important increase was observed, as reflected by catch per unit of effort. For instance, mean summer fishing yield, in Robert-Bourassa reservoir, went from 7.8 kg net^{-1} day^{-1} to 2.6 kg net^{-1} day^{-1} during filling, then to 21.3 kg net^{-1} day^{-1} eight years later (Deslandes et al. 1993). Twelve years after filling, with gradual return to pre-impoundment primary (phytoplankton, bacteria) and secondary (zooplankton) productivity conditions, fish biomass had returned to original levels (6.5 kg net^{-1} day^{-1}). Fish communities underwent significant changes after impoundment. Coregonids, namely lake whitefish (*Coregonus clupeaformis*), and northern pike (*Esox lucius*) increased their relative importance, while longnose sucker (*Catostomus catostomus*), white sucker (*Catostomus commersoni*) and walleye (*Stizostedion vitreum*) were generally less abundant and concentrated in particular areas (Deslandes et al. 1994). For instance, lake whitefish and pike, which composed 16% and 9% of the catch respectively before impoundment of the Robert-Bourassa reservoir, accounted for 45% and 16% of fish captured thirteen years after flooding. Increased productivity of reservoirs also resulted in increased growth rates for a few years after impoundment (Deslandes et al. 1994; Deslandes et al. 1995).

Feeding habits of fish have been studied in different environments of the La Grande complex, such as reservoirs (Robert-Bourassa, Opinaca, Caniapiscau, Desaulniers), increased flow areas (Boyd-Sakami diversion, La Grande river) and natural lakes (Detcheverry, Rond-de-Poêle, Hazeur, Sérigny, Des Vœux, Jobert) (Figure 10.1). The diet of dominant non-piscivorous and piscivorous species were monitored a few years after filling, from 1980 to 1982 (SAGE 1983) and several years later, from 1992 to 1994 (Doyon 1995ab; Doyon et al. 1996). Hg determinations of fish stomach contents were carried out in 1993 and 1994 (Doyon et al. 1996).

Although direct absorption of dissolved MeHg by gills during respiration is a possible pathway of accumulation, food is recognized as the major entry of MeHg for fish (Cope et al. 1990; Meili 1991a). It has also been demonstrated that food chain length affects Hg accumulation in fish (Cabana et al. 1994). Thus, a change either in the food web or in the MeHg level of the ingested food could affect the Hg accumulation in fish in reservoirs after flooding. In this chapter, we will examine the influence of the fish diet and the Hg level of their prey on the Hg concentration in fish from reservoirs.

10.2
Materials and Methods

Fish were captured with gill nets, measured and weighed, as part of a more extensive sampling program aimed at the monitoring of fish populations of the La Grande complex (Figure 10.1). The non-piscivorous species examined were

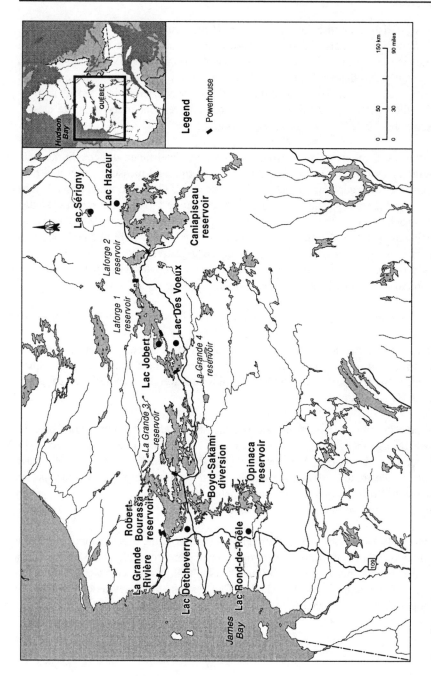

Fig. 10.1. Location of the study area, Québec, Canada

lake whitefish, longnose sucker and cisco (*Coregonus artedii*). Although sympatric populations of normal and dwarf lake whitefish occur in the eastern part of the La Grande complex, only data for the normal lake whitefish were considered in this chapter. Northern pike, burbot (*Lota lota*), walleye and lake trout (*Salvelinus namaycush*) were the most common piscivorous species sampled. In order to examine the variation of the diet with fish size, fish were separated in 3 size classes (see Figures 10.2, 10.3 and 10.4) (Doyon et al. 1996).

Stomach contents of piscivorous species were determined directly in the field laboratory by opening the stomach and by counting and measuring ingested fish. Other types of organisms (insects, rodents, etc) were identified when possible. For all non-piscicorous species, the stomachs were incised to assign a repletion index of stomach fullness (trace of food, filled to 25%, 50%, 75% and 100%). Only the specimens with a repletion index between 25% and 100% were retained for stomach content analysis. For suckers, the entire digestive tract was systematically preserved for prey identification in the laboratory, however, only the first third was retained for the quantitative inventory of ingested organisms.

All organisms in the stomach of non-piscivorous fish were identified to the lowest possible taxonomic level. The inventory of the organisms was done by successive dilution in distilled water which allowed a clear discrimination between organisms and organic and inorganic debris. A manual separation and a grouping of organisms was performed to obtain a minimum weight of material (0.2 mg dw) for the Hg determination.

The occurrence of organisms was obtained by dividing the number of stomachs in which organisms were present by the total number of stomachs analyzed. The volume of a given prey for non-piscivorous species was evaluated by volumetric displacement in alcohol, and the weight of fish in stomach of piscivorous species was estimated by weight-length relationships, to obtain the contribution of a given prey to the total volume or total biomass ingested.

Hg analyses in fish were performed on a sample of the dorsal lateral muscle by cold vapor atomic absorption spectrophotometry and total Hg and MeHg in prey ingested by non-piscivorous fish (insects, plankton, plants, etc) were determined by atomic fluorescence as described in Chapter 2.0. More details concerning the manipulations and the analyses are available in Doyon et al. (1996).

Lake whitefish

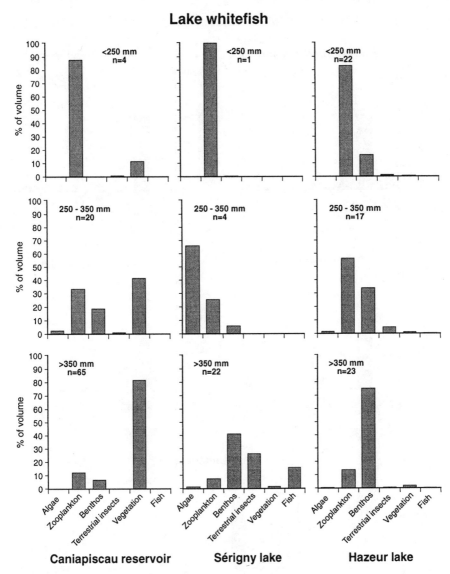

Fig. 10.2. Main prey groups identified in stomachs of normal lake whitefish, for different size classes, in Caniapiscau reservoir, Sérigny lake and Hazeur lake, in 1993

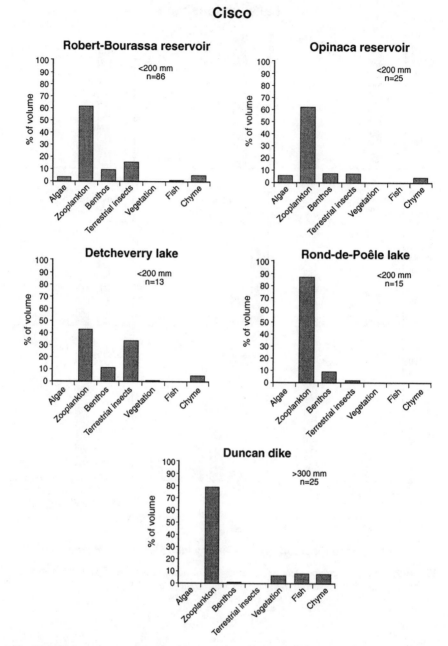

Fig. 10.3. Main prey groups identified in stomachs of cisco, for different size classes, in Robert-Bourassa and Opinaca reservoirs, Detcheverry lake, Rond-de-Poêle lake, and near Duncan dyke (Robert-Bourassa reservoir) in 1994

Longnose sucker

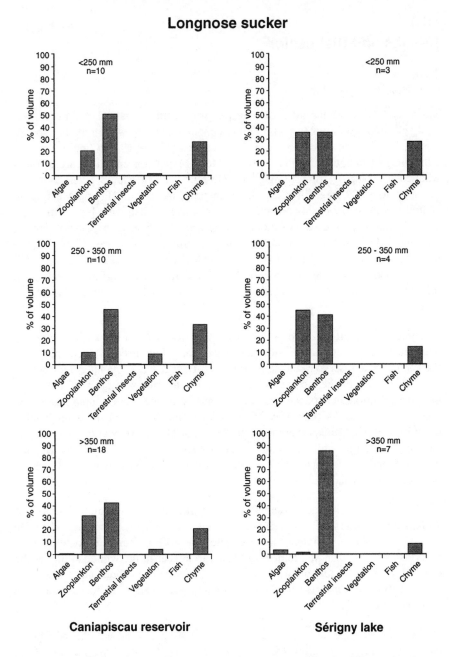

Fig. 10.4. Main prey groups identified in stomachs of longnose sucker, for different size classes, in Caniapiscau reservoir and Sérigny lake, in 1993

10.3
Results and Discussion

10.3.1
Non-Piscivorous Fish

Mean total Hg concentrations of ingested prey from natural lakes ranged from 42 ng g^{-1} dw in molluscs to 845 ng g^{-1} in ephemeropterans, while the corresponding values for the reservoirs were respectively 172 ng g^{-1} dw (molluscs) and 1 856 ng g^{-1} (trichopterans) (Table 10.1). Mean MeHg concentrations were somewhat lower, ranging from 13 ng g^{-1} to 718 ng g^{-1} dw in natural lakes and from 68 ng g^{-1} to 1 709 ng g^{-1} in reservoirs. Generally, mean total Hg and MeHg concentrations in the stomach contents of non-piscivorous fish were higher in reservoirs than in natural lakes by factors varying from 2 to 5. Similarly, maximum Hg concentration in non-piscivorous fish of standardized length were 2 to 5 times higher in reservoirs (0.34 to 0.73 mg kg^{-1} ww for suckers and 0.37 to 0.53 mg kg^{-1} for whitefish) than those from natural lakes (0.07 to 0.22 mg kg^{-1} ww and 0.05 to 0.3 mg kg^{-1} for suckers and whitefish respectively) (Schetagne et al. 1996; Chapter 11.0).

The diet of small (< 250 mm) lake whitefish from natural lakes and reservoirs, between 11 and 15 years after filling, was generally dominated by zooplankton. In the Caniapiscau reservoir, and in Sérigny and Hazeur lakes, plankton represented more than 80% of the stomach content volume for this size class of fish (Figure 10.2). Chironomid pupae and cladocerans were the main zooplanktonic organisms ingested. As they grow older, lake whitefish preyed more on benthic organisms. In the Caniapiscau reservoir, the importance of vegetation debris in the stomachs of large whitefish indicates that they are bottom dwellers. Benthic groups found in stomachs were mainly composed of trichopterans, dipterans, ephemeropterans and molluscs.

Fish, mainly sticklebacks, represented a small fraction of the diet of large lake whitefish in natural lakes and reservoirs. They represented generally less than 5% of the prey volume, although they accounted for 10% in Rond-de-Poêle lake for individuals larger than 450 mm. Fish tended to appear in the whitefish diet at the end of the summer, possibly as a consequence of a reduced abundance of invertebrates, particularly insects for which emergence is reduced, or increased availability of small fish.

The relative importance of taxa found in fish stomachs did not change significantly after reservoir filling (SAGE 1983; Doyon 1995ab; Doyon et al. 1996; Doyon and Tremblay 1997). During the first years following impoundment of the Robert-Bourassa and Opinaca reservoirs, the aquatic insects, mainly dipterans and ephemeropterans, were the main prey items found in lake whitefish during the summer, while zooplankton, mainly cladocerans, was dominant during the fall.

Table 10.1. Total mercury and methylmercury concentrations (mean ± standard deviation (n), ng g⁻¹ dw) and the proportion of methylmercury to total mercury in different food type ingested by non-piscivorous fish collected in 1993 and 1994 in the La Grande complex. A missing standard deviation indicates that only one homogenate was analyzed

| Food type | | Reservoirs or diversion | | | | Natural lakes | | | | |
| | | Western part of the complex | | | Eastern part of the complex | Western part of the complex | | | Eastern part of the complex | |
		Robert-Bourassa	Opinaca	Boyd-Sakami	Caniapiscau	Detcheverry	Rond-de-Poêle	Des Vœux	Sérigny	Jobert
Molluscs	Total Hg	466	172±35 (7)	202±77 (6)	1016±25 (2)	78±22 (3)	92±18 (2)	42	354	62
	MeHg	332	85±15 (7)	161±77 (6)	68±43 (3)	25±9 (3)	13±2 (2)	32		40±6 (2)
	% MeHg	71	52	78	69	32	14	77		63
Dipterans	Total Hg	645±333 (10)	857±265 (6)	973	891±521 (9)	495±260 (15)	403±100 (2)	227±120 (5)	532±284 (4)	441±209 (6)
	MeHg	480±356 (10)	631±294 (6)	868	374±213 (9)	101±36 (18)	175±75 (2)	157±109 (5)	318±308 (4)	382±151 (6)
	% MeHg	72	71	89	48	21	44	66	51	78
Ephemeropterans	Total Hg	421±149 (5)				275±108 (2)	207±133 (4)	200	845	
	MeHg	331±130 (5)				205±141 (3)	185±145 (5)	98	718	
	% MeHg	77				74	86	48	85	

Table 10.1. Total mercury and methylmercury concentrations (mean ± standard deviation (n), ng g^{-1} dw) and the proportion of methylmercury to total mercury in different food type ingested by non-piscivorous fish collected in 1993 and 1994 in the La Grande complex. A missing standard deviation indicates that only one homogenate was analyzed (continued)

| Food type | | Reservoirs or diversion | | | | Natural lakes | | | | |
| | | Western part of the complex | | | Eastern part of the complex | Western part of the complex | | | Eastern part of the complex | |
		Robert-Bourassa	Opinaca	Boyd-Sakami	Caniapiscau	Detcheverry	Rond-de-Poêle	Des Vœux	Sérigny	Jobert
Trichopterans	Total Hg	530±215 (27)	971±288 (8)	1150±189 (2)	1856	305±98 (12)	232±75 (4)	124	390±90 (2)	
	MeHg	468±224 (27)	964±266 (9)	1123±362 (3)	1709	214±70 (13)	153±63 (4)	97	151±39 (2)	
	% MeHg	88	99	98	92	71	70	78	38	
Heteropterans	Total Hg	566±52 (2)				140				
	MeHg	525±15 (2)				139				
	% MeHg	93				100				
Plants debris	Total Hg	516±109 (4)	579±95 (4)	385±210 (3)	809±547 (7)	74	199	422	645	
	MeHg	296±55 (4)	436±414 (4)	333±152 (3)	367±319 (7)	52	132	370	455	
	% MeHg	61	74	91	47	69	67	88	71	
Zooplankton	Total Hg	945			868±363 (4)	595±234 (4)		106	336±37 (2)	308±30 (2)
	MeHg	617			524±107 (4)	104±27 (4)		94	136	183±83 (2)
	% MeHg	65			62	18		89	45	57

The diet of cisco, even for large individuals, was largely dominated by zooplankton (Figure 10.3). Ostracods, cladocerans and copepods were the major groups ingested. The stomach content of cisco reflects its pelagic niche. Experimental fishing with pelagic gill nets in Robert-Bourassa and Opinaca reservoirs, and in Boyd-Sakami diversion showed that cisco accounted generally for more than 80% of the catch. In cisco larger than 300 mm, fish represented a significant portion of the diet, up to 8% at Duncan dyke station, in the Robert-Bourassa reservoir; sticklebacks (average length: 28 mm) were most commonly ingested. As for lake whitefish, the importance of fish in cisco diet increased as summer progressed.

Longnose sucker fed mainly on benthic organisms in lakes and reservoirs, although zooplankton represented an important part of their diet, particularly for smaller size fish (Figure 10.4). Dipterans, ephemeropterans and trichopterans represented the dominant prey among benthic organisms, while cladocerans were the zooplanktonic organisms most commonly preyed upon. This situation was observed just after filling, as well as 11 and 13 years after filling, although pelecypod molluscs accounted for up to 38% of the stomach content volume during filling of the Opinaca reservoir. In the Robert-Bourassa reservoir, stomach contents of large suckers (> 450 mm) included important quantities of vegetation, up to 17.6% of the volume. As for lake whitefish, this phenomenon was typical of bottom feeders in the littoral zone. Unlike benthic organisms, the importance of zooplankton, plants and adult insects decreases in sucker diet as summer progresses.

The diet of the non-piscivores were generally similar between reservoirs and natural lakes. Thus, the higher level of total Hg in fish from reservoirs, in comparison to natural lakes, is probably related to higher concentrations in their ingested prey.

Another argument supporting the importance of MeHg concentrations in the ingested food is provided by the differences of MeHg concentrations in ingested organisms between natural lakes. For instance, MeHg concentrations in invertebrates of the Des Vœux lake are generally lower than those from Jobert or Sérigny lakes, in the same area (Chapter 4.0; Tremblay et al. 1996a; Tremblay et al. 1998a). Total Hg levels in non-piscivorous fish from these lakes show a trend similar to that of their ingested prey (Schetagne et al. 1996).

In reservoirs, mean MeHg concentrations in pelagic zooplankton increased rapidly after impoundment, reaching a maximum 4 to 5 years after flooding. Concentrations then declined gradually to reach concentrations measured in natural lakes of the area, 8 years after flooding (Tremblay et al. 1998a; Chapter 9.0). The same temporal trend was observed for whitefish and suckers that prey upon invertebrates, although it was somewhat delayed as they are at a higher level in the food chain and as they have longer life cycle (Schetagne et al. 1996; Chapter 11).

The accumulation of total Hg in fish can also be affected by a change in their diet. For instance, downstream from generating stations, large lake whitefish take advantage of the entrainment of small stunned fish, which are an easy prey and become an important component of their diet. Below the Robert-Bourassa generating station, fish represent 84.2% of the stomach content volume of lake whitefish > 450 mm, compared to 1.3% above the dam (Table 10.2) (Brouard et al. 1994). In that case, small cisco were the most important prey. As a consequence, whitefish below the generating station have much higher Hg levels than those captured above it (Chapter 11).

Table 10.2. Importance of fish in the diet and mean mercury concentration (stomach content and flesh) for lake whitefish of two length classes caught above (reservoir) and below (river) the Robert-Bourassa generating station (number of samples is given in parentheses) (adapted from Brouard et al. 1994)

	Robert-Bourassa upstream station[a]		Robert-Bourassa downstream station[b]	
	< 450 mm	> 450 mm	< 450 mm	> 450 mm
Occurrence of fish in stomach (%)	9.23 (65)	4.16 (48) NS	32.36 (380)	87.50 (80)*
Volume of fish in stomach contents (%)	0.84 (65)	1.27 (48) NS	21.77 (380)	84.15 (80)**
Total Hg in stomachs (mg kg^{-1} ww)	0.12 (1)	0.20±0.05 (14) NS	0.71±0.31 (11)	1.75±0.33 (2)**
Total Hg in flesh (mg kg^{-1} ww)	0.29±0.08 (16)	0.54±0.21 (13)**	1.07±0.23 (11)	2.95±0.76 (8)**

[a]   Immediately above the Robert-Bourassa powerhouse.
[b]   Immediately below the Robert-Bourassa powerhouse.
Note: (*) indicate significant difference between size for a given station using Chi square test ($p < 0.05$); (**) indicate significant differences between size using Mann-Whitney U test ($p < 0.05$); NS = no significant difference.

10.3.2
Piscivorous Fish

Hg concentrations in small fish (< 120 mm) ingested by piscivorous fish were, as for the prey of non-piscivorous species, more elevated in reservoirs (0.17 mg kg^{-1} ww to 0.68 mg kg^{-1}) as compared to natural lakes (0.06 mg kg^{-1} ww to 0.22 mg kg^{-1}) (Table 10.3). Maximum total Hg concentrations in larger specimens of standardized length show similar trends with levels being 3 to 6 times higher in reservoirs (1.65 mg kg^{-1} ww to 3.34 mg kg^{-1} for pike and 1.03 mg kg^{-1} ww to 2.72 mg kg^{-1} for walleye) than their counterparts in natural lakes (0.3 mg kg^{-1} ww to 0.93 mg kg^{-1} for pike and 0.30 mg kg^{-1} to 1.02 mg kg^{-1} ww for walleye) (Schetagne et al. 1996).

Table 10.3. Total mercury concentrations (mean ± standard deviation (n, length in mm), mg kg^{-1} ww) in different small individual prey species ingested by piscivorous fish collected in 1994 in the western part of the La Grande complex. A missing standard deviation indicates that only one individual was analyzed

Fish species	Reservoirs		Natural lakes	
	Robert-Bourassa	Opinaca	Detcheverry	Rond-de-Poêle
Young fish	0.28±0.16 (5, 39)	0.32±0.16 (4, 39)	0.06 (1, 35)	0.18±0.16 (3, 42)
Sculpins[1]	0.37 (1, 75)	0.28±0.1 (3, 77)	0.06±0.01 (3, 76)	
Cisco	0.33±0.09 (5, 89)		0.11 (1, 115)	0.16 (1, 88)
Lake whitefish		0.38±0.06 (3, 105)		
Walleye	0.27±0.08 (4, 48)	0.17 (1, 50)		0.22 (1, 192)
Sticklebacks[2]	0.24±0.08 (4, 48)		0.15±0.22 (2, 40)	
Burbot		0.20 (1, 230)		0.17 (1, 357)
Trout perch[3]	0.68 (1, 110)		0.17±0.03 (4, 56)	0.11 (1, 65)

[1] *Cottus* sp.

[2] Gasteroiteidae

[3] *Percopsis omiscomaycus*

With the exception of pike in Robert-Bourassa reservoir, the diet of piscivorous fish of the La Grande complex were similar between reservoirs and natural lakes (Tables 10.4 and 10.5). As for the non-piscivorous fish species, the higher level of Hg in fish from reservoirs is generally related to the concentration of that metal in the ingested food rather than a change in diet.

In the eastern part, and particularly in the Caniapiscau reservoir, the diet of predatory fish was highly dominated by lake whitefish. More than 80% of pike

Table 10.4. Percentage of occurence of prey in piscivorous fish stomachs from the eastern part of the La Grande complex in 1993

	n	Longnose sucker	Lake whitefish	Round* whitefish	Northern pike	Burbot	Sculpins	Catostomids	Debris	Unidentified fish	Other than fish
CANIAPISCAU RESERVOIR											
Lake trout											
<600 mm	1									100	
600 - 750 mm	2		100								
>750 mm	7	14.2	14.2						14.2	71.4	14.2
Pike											
400 - 600 mm	6		83.3							16.7	
>600 mm	31		90.3			6.3			3.2	19.4	3.2
Burbot											
400 - 600 mm	9		22.2		11.1					55.6	11.1
HAZEUR LAKE											
Lake trout											
<600 mm	19		31.6				5.3	5.3		47.5	10.5
600 - 750 mm	4					25.0			25.0	75	
>750 mm	1								8.3	100	
SÉRIGNY LAKE											
Lake trout											
<600 mm	18	5.6	22.2	5.6		25.0	8.3			27.8	38.9
600 - 750 mm	12	8.3	8.3	8.3						33.3	16.7
>750 mm	1									100	

* *Prosopium cylindraceum*

Table 10.5. Percentage of occurence of prey in piscivorous fish stomachs from the western part of the La Grande complex in 1994

	n	White sucker	Cisco	Lake whitefish	Coregonus (undetermined)	Walleye	Northern pike	Burbot	Lake chub	Sticklebacks	Trout perch	Yellow perch	Longnose sucker	Sculpins	Unidentified fish	Other than fish
ROBERT-BOURASSA RESERVOIR																
Pike <400 mm	12		16.7	8.3	16.7				8.3	8.3		8.3			25.0	
400-750 mm	64	3.1	34.4		6.3	6.3	3.1	3.1	1.6	10.9					28.1	6.3
>750 mm	18	16.7		5.6		22.2	11.1	5.6					5.6		33.3	5.6
Walleye <250 mm	33		3.0												63.6	30.3
250-450 mm	58		1.7												86.2	13.8
>450 mm	17		11.8		5.9										58.9	35.3
Burbot 400-750 mm	10			20.0					10.0	10.0				70.0		
OPINACA RESERVOIR																
Pike <400 mm	8		12.5						12.5		25.0				37.5	12.5
400-750 mm	38		26.3		10.5		7.9	7.9	5.3			5.3			36.8	
>750 mm	7			28.6				14.3				28.6			28.6	
Walleye <250 mm	8														87.5	12.5
250-450 mm	5				40.0										60.0	
>450 mm	14		7.1		7.1			7.1							78.6	

Table 10.5. Percentage of occurence of prey in piscivorous fish stomachs from the western part of the La Grande complex in 1994 (continued)

	n	White sucker	Cisco	Lake whitefish	Coregonus (undetermined)	Walleye	Northern pike	Burbot	Lake chub	Sticklebacks	Trout perch	Yellow perch	Longnose sucker	Sculpins	Unidentified fish	Other than fish
BOYD-SAKAMI DIVERSION																
Pike <400 mm	1				100											100
400-750 mm	4		50	25.0											25.0	
>750 mm	1			100												
Walleye <250 mm	16														93.8	6.3
250-450 mm	9				11.1			11.1							66.7	11.1
Burbot 400-750 mm	1															100
ROND-DE-POÊLE LAKE																
Pike <400 mm	2										50.0			50.0		
400-750 mm	18	11.1	5.6			5.6		5.6	5.6		5.6	5.6	11.1		50.0	11.1
>750 mm	9	22.2	11.1	22.2	11.1										33.3	
Walleye <250 mm	8		12.5												37.5	50.0
250-450 mm	8		12.5												37.5	50.0
>450 mm	26		7.7		7.7							3.8			53.8	26.9
DETCHEVERRY LAKE																
Pike <400 mm	1													100		100
400-750 mm	4		50								25.0				25.0	
Walleye <250 mm	15									26.7					73.3	
250-450 mm	21		19.0		33.3										57.1	
>450 mm	4		25.0												75.0	
Burbot 400-750 mm	6				33.3										50.0	16.7

had ingested whitefish and this species accounted for more than 95% of the stomach content biomass. The ingested whitefish had an average length varying from 160 mm to 187 mm in the Caniapiscau reservoir, but were much smaller in the two natural lakes in the same region (between 80 mm and 151 mm). Longnose sucker, pike and burbot were, to a lesser extent, also present in the piscivores stomachs. However, the small number of the latter make it difficult to quantify the importance of these species in the diet of piscivorous fish. This picture also reflects the situation in natural lakes of this region. However, burbot seems more abundant in lake trout stomachs from natural lakes than from reservoirs. They were observed in 25% of intermediate size (600 - 750 mm) lake trout caught in Hazeur and Sérigny lakes. Prey items in natural lakes were also more diversified than in the Caniapiscau reservoir, since round whitefish and sculpins were found only in fish caught in these lakes (Table 10.4).

In the western part of the La Grande complex, cisco is the prey which were most commonly found in predator stomachs. The size of cisco present in stomachs was generally small, between 87 mm and 152 mm in reservoirs and the Boyd-Sakami diversion, and between 88 mm and 175 mm in natural lakes. Cisco are only present in this part of the territory and are typically pelagic.

In most of the studied environments, pike fed on a large diversity of prey. In addition to cisco, which represent the most often ingested prey in the Robert-Bourassa reservoir, they preyed on suckers, lake whitefish, walleye, pike, burbot, lake chub (*Couesius plumbeus*), sticklebacks, trout-perch (*Percopsis omiscomaycus*), yellow perch (*Perca flavescens*) and sculpins (Table 10.5). Lake whitefish ingested by pike were much larger than cisco, ranging from 160 mm to 400 mm. As they grew in size, pike fed less on coregonids and more on piscivorous fish, such as walleye, pike and burbot. These piscivorous fish represented between 35% and 59% of the biomass ingested by large pike (> 400 mm) in the Robert-Bourassa reservoir, 15 years after impoundment and their average size varied between 288 mm and 354 mm. This feeding behavior puts them in a situation where they would ingest large amounts of MeHg. The predominance of coregonids in pike stomachs, as the occurrence of burbot, was also observed in 1980 and 1982 in the Robert-Bourassa reservoir, during the second and third year after the beginning of impoundment (SAGE 1983).

Pike from reservoirs in the western part of the La Grande complex and from Rond-de-Poêle lake had a diet that was generally more diversified than in the Caniapiscau reservoir. The latter being located in a harsher environment, many species present in the west are absent in this reservoir, namely cisco, walleye, perch-trout and several species of minnows. In addition, pike from reservoirs in the western part of the complex fed more on piscivorous fishes (~ 50% of the biomass) than those from the Caniapiscau reservoir (~ 4%).

The diet of walleye caught in 1994 was dominated by coregonids (cisco and lake whitefish), regardless of their size or sampling site. A few burbot were also observed in walleye stomachs caught downstream from the Opinaca reservoir. This situation is similar to what was observed in 1980 and 1982, where walleye in the Robert-Bourassa reservoir ingested mainly coregonids and sticklebacks (SAGE 1983).

Burbot fed essentially on coregonids (79% to 100% of the biomass). They may also occasionally ingest piscivorous fish, as a pike was found in a burbot from the Caniapiscau reservoir in 1993. Data from 1980 to 1982 showed that burbot fed principally on coregonids, sticklebacks and suckers.

Fish Hg concentrations increased significantly in all species after impoundment of the La Grande complex reservoirs with peak concentrations in pike occurring 10 to 13 years after flooding (Schetagne et al. 1996; Chapter 11). The temporal evolution of the Hg concentrations in fish from reservoirs are related to the physical and hydrologic characteristics of the reservoirs and the temporal changes in the concentration of MeHg in the organisms ingested. The abiotic factors influence the transfer of MeHg from the flooded soils to the base of the food chain, mainly the invertebrates (Tremblay and Lucotte 1997; Chapter 9). This transfer of MeHg to the invertebrates is more active in reservoirs (mostly the first 2 to 4 years after flooding), than in natural lakes, leading to higher Hg accumulation in the food chain of reservoirs.

However, as shown for the non-piscivorous fish species, a shift in diet can also contribute to increased Hg accumulation in piscivorous species. For instance, the higher proportion of piscivorous fish ingested by the pike from the Robert-Bourassa reservoir (50%), in comparison to those of Canispiscau reservoir (4%), increases the length of the food chain and may have contributed, with the physical and hydrologic characteristics of the reservoirs, to a higher Hg accumulation (maximum of 3.3 mg kg^{-1} ww for pike of the Robert-Bourassa reservoir and of 2.3 mg kg^{-1} ww for the pike of the Caniapiscau reservoir).

10.4
Conclusions

MeHg is biomagnified from invertebrates to non-piscivorous and to piscivorous fish. An increase of the MeHg concentration with trophic level occurs both in natural lakes and in reservoirs, but our results indicate that Hg levels are 2 to 5 times higher in reservoirs than in natural lakes. Generally, fish diet were similar in both environments and could not explain the differences in fish Hg accumulation after flooding. Differences in fish Hg concentration between natural lakes and reservoirs, and the variations over time in reservoirs, are more related to the temporal changes in the concentration of MeHg in ingested prey than changes in feeding habits. However, piscivores of reservoirs eat more of their congeners or other piscivorous species and this change in the structure of the food web would lead to differences in Hg accumulation; the longer food chain for pike of Robert-Bourassa reservoir, in comparison to the ones of Caniapiscau and Opinaca reservoirs, leads to greater Hg accumulation. Furthermore, downstream of the hydroelectric turbines, typical non-piscivorous fish shift their diet to small fish stunned by their passage through turbines, thus leading to higher Hg levels.

Acknowledgements

This study was funded by Hydro-Quebec. We are thankful to the Groupe-conseil Génivar for their assistance in the realization of this study. The participation of Isabelle Rheault from the Université du Québec à Montréal is also gratefully acknowledged.

11 Post-Impoundment Evolution of Fish Mercury Levels at the La Grande Complex, Québec, Canada (from 1978 to 1996)

Roger Schetagne and Richard Verdon

Abstract

At the La Grande hydroelectric complex, the evolution of fish mercury (Hg) levels was monitored in reservoirs, along river diversion routes, and in rivers with modified flow, from 1978 to 1996. Five fish species were considered: two non-piscivorous, lake whitefish (*Coregonus clupeaformis*) and longnose sucker (*Catostomus catostomus*), and three piscivorous, northern pike (*Esox lucius*), walleye (*Stizostedion vitreum*) and lake trout (*Salvelinus namaycush*). Total Hg concentrations were measured by standard cold vapour atomic absorption spectrophotometry, and expressed in mg kg^{-1}, wet weight. In reservoirs, concentrations in all species increased rapidly after impoundment, peaking after 5 to 9 years in non-piscivorous fish, and after 10 to 13 years in piscivorous species, at levels 3 to 7 times those measured in surrounding natural lakes, then significantly and gradually declined. Data from the La Grande complex strongly suggest that concentrations return to natural levels after 10 to 25 years for non-piscivorous species, and after 20 to 30 years for piscivorous ones. This duration of the phenomenon is corroborated by results from other reservoirs in Canada and Finland. Monitoring of Hg levels in fish and studies of drifting organisms also show that Hg is exported downstream from reservoirs, probably mostly by reservoir Hg-rich organic debris as well as by plankton, aquatic insects or small fish. This export may affect Hg levels in fish over long distances downstream from reservoirs in the absence of large deep bodies of water allowing sedimentation of organic debris or biological uptake of Hg-rich organisms originating from reservoirs. Along a series of 4 large reservoirs emptying into one another, the highest Hg levels were systematically measured in fish caught immediately downstream from reservoir outputs, but no cumulative effect in fish Hg levels was observed from the first to the last reservoir. In a river which sustained a reduction of flow of over 90%, fish Hg levels remained within the range of concentrations measured in fish collected from natural surrounding lakes.

11.1
Introduction

The La Grande hydroelectric complex is located on the eastern side of James Bay, approximately at latitude 54°N. The whole region is part of the Canadian Shield. At the onset of the project, the territory, composed of scattered coniferous forest with numerous peat bogs, was free of any industrial activity and sparsely occupied by native Cree. Most common fish species in the region include longnose sucker, white sucker (*Catostomus commersoni*), lake whitefish, cisco (*Coregonus artedii*), northern pike, walleye and lake trout. The La Grande complex resulted in the creation of several large reservoirs, each flooding land areas ranging from 200 to 3 400 km^2, as well as in the diversion of 3 main rivers, the Caniapiscau, from the east, and the Eastmain and Opinaca, from the south. Fish Hg concentrations were monitored before and after the impoundment of these reservoirs, as a significant proportion of the native Cree still practice traditional subsistence hunting and fishing.

11.2
Materials and Methods

Each monitored environment (reservoirs, rivers with reduced or increased flow, diverted rivers and coastal areas) was sampled at 3 to 5 stations, before and after modification, usually every second year from 1978 to 1996 (Figure 11.1). Gill nets with mesh size ranging from 2.5 to 10.2 cm were used at each site, with the aim of obtaining 30 specimens of the following target species: longnose sucker, lake whitefish, northern pike, walleye and lake trout. For every species, these 30 specimens were evenly distributed in pre-selected length classes. Over all, more than 18 000 fish were analyzed for total Hg during the 1978-1996 period. Fish flesh samples consisting of approximately 10 g were conserved frozen until total Hg analysis by the standard cold vapour atomic absorption spectrophotometry method, as described in Chapter 2.0. The standard quality control procedures applied by the analytical laboratory were enhanced by the addition of blind triplicates taken on ten percent of all fish sampled. The mean annual variation coefficients for these blind triplicates ranged from 5.2 to 9.2% over the 1986 to 1996 period. The statistical approach used to follow the spatio-temporal evolution of fish muscle Hg concentrations is based on the polynomial regression analysis with indicator variables, as described in Tremblay et al. (1998c).

This method allows the calculation of linear or curvilinear models to express muscle total Hg concentration (mg kg^{-1} wet weight) as a function of a polynomial of total length (usually of the second order) such as:

$$[Hg] = a + b \times length + c \times length^2, \text{ where a, b and c are constants.}$$

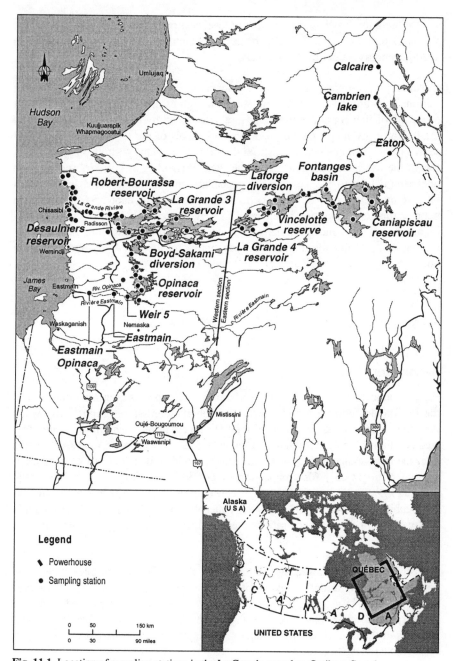

Fig. 11.1. Location of sampling stations in the La Grande complex, Québec, Canada

Interpretation was based on mean Hg levels estimated by these polynomials for standardized length specific to each species, which corresponds approximately to average lengths caught by our gill nets.

Although pre-impoundment fish Hg levels were available at several sampling stations, background levels used in tables and figures depicting post-impoundment evolution of fish Hg concentrations were determined by treating the values of all the fish of a given species sampled in natural lakes of the region as a single data set (as if they had been captured from the same lake). This permits one to account for all the available data but masks the great variability in levels observed from one natural lake to the next (mean concentrations at standardized length often varying by factors of 3 to 4 between lakes, as shown in Chapter 5.0). In order to better estimate the duration of elevated fish Hg concentrations in reservoirs, post-impoundment Hg concentrations will be compared to the range of mean concentrations obtained in natural lakes, thus accounting for the variability observed from one lake to another (see figures of this chapter).

11.3
Results

11.3.1
Reservoirs

Fish Hg concentrations increased significantly in all species after impoundment of the La Grande complex reservoirs. In the Robert-Bourassa reservoir[1], for which the longest time series is available, total Hg levels in lake whitefish of standardized length (400 mm) increased by a factor of 5, from 0.11 to 0.53 mg kg^{-1}, 5 years after the impoundment (Figure 11.2a). Concentrations then declined gradually and significantly ($p < 0.05$, for comparison of confidence limits of means) to reach 0.21 mg kg^{-1} after 17 years. Although concentrations may still be decreasing and may stabilize at lower levels, the latter concentration (0.21 mg kg^{-1}) is not significantly different from concentrations measured in 5 neighboring natural lakes (see Figure 11.3).

Fish Hg levels followed similar trends in all La Grande complex reservoirs, with peak concentrations in lake whitefish occurring 5 to 9 years after impoundment, followed by a gradual decline towards concentrations measured in neighboring natural lakes, reaching the range of background levels after 10, 11 and 17 years in the case of the La Grande 4, Caniapiscau and Robert-Bourassa reservoirs respectively (Figure 11.2b).

---

[1] Previously called the La Grande 2 reservoir, this reservoir has been renamed the Robert-Bourassa reservoir. The same applies to the La Grande-2 generating station.

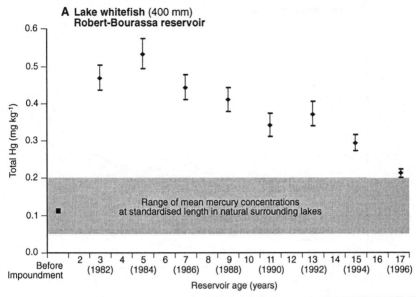

A Lake whitefish (400 mm)
   Robert-Bourassa reservoir

Range of mean mercury concentrations
at standardised length in natural surrounding lakes

Comparison of mean mercury concentrations at standardised length (0.05 significance level)									
Reservoir age (years)	Natural lakes g	3 (1982) ab	5 (1984) a	7 (1986) b	9 (1988) bc	11 (1990) de	13 (1992) cd	15 (1994) e	17 (1996) f
Mean level	0.112	0.469	0.533	0.444	0.411	0.342	0.371	0.294	0.210
Lower limit	0.107	0.436	0.494	0.411	0.380	0.312	0.340	0.273	0.232
Upper limit	0.117	0.504	0.575	0.478	0.444	0.374	0.406	0.316	0.199
N	503	150	134	148	140	101	102	164	156

Note: Bars indicate 95% confidence intervals.
   Sampling years with different letters indicate significantly different mean mercury concentrations as 95% confidence
   intervals do not overlap.

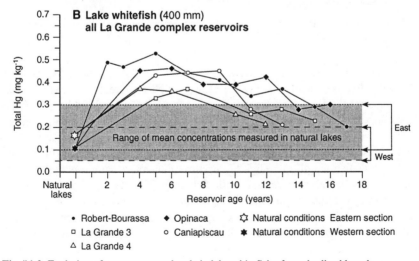

B Lake whitefish (400 mm)
   all La Grande complex reservoirs

Range of mean concentrations measured in natural lakes

East

West

• Robert-Bourassa   ♦ Opinaca   ☆ Natural conditions Eastern section
□ La Grande 3   ○ Caniapiscau ✱ Natural conditions Western section
△ La Grande 4

Fig. 11.2. Evolution of mean mercury levels in lake whitefish of standardized length

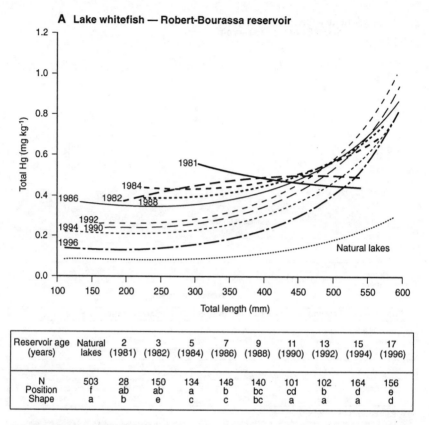

Reservoir age (years)	Natural lakes	2 (1981)	3 (1982)	5 (1984)	7 (1986)	9 (1988)	11 (1990)	13 (1992)	15 (1994)	17 (1996)
N	503	28	150	134	148	140	101	102	164	156
Position	f	ab	ab	a	b	bc	cd	b	d	e
Shape	a	b	e	c	c	bc	a	a	a	d

Note: Sampling years with different letters indicate significant differences (p<0.05) in the shape or position of the mercury-to-length relationship.

Fig. 11.3. Temporal evolution of the relationship between mercury and length after impoundment of the Robert-Bourassa reservoir

The shape of the relationship between total Hg and fish length also changed a few years after impoundment. Figure 11.3a, which represents the case of lake whitefish of the Robert-Bourassa reservoir, shows the typical evolution of this relationship after impoundment. Two to three years after flooding, as Hg concentrations increase more rapidly in younger, smaller fish than in older, larger fish, the relationship between Hg and length shifts, from concentrations typically increasing with fish length in a slightly curvilinear manner (natural conditions), to concentrations actually decreasing with fish length (1981). In young fish, the annual gain in flesh is proportionately higher than in old ones, with the result that their Hg concentration increases more rapidly, as prey organisms are more contaminated immediately after impoundment. In older fish, the large proportion of flesh produced before impoundment (with less Hg) serves as a buffer, so that their levels increase more slowly. After 7 to 9 years, as concentrations also increase in

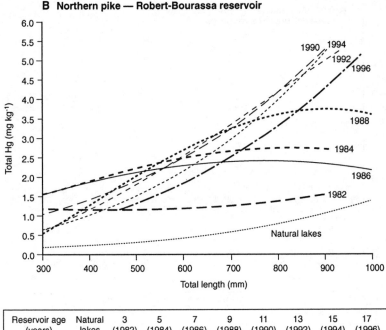

B Northern pike — Robert-Bourassa reservoir

Reservoir age (years)	Natural lakes	3 (1982)	5 (1984)	7 (1986)	9 (1988)	11 (1990)	13 (1992)	15 (1994)	17 (1996)
N	373	147	148	150	139	114	85	176	146
Position	e	d	bc	c	ab	a	a	b	c
Shape	b	c	de	e	d	a	a	de	d

Note: Sampling years with different letters indicate significant differences (p<0.05) in the shape or position of the mercury-to-length relationship.

Fig. 11.3. Temporal evolution of the relationship between mercury and length after impoundment of the Robert-Bourassa reservoir (continued)

larger fish and begin to decrease in smaller ones (as the Hg concentrations in prey decreases), the relationship between Hg and length shifts back towards its original shape, but at higher concentrations (1990). Subsequently, as Hg concentrations decrease in both small and large fish, the whole curve shifts downward towards natural concentrations (1996). The same temporal evolution in the relationship between Hg and length was observed for northern pike, although it was somewhat delayed as they are at a higher level in the food chain (Figure 11.3b).

In the Robert-Bourassa reservoir, mean Hg levels in northern pike of standardized length also increased by a factor of about 5, from 0.58 to 3.34 mg kg^{-1}, but peaked 9 to 13 years after impoundment (Figure 11.4a). Concentrations in northern pike measured after 15 and 17 years had also declined significantly (p < 0.05) reaching 2.46 mg kg^{-1}. The same holds true for the other

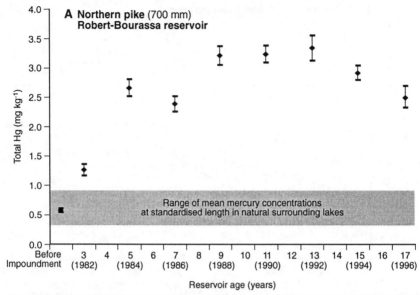

Comparison of mean mercury concentrations at standardised length (0.05 significance level)								
Reservoir age (years)	Natural lakes e	3 (1982) d	5 (1984) c	7 (1986) c	9 (1988) a	11 (1990) a	13 (1992) a	15 (1994) b	17 (1996) c
Mean level	0.575	1.265	2.661	2.389	3.213	3.243	3.343	2.922	2.460
Lower limit	0.544	1.170	2.521	2.266	3.051	3.101	3.129	2.802	2.580
Upper limit	0.607	1.363	2.806	2.515	3.379	3.389	3.565	3.044	2.340
N	373	147	148	150	139	114	85	167	146

Note: Bars indicate 95% confidence intervals.
Sampling years with different letters indicate significantly different mean mercury concentrations as 95% confidence intervals do not overlap.

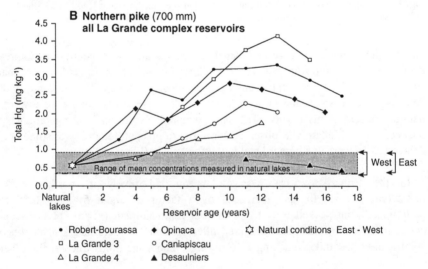

Fig. 11.4. Evolution of mean mercury levels in northern pike of standardized length

reservoirs, as mean Hg levels in northern pike peaked after 10 to 13 years depending on the reservoir and then also gradually and significantly decreased ($p < 0.05$) as shown in Figure 11.4b.

11.3.2
Dwarf Lake Whitefish

In the eastern part of the La Grande complex, populations of dwarf and normal lake whitefish coexist in both reservoirs and natural lakes (Doyon et al. 1998a). Figure 11.5 shows that dwarf specimens bioaccumulate Hg more rapidly than normal individuals. Eleven years after the beginning of the impoundment of the Caniapiscau reservoir, 6 and 7-years old dwarf specimens showed total Hg concentrations 3 times higher than those of normal individuals of similar ages (0.74 vs 0.27 mg kg^{-1}). As suggested by Doyon et al. (1998b), higher bioaccumulation rates in dwarf specimens may be caused by the fact that they begin to mature at an earlier age (usually at age 2 or 3 compared to age 6 or 7 for normal individuals), thus producing proportionally less flesh to dilute the assimilated Hg. Total Hg analyses indicating higher concentrations in flesh (0.42 mg kg^{-1}) than in gonads (0.17 mg kg^{-1}) support this hypothesis.

11.3.3
Rivers with Reduced Flow

In the reduced flow sections of the Eastmain river (over 90% reduction), mean Hg concentrations in standardized-length fish, measured after cut-off, were not significantly different from the range of concentrations measured in fish of natural neighboring lakes (Table 11.1).

11.3.4
Diversion Routes

Along the 250-km long Laforge diversion route, which diverts the upper Caniapiscau river into the La Grande river basin, water sequentially flows through the Caniapiscau, Fontanges, Vincelotte and La Grande 4 reservoirs. Figure 11.6 shows mean Hg levels obtained 9 years after the beginning of the diversion, for standardized-length lake whitefish and northern pike. They show that Hg is exported downstream from reservoirs, higher concentrations consistently being measured immediately below the different reservoirs. For example, the mean Hg concentration for lake whitefish of standardized length caught immediately below the Fontanges reservoir was 0.54 mg kg^{-1} compared to 0.35 mg kg^{-1} for lake whitefish caught within this reservoir. These data also show no cumulative effect on fish Hg concentrations passing from the Caniapiscau reservoir (1.97 mg kg^{-1} for northern pike of standardized length) to the La Grande 4 reservoir (corresponding concentration of 1.40 mg kg^{-1}) via the Fontanges and Vincelotte reservoirs.

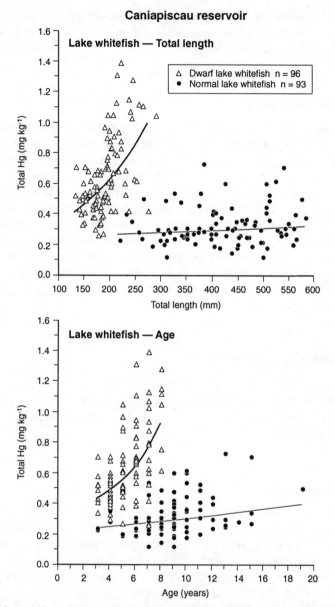

Fig. 11.5. Mercury bioaccumulation rates, as a function of length and age, for dwarf and normal lake whitefish of the Caniapiscau reservoir (11 years after impoundment)

Table 11.1. Mean total mercury levels (mg kg^{-1} ww) in fish of standardized length in reduced flow sections of the Eastmain river

Sampling station		Lake whitefish (400 mm) (mg kg^{-1})			Northern pike (700 mm) (mg kg^{-1})		
Range of mean Hg levels in natural lakes		0.05-0.20			0.30-0.93		
		N	[Hg]	C.I.[2]	N	[Hg]	C.I.[2]
Eastmain	Year 4[1]	25	0.22	(0.18-0.27)	29	0.85	(0.75-0.96)
	Year 6	-	-	-	28	0.84	(0.73-0.95)
	Year 8	11	0.15	(0.13-0.18)	22	0.83	(0.72-0.95)
	Year 12	11	0.12	(0.09-0.15)	26	0.75	(0.65-0.85)
Weir 5	Year 6	11	0.10	(0.08-0.13)	30	0.86	(0.76-0.98)
Eastmain-Opinaca	Year 4	-	-	-	30	0.62	(0.48-0.78)
	Year 6	-	-	-	11	0.96	(0.74-1.22)
	Year 8	-	-	-	19	0.64	(0.57-0.81)
	Year 12	-	-	-	11	0.80	(0.64-0.99)

[1] Years after flow reduction
[2] Confidence interval of estimated mean (95%)

11.3.5
Downstream from Reservoirs

11.3.5.1
The Caniapiscau River

During two consecutive summers (1984 and 1985), water from the Caniapiscau reservoir was spilled downstream in the Caniapiscau river. Figure 11.7 illustrates the changes in Hg levels in lake whitefish and lake trout of standardized length, before and after these spills. It shows that Hg was exported downstream, as Hg levels in fish caught at Eaton and Cambrien lake stations (respectively 100 and 275 km downstream) increased as much as in those of the Caniapiscau reservoir itself. At Eaton station, mean concentrations in standardized-length lake whitefish peaked two years after spilling at 0.58 mg kg^{-1}, which corresponds to 3 times the overall mean concentration measured in natural lakes of the eastern section of the La Grande complex (0.17 mg kg^{-1}). Hg concentrations in fish rapidly returned to natural levels (4 to 10 years after spilling depending on species and downstream station). At Calcaire station, below the large and deep Cambrien lake, Hg concentrations in these two species were never significantly different (p < 0.05) from values measured in natural lakes of the area.

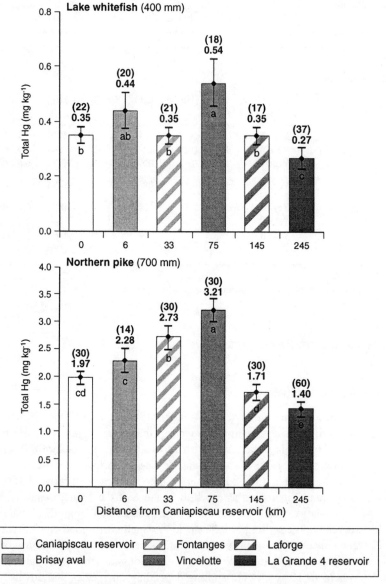

Fig. 11.6. Mean mercury levels measured in fish of standardized length along the Laforge diversion 9 years after diversion of the Caniapiscau waters towards the La Grande 4 reservoir (1993)

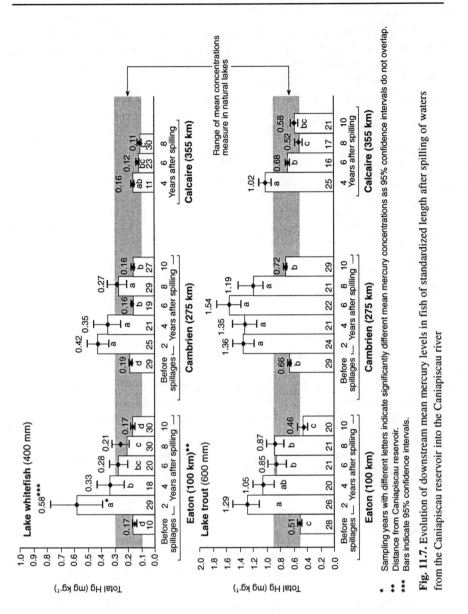

Fig. 11.7. Evolution of downstream mean mercury levels in fish of standardized length after spilling of waters from the Caniapiscau reservoir into the Caniapiscau river

* Sampling years with different letters indicate significantly different mean mercury concentrations as 95% confidence intervals do not overlap.
** Distance from Caniapiscau reservoir.
*** Bars indicate 95% confidence intervals.

11.3.5.2
The La Grande River

Large lake whitefish caught immediately downstream from the Robert-Bourassa reservoir bioaccumulate Hg much more rapidly than those of comparable size caught within the reservoir (Figure 11.8). A study of stomach contents carried out in 1990 revealed that downstream lake whitefish (usually benthic feeders) became piscivorous, feeding on small ciscos (total length of 20 to 130 mm) stunned by their passage through the turbines (Brouard et al. 1994). As a result, mean concentration for specimens >450 mm of total length was 0.54 mg kg^{-1}, within the reservoir compared to 2.95 mg kg^{-1}, immediately downstream.

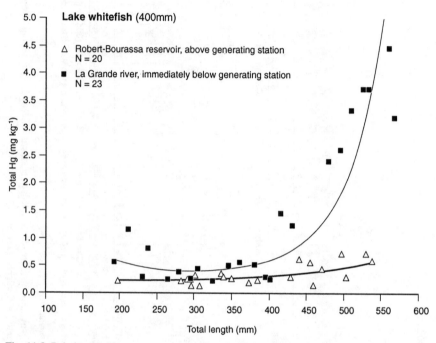

Fig. 11.8. Relationship between mercury and length for lake whitefish captured above and below the Robert-Bourassa generating station (1992)

11.3.5.3
The Coast of James Bay

As shown in Table 11.2, the effect of the La Grande complex reservoirs on Hg levels in fish of James Bay is limited to the area influenced by the summer fresh water plume of the La Grande river. Sedimentation of Hg-rich suspended organic particles, dilution and flocculation mechanisms at the salt and fresh water interface are believed to limit the export of reservoir-produced methylmercury (MeHg) in coastal areas.

Table 11.2. Estimated mean mercury levels (mg kg^{-1} ww) in fish of standardized length on the coast of James Bay

Area	Species	Cisco (300 mm)	Lake whitefish (400 mm)	Brook trout (300 mm)	Fourhorn sculpin (250 mm)	Greenland cod (400 mm)
North of the	Number of fish	231	58	157	29	14
La Grande summer	Number of stations	2	2	2	1	1
plume	Range of mean [Hg] (mg kg^{-1})	(0.09-0.10)	(0.08-0.15)	(0.09-0.14)	(0.10)	(0.14)
La Grande summer	Number of fish	279	213	221	106	58
plume	Number of stations	4	3	3	3	3
	Range of mean [Hg] (mg kg^{-1})	(0.10-0.13)	(0.14-0.21)	(0.19-0.27)	(0.31-0.55)	(0.26-0.42)
South of the	Number of fish	129	203	217	31	60
La Grande summer	Number of stations	4	5	3	1	2
plume	Range of mean [Hg] (mg kg^{-1})	(0.10)	(0.06-0.15)	(0.12-0.14)	(0.22)	(0.18-0.42)*

* High value from one unaffected station exhibiting naturally high concentrations.
Brook trout (*Salvelinus fontinalis*), Fourhorn sculpin (*Myoxocephalus quadricornis*), Greenland cod (*Gadus ogac*).

11.4
Discussion

11.4.1
Reservoirs

11.4.1.1
Increases of Fish Mercury Levels in Reservoirs

The monitoring of fish Hg levels at the La Grande complex clearly shows important increases in all species after impoundment of all the reservoirs. The cause-and-effect relationship between reservoir creation and elevated Hg concentrations in fish has been established by numerous authors since the mid 1970's (Potter et al. 1975; Abernathy and Cumbie 1977; Bruce et al. 1979; Lodenius et al. 1983; Bodaly et al. 1984; Messier and Roy 1987; Verdon et al. 1991).

Cox et al. (1979) inferred that Hg entered the food chain from the flooded soils. Since then, it has been well demonstrated that the flooding of vegetation and organic matter in soils enhances bacterial activity which produces MeHg from inorganic Hg (Furutani and Rudd 1980; Ramlal et al. 1987; Hecky et al. 1991). Experiments in limnocorrals demonstrated that organic materials consistently stimulated bioaccumulation of MeHg in perch muscle (Hecky et al. 1987). Studies at the La Grande complex showed that the proportion of MeHg to total Hg in flooded soils increased gradually after the impoundment of the La Grande complex reservoirs (Grondin et al. 1995; Bégin 1997; Tremblay and Lucotte 1997).

A certain proportion of the MeHg produced at the flooded soils and vegetation levels is rapidly transferred through the reservoir food chain up to the fish (Chapter 9.0). This transfer may be attributed to a number of processes: (1) diffusion to the water column and rapid adsorption to suspended particles (Morrison and Thérien 1991b; Chapter 7.0); (2) erosion and resuspension of shoreline flooded organic matter by wave action (Mucci et al. 1995); (3) periphyton and aquatic insects may also play an important role in the biological transfer from shallow flooded soils to fish (Lebeau 1996; Tremblay et al. 1996ab; Chapter 8.0).

11.4.1.2
Factors Explaining Differences Observed Between Reservoirs

Monitoring at the La Grande complex indicates that all reservoirs show a similar trend in fish Hg levels, although they exhibit differences in the magnitude of the after impoundment increases, as well as in the rates of increase and subsequent decrease. The following discussion will try to relate these differences to reservoir-specific physical and hydrologic characteristics such as: land area flooded, annual volume of water flowing through, filling time (duration) and proportion of flooded land area located in the drawdown zone. The ratio between the land area flooded (in km^2) and the annual volume of water (in km^3) flowing through the reservoir (LAF/AVW ratio) should be a good indicator of the magnitude of increase of Hg levels in fish. The land area flooded indicates the amount of organic matter stimulating bacterial methylation of Hg. A number of authors have suggested the extent of flooding as a key factor in determining reservoir fish Hg concentrations (Jones et al. 1986; Messier and Roy 1987; Johnston et al. 1991; Verdon et al. 1991; Kelly et al. 1997). The annual volume of water flowing through a reservoir is also considered a key factor because it indicates the diluting capacity of the Hg released into the water column, it plays a role in the extent of oxygen depletion (as Gilmour and Henry (1991) have shown that anoxic conditions promote methylation), and it determines the extent of export of Hg downstream (see further). The annual volume of water flowing through a reservoir is preferred to the static volume of a reservoir as an indicator of the diluting capacity because it integrates the notion of water residence time, thus allowing one to distinguish between two reservoirs having equal volumes but different water residence times. The LAF/AVW ratio has also been identified by Schetagne (1994) as a good indicator of the potential for post-impoundment water quality modifications in reservoirs (such as dissolved oxygen depletion or CO_2 and total phosphorus increases), which are also related to bacterial decomposition of flooded organic matter. Thus, the higher this ratio, the greater would be the increase of Hg levels in fish.

Table 11.3, which relates a number of reservoir characteristics to the maximum increase factors of Hg concentrations in fish obtained after impoundment for lake whitefish and northern pike, shows that this ratio alone cannot explain all the observed differences, although it allows a good differentiation between the maximum increase factors obtained for the Robert-Bourassa, Opinaca and La Grande 4 reservoirs. Indeed, the LAF/AVW ratio of these 3 reservoirs decreases from 31 to 14, while the maximum Hg increase factor in fish (average for both species) decreases from 5.3 to 2.3.

The time (in months) it takes to fill a reservoir is also considered an important factor in determining peak Hg levels in fish after impoundment, as a number of authors have shown that the release of Hg from flooded organic material to the water column is very rapid (Morrison and Thérien 1991b; Kelly et al. 1997). In the case of a reservoir flooded over a number of years, the land area flooded first would have already released its Hg when the land area flooded last would begin to release it. Chartrand et al. (1994) showed that water quality modifications, indicative of bacterial decomposition, peaked after 2 to 3 years in reservoirs filled within a year (Robert-Bourassa and Opinaca), but peaked after 6 to 10 years in the Caniapiscau reservoir, filled in 3 years. Thus, considering all other factors equal, the longer the filling time, the lower would be the peak Hg levels in fish, but the longer would be the period necessary to return to concentrations typical of neighboring lakes.

At the La Grande 3 reservoir, the lower increase factor in lake whitefish Hg levels, compared to the Robert-Bourassa reservoir, as well as the peak concentration occurring later (after 7 years compared to 5 years, Figure 11.4), may be related to the longer filling time (38 months compared to 13 months), other characteristics such as LAF/AVW ratio, being similar (Table 11.3). For pike of this reservoir, the increase factor (7.2) may be overestimated because a large proportion of pike analyzed for Hg were captured at a sampling station located in a shallow area, where actual LAF/AVW ratio is higher. Over the monitoring period, concentrations in pike have systematically been much higher at this station than at the other stations of this reservoir (from one to two mg kg^{-1} higher). This would also explain the discrepancy in the increase factors for pike (7.2) compared to lake whitefish (3.4) in this reservoir, while increase factors between species are similar for all the other La Grande complex reservoirs.

The proportion of the total land area flooded located in the drawdown zone would be a good indicator of the importance of the active biological transfer of

Table 11.3. Comparison of physical and hydrologic characteristics of reservoirs with post-impoundment increases in fish mercury concentrations for the La Grande complex[1]

Reservoir	Land area flooded (in km²)	Reservoir water volume (in km³)	Annual volume of water (in km³)	Filling time (in months)	Land area flooded to annual volume of water ratio (LAF/AVW)	Proportion of flooded area in drawdown zone	Initial mean fish [Hg] at standardized length[2] (mg kg⁻¹)	Maximum after impoundment fish [Hg] at standardized length (mg kg⁻¹)	Maximum after impoundment fish [Hg] increase factor
Robert-Bourassa	2 478	57	81	13	31	29 %	Whitefish = 0.11 Pike = 0.58	Whitefish = 0.53 Pike = 3.34	Whitefish = 4.8 Pike = 5.8
Opinaca	622	7	27	6	23	28 %	Whitefish = 0.11 Pike = 0.58	Whitefish = 0.46 Pike = 2.85	Whitefish = 4.2 Pike = 4.9
La Grande 3	1 923	48	60	38	32	29 %	Whitefish = 0.11 Pike = 0.58	Whitefish = 0.37 Pike = 4.16 (3)	Whitefish = 3.4 Pike = 7.2[3]
La Grande 4	596	17	43	9	14	36 %	Whitefish = 0.17 Pike = 0.58	Whitefish = 0.37 Pike = 1.65	Whitefish = 2.2 Pike = 2.8
Caniapiscau	2 808	40	24.5	35	115	61 %	Whitefish = 0.17 Pike = 0.58	Whitefish = 0.45 Pike = 2.29	Whitefish = 2.6 Pike = 3.9

[1] Average values of physical and hydrologic characteristics calculated for the first 10 years after impoundment.

[2] Standardized length of 400 mm for lake whitefish and of 700 mm for northern pike.

[3] Northern pike [Hg] is overestimated because of sampling in an area not representative of the reservoir as a whole.

MeHg from the flooded soils to fish. According to Tremblay and Lucotte (1997) and to the findings in Chapter 9.0, this biological transfer could play a significant role during a prolonged period (at least 14 years) in shallow areas, protected from wave action, where organic matter has not been eroded. At the La Grande complex reservoirs, the flooded soils are generally very thin and are rapidly eroded and subsequently deposited in deeper colder areas, less favorable to bacterial methylation. This erosion reduces the surface of flooded soils, still containing organic matter after a few years of water level fluctuations, where biological transfer of Hg (by periphyton or insects) still takes place. Thus, for reservoir such as those found in the La Grande complex where the organic layer of flooded soils is quickly removed by wave action, the greater the proportion of land area flooded located in the drawdown zone, the lower would be the magnitude of increase of fish Hg levels and the quicker the return to background levels.

As shown in Table 11.3, comparatively lower increase factors were observed in the Caniapiscau reservoir, compared to the Robert-Bourassa and Opinaca reservoirs, in spite of a greater LAF/AVW ratio (115 vs 31 and 23). This may be attributed, to a longer filling time (35 months compared to 13 and 6 months), as well as to a much greater proportion of flooded land area in the drawdown zone (61% vs 29 and 28%). By greatly reducing the biological transfer of Hg (by benthic organisms) from the flooded organic matter to fish, this latter characteristic may also have contributed to a shorter return time to lake whitefish concentrations equivalent to those of natural lakes of the area (after 11 years in the Caniapiscau reservoir compared to 17 or more years for the Robert-Bourassa and Opinaca reservoirs, see Figure 11.4).

Other factors may also explain the differences observed between reservoirs. At the Opinaca reservoir, higher water temperatures (Schetagne and Roy 1985), favoring methylation (Bodaly et al. 1993; Kelly et al. 1997), as well as denser flooded vegetation and flooded soils richer in organic matter (Poulin-Thériault–Gauthier-Guillemette 1993), may explain why peak fish Hg levels are a little higher than could be expected from the LAF/AVW ratio alone, when compared to the value obtained for the Robert-Bourassa reservoir. Furthermore, lower water temperatures (Schetagne and Roy 1985) and flooded soils poorer in organic carbon (Poulin-Thériault–Gauthier-Guillemette 1993) may also explain the lower fish Hg levels measured in the Caniapiscau reservoir.

Although increasing fish Hg levels in reservoirs is a very complex phenomenon in which numerous other factors must be involved, including biological factors such as the structure of the fish community, fish growth rates and diet (Chapter 10.0), the results presented in Table 11.3 suggest that a limited number of physical and hydrologic factors may be used to predict the order of magnitude of the increase in fish Hg levels.

11.4.1.3
Duration of the Phenomenon in Reservoirs

In reservoirs of northern Québec and Labrador, Hg levels in lake whitefish return to natural levels 10 to 25 years after impoundment (Figure 11.9a). In the case of the Baskatong reservoir, the still elevated concentrations in lake whitefish, after over 50 years, may be attributed to log floating operations which, until recently, brought continuous allochtonous organic material stimulating bacterial methylation.

For piscivorous species, such as northern pike, Figure 11.9b shows that concentrations have returned within the range of concentrations measured in natural lakes in all reservoirs older than 20 to 30 years. A similar duration has also been observed in Finland and northern Manitoba (Verta et al. 1986b; Strange 1993, 1995). Although the age of the La Grande complex reservoirs varied only from 10 to 17 years at the time of the latest monitoring campaign, the general shape of the curves obtained for piscivorous species in these reservoirs follows the same time trend.

A number of studies and field observations show that the factors responsible for the increase of fish Hg levels in reservoirs are temporary. Water quality monitoring at the La Grande complex has shown that temporary modifications due to decomposition of flooded organic matter, such as the increases of CO_2 and phosphorus, are virtually over, 8 to 10 years after impoundment in reservoirs filled within a year and after 14 years in the Caniapiscau reservoir, filled during a 3 year period (Schetagne et al. 1996). Schetagne (1994) suggested that this short period of modification was due to the rapid depletion of readily decomposable flooded organic matter. According to Garzon (1984), only the green part of flooded vegetation biomass decomposes readily. The remainder is composed of substances resistant to biochemical degradation. Van Collie et al. (1983) showed that ligneous components of flooded vegetation remained virtually unaffected after many decades, as spruce tree trunks had lost less than 1% of their biomass after 55 years of flooding in the Gouin reservoir located in Québec. Field observations at the La Grande complex reservoirs have shown that organic material was rapidly eroded on a large proportion of the shorelines (within a 5 to 10 year period), thus rapidly reducing active biological transfer of MeHg from flooded soils to water column by benthic organisms (Poulin-Thériault 1983, 1987, 1994). Zooplankton and benthic organisms collected from stomach contents of white suckers of the Desaulniers reservoir, located close to the Robert-Bourassa reservoir, had MeHg concentrations similar to those of natural lakes, 17 years after impoundment (Doyon et al. 1996). Hg concentrations in zooplankton collected in the pelagic zone of La Grande complex reservoir also had MeHg levels equivalent to those of zooplankton of surrounding natural lakes 8 to 10 years after impoundment

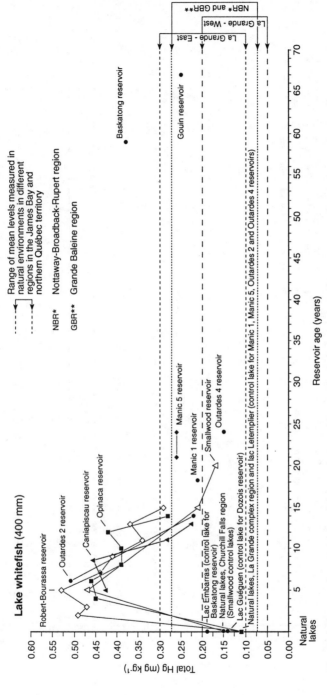

Fig. 11.9. Evolution of mean mercury levels in fish of standardized length in young and old reservoirs of Québec and Labrador

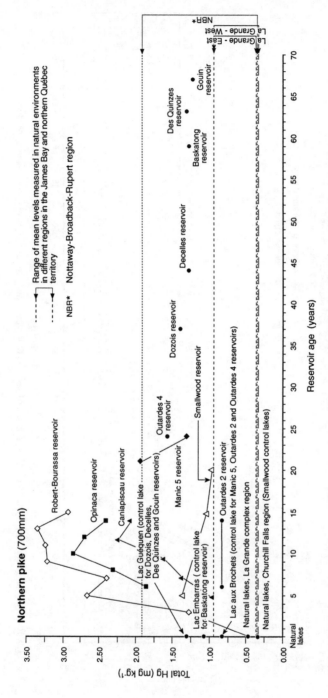

Fig. 11.9. Evolution of mean mercury levels in fish of standardized length in young and old reservoirs of Québec and Labrador (continued)

(Tremblay et al. 1998a; Chapter 9.0). Thus, after a rapid upsurge of Hg through the reservoir food chain, Hg concentrations in fish gradually return to levels typical of surrounding natural lakes.

11.4.2
Downstream from Reservoirs

Monitoring of fish Hg levels at the La Grande complex has also shown that Hg is exported downstream from reservoirs (Figures 11.6, 11.7 and 11.8). This has also been acknowledged in other studies (Johnston et al. 1991; Kelly et al. 1997; LGL Ltd. 1993). Studies at the La Grande complex indicate that Hg is mostly exported by reservoir Hg-rich organic debris as well as by plankton, aquatic insects or small fish (Brouard et al. 1994; Montgomery et al. 1996; Brouard et al. 1990). Schetagne et al. (1996) have suggested that the magnitude of increase in downstream fish Hg levels, as well as the distance to which it occurs, depends on: (1) the importance of tributaries diluting the reservoir Hg-rich debris or organisms (food source for fish) and (2), the presence or absence downstream of large bodies of water (lakes or other reservoirs) permitting the sedimentation of organic material or the biological uptake of the Hg-rich organisms. The fact that fish Hg levels did not increase at Cambrien station (Figure 11.7), located below large and deep Cambrien lake, supports this hypothesis. LGL Ltd (1993) also showed an increase of fish Hg levels downstream from the Smallwood reservoir (Labrador), as well as an important decrease in concentrations below Winokapau lake, located 105 km downstream.

The importance of large downstream bodies of water is further corroborated by the absence of a cumulative effect in fish Hg levels passing from the Caniapiscau reservoir to the La Grande 4 reservoir, via the Fontanges and Vincelotte reservoirs. The above mentioned results strongly suggest that in the case of a chain of large deep reservoirs, the downstream effect of one reservoir is limited to only part of the first reservoir located immediately below. Indeed, spatial distributions of fish Hg levels within a given receiving reservoir show that this effect, in the case of large reservoirs, is limited to the area located near the input from the upstream reservoir (Schetagne et al. 1996).

11.4.3
Rivers with Reduced Flow

In the reduced flow sections of the Eastmain and Opinaca rivers, fish Hg levels remained within the range of concentrations observed in surrounding natural lakes, since there was no flooding of organic material stimulating methylation of Hg by decomposing bacteria.

11.5
Conclusions

The creation of reservoirs, flooding large tracts of land, causes important but temporary increases of Hg levels in fish, to levels 3 to 7 times those measured in neighboring natural lakes, as observed in the La Grande complex reservoirs. Monitoring at the La Grande complex indicates that all reservoirs show a similar trend in fish Hg levels, although they exhibit differences in the magnitude of the post-impoundment increases, as well as in the rates of increase and subsequent decrease. Although the increase in fish Hg levels in reservoirs is a complex phenomenon, our results suggest that these differences between reservoirs may be explained, at least in part, by a limited number of physical and hydrologic characteristics such as: the extent of land area flooded, the annual volume of water flowing through, the filling time and the proportion of flooded area in the drawdown zone. In developing models to predict the evolution of fish Hg levels in future reservoirs, these characteristics should be taken into account. Water temperature, the density of flooded vegetation and organic content of flooded soils should also be incorporated.

The results of the La Grande complex monitoring program also suggest that, in order to reduce health risks related to the temporary increase in Hg levels in fish resulting from the creation of hydroelectric reservoirs, future development plans should favour schemes that result in lower ratios of land area flooded to annual volume of water flowing through a reservoir, as well as development schemes in northern regions where temperature, density of vegetation and organic content of soils are usually lower. Although reservoirs with significant drawdown may lead to lower Hg levels in fish, they should not be favoured because of their potential negative effects on fish production, particularly considering the temporary nature of the increase in fish Hg levels.

Acknowledgements

Funding for this research was provided by Hydro-Québec, the Société d'énergie de la Baie James and the James Bay Mercury Committee. We are grateful to Judit Vass of Philip Services Corp. for the quality control of the laboratory analyses and several employees of Groupe-conseil Génivar Inc. who carried out the field work and the statistical analyses.

12 Calculated Fluxes of Mercury to Fish in the Robert-Bourassa Reservoir

Normand Thérien and Ken Morrison

Abstract

A detailed analysis of existing fish and plankton data allowed estimation of bio-mass and mercury (Hg) fluxes in fish during the years 1978-1984 immediately following the flooding of the reservoir. Fish were divided into two arbitrary groups: piscivores and prey (non-piscivores) accounting globally for more than 90% of all fish biomass captured during the 7 years period. Extensive catch-per-unit-effort data available over time was used to estimate fish yield based on pro-ductivity and to calculate standing stocks allocated to both groups. A strictly pe-lagic food chain (from plankton to non-piscivorous fish to piscivores) was as-sumed for the computation of the biomass and Hg fluxes from one trophic level to another. Using a verified plankton model for the reservoir, "available" biomass and Hg fluxes to prey fish (non-piscivores) were calculated. From the field data and calculated standing stocks, the "available" biomass and Hg fluxes from the prey fish to the piscivorous fish were also calculated. Conversely, from field data and the calculated standing stocks for both groups of fish, fluxes "required" to explain the fish biomass and Hg concentration of fish, were also computed. The magnitude of the deficits between the "available" and "required" fluxes showed that an entirely pelagic food chain could not account for observed Hg in fish, but that the greatest vector of Hg transfer to fish must have been the benthos. This conclusion was supported independently by analyses of fish stomach contents and recent data on Hg concentrations in benthos. The fluxes of total Hg, as dissolved Hg but more likely Hg associated with particulates, from decomposing vegetation and soil to the water column were also calculated in order to estimate the global fluxes of Hg within several compartments of the reservoir in the first few years after flooding.

12.1
Introduction

Following construction of the large reservoirs in northern Québec, marked in-creases in fish Hg concentrations have been measured. The evolution of Hg levels in fish were followed (Chapter 11.0), as were numerous aspects of the aquatic environment in general (phytoplankton and zooplankton populations, fish populations, dissolved nutrients, dissolved oxygen, water temperature, etc). Unfortunately, at the time of flooding of the largest of these reservoirs (1978-1983), analytical techniques were not sufficiently sensitive to measure changes in dissolved Hg or Hg concentrations in very small samples (e.g., mg of plankton). Studies were carried out *in vitro* to quantify Hg release from vegetation and humus once analytical techniques were adequate (Chapter 7.0). Also, several studies have been carried out on Hg levels in plankton, benthos, and fish stomach contents (Chapters 3.0, 9.0 and 10.0), but these were done long after the initial flooding, when the greatest changes in fish Hg levels occurred. Therefore, the actual processes leading to the increased fish Hg levels in reservoirs have not been clearly quantified. A better identification of the most significant Hg pathways to fish thus appeared a prerequisite before predicting Hg concentration in fish of reservoirs.

Since fish are at the top of the aquatic food web, they may reflect an integration of the processes occurring lower down. Therefore, it should be possible to infer the fluxes through fish from changes in fish biomass and fish Hg concentrations, and thus estimate some of these processes. This is what we have done in this study. There were three phases to the study. The first was a detailed analysis of existing fish and plankton data to estimate biomass and Hg fluxes in fish and to identify information deficits. In effect, it appeared important to test the validity of the pelagic food chain (Zooplancton $\rightarrow$ Non-piscivorous fish $\rightarrow$ Piscivorous fish) as the dominant pathway explaining the biomass and Hg concentration of fish in the years following flooding of the reservoir. The approach was similar to the testing of an hypothesis. To do this, mathematical relationships would be required to calculate standing stocks of fish and biomass and Hg fluxes between trophic levels. However, the emphasis was not to use or develop specific bioenergetic models for biomass nor to use or develop specific predictive models for Hg in fish but to make the best use of the available field data. The field data was used to compute "available" biomass and Hg fluxes from zooplankton to the non-piscivorous fish and, similarly, to compute "available" fluxes from non-piscivorous fish to the piscivores. Conversely, field data was used to compute biomass and Hg fluxes that would be "required" from zooplankton (under the assumption that it is the unique source of food to non-piscivorous fish) to explain observed prey fish (non-piscivores) biomass and Hg concentration. Computation was also done of the biomass and Hg fluxes "required" from prey fish to explain the observed piscivorous fish biomass and Hg concentration. Deficits of fluxes between trophic levels were then calculated as the difference between the "available" and "required" biomass and Hg fluxes. The second phase of the study was to estimate the fluxes of Hg coming from flooded vegetation since data was recently available (Chapter 7.0). The final phase was an integration of the first two phases to estimate the global fluxes of Hg within the reservoir in the first few years after flooding.

12.2
Materials and Methods

Fish were divided into two arbitrary groups: piscivores and prey (non-piscivores). The first group included northern pike (*Esox lucius*) and walleye (*Stizostedion vitreum)* constituting 90% of all piscivores species caught during 1978-1984. The second group included lake whitefish (*Coregonus clupeaformis*), longnose sucker (*Catostomus catostomus*) and white sucker *(Catostomus commersoni)* constituting more than 95% of all non-piscivorous species caught during that time. These two groups of fish accounted globally for 90-95% of all fish biomass captured during the 7 year period (Doyon and Belzile 1998).

12.2.1
Estimation of Fish Standing Stocks

No estimates of standing stocks were available, but there were extensive catch-per-unit-effort data for the various stations over time. Using Ryder's morpho-edaphic index (Ryder 1965; Ryder et al. 1974) as a first approximation, fish yield was estimated at two lake stations as: $Y = 0.966 \, MEI^{0.5}$ with $Y =$ fish yield (kg ha^{-1} yr^{-1}) and $MEI =$ morphoedaphic index. This index is simply the total dissolved solids divided by mean depth (TDS/ $\overline{Z}$). Standing stock was then estimated as 10 x yield, since productivity of the region was evaluated at 10% (Power and Le Jeune 1976). Reservoir standing stocks were estimated after flooding in proportion to catch-per-unit-effort. Standing stock was allocated to piscivore or prey (non-piscivore) groups proportional to catches. Hg in standing stocks was then estimated using appropriate weighted mean concentrations, the weighted means being calculated based on Hg-age relationships (Morrison and Thérien 1995b) and estimated population age structure.

12.2.2
Estimation of Biomass Fluxes

The approach chosen in this work was not to use or develop a detailed bioenergetic predictive model for biomass over time as presented by Kitchell et al. (1977) or Hewett and Johnson (1992), but rather to rely on mass balance relationships to compute estimates of annual biomass fluxes from fish biomass data. To this end, different equations were written for biomass changes for the four trophic levels (phytoplankton, zooplankton, prey fish, piscivorous fish) based on classical mass-balance principles (Jorgensen 1980):

$$\frac{\Delta PHYTO}{\Delta t} = HYDR_{phyto} + GROWTH_{phyto} - RESP_{phyto} - SED_{phyto} - PRED_{phyto}$$

$$\frac{\Delta ZOO}{\Delta t} = HYDR_{zoo} + PRED_{phyto} - DEF_{zoo} - RESP_{zoo} - SED_{zoo} - PRED_{zoo}$$

$$\frac{\Delta PREY}{\Delta t} = PRED_{zoo} + PRED_{benthos} - DEF_{prey} - RESP_{prey} - PRED_{prey}$$

$$\frac{\Delta PISC}{\Delta t} = PRED_{prey} + PRED_{auto} - DEF_{pisc} - RESP_{pisc} - PRED_{auto} - PRED_{pisc}$$

where:

$HYDR_{phyto,zoo}$	- hydraulic effects (net rate of change due to inflows and outflows) on phytoplankton and zooplankton
$GROWTH_{phyto}$	- growth of phytoplankton via photosynthesis
$RESP_{phyto,zoo,prey,pisc}$	- respiration (metabolic losses) for each trophic level
$SEDp_{hyto,zoo}$	- sedimentation of phytoplankton and zooplankton
$PRED_{phyto,zoo,benthos,prey,pisc}$	- predation (biomass transfer) from each trophic level to the next
$PRED_{auto}$	- auto-predation of piscivores (piscivores eating other piscivores)
$DEF_{zoo,prey,pisc}$	- defecation by zooplankton, prey fish and piscivores

Rate formulations and constants were consistent with basic bioenergetics and observed population structures (Thérien and Morrison 1994). The rates for the two plankton levels were obtained from a previously-verified model of plankton dynamics for the Robert-Bourassa reservoir (Thérien et al. 1982; Morrison et al. 1987; Morrison and Thérien 1987). To be conservative, all mortality of piscivores was considered to be auto-predation, and thus the $PRED_{pisc}$ rate was assumed to be zero. In addition, no information was available on benthic populations, and assuming a strictly pelagic food chain, $PRED_{benthos}$ was assumed to be zero. Although $PRED_{auto}$ cancels out in the equation for piscivores, it is included explicitly because defecation is proportional to the sum of this and $PRED_{prey}$.

Since standing stocks had already been estimated, the biomass changes were calculated. Therefore, the different equations for zooplankton and prey fish could be solved separately to estimate the amount of predation possible on the observed populations. For comparison, the different equations were also solved individually for the two fish trophic levels to estimate the amount of predation necessary to have the observed biomass levels. Finally, this process was also extended to estimate the amount of predation necessarily flowing into prey fish to support a standing stock capable of subsequently supporting the estimated piscivores standing stocks.

12.2.3
Estimation of Hg Fluxes

Again, the approach chosen in this work was not to use or develop a detailed bioenergetics-based predictive model for Hg in fish as presented by Norstrom et al. (1976), Rodgers (1994), or Harris and Spencer (1998) to name a few, but rather to rely on Hg mass balance relationships to compute estimates of annual Hg fluxes from fish biomass and Hg data. In the same manner as for biomass, different equations were written for Hg changes for the four trophic levels:

$$\frac{\Delta MERC_{phyto}}{\Delta t} = ABSORB_{phyto} - EXCR_{phyto} - SEDHG_{phyto} - TRANS_{phyto}$$

$$\frac{\Delta MERC_{zoo}}{\Delta t} = TRANS_{phyto} - DEFHG_{zoo} - EXCR_{zoo} - SEDHG_{zoo} - TRANS_{zoo}$$

$$\frac{\Delta MERC_{prey}}{\Delta t} = TRANS_{zoo} + TRANS_{benthos} - DEFHG_{prey} - EXCR_{prey} - TRANS_{prey}$$

$$\frac{\Delta MERC_{pisc}}{\Delta t} = TRANS_{prey} + TRANS_{auto} - DEFHG_{pisc} - EXCR_{pisc} - TRANS_{auto} - TRANS_{pisc}$$

where:

$ABSORB_{phyto}$ - absorption of Hg due to growth of phytoplankton

$EXCR_{zoo,prey,pisc}$ - excretion of Hg from each trophic level respectively

$SEDHG_{phyto,zoo}$ - sedimentation of phytoplankton and zooplankton Hg

$TRANS_{phyto,zoo,benthos,prey,pisc}$ - transfer of Hg via predation from each trophic level to the next

$TRANS_{auto}$ - Hg transfer due to auto-predation of piscivores

$DEFHG_{zoo,prey,pisc}$ - Hg in feces of zooplankton, prey fish and piscivores

Rate formulations and constants were again consistent with basic bioenergetics and observed population structures (Thérien and Morrison 1994). In fact, most of these rates were equal to the corresponding biomass rates multiplied by the appropriate Hg concentrations. Due to a lack of data for plankton Hg concentrations, it was assumed that plankton contained 50 ng Hg g^{-1} wet weight. Assimilation efficiency for Hg was set at 90% (e.g., $DEFHG_{pisc} = 0.10 \times [TRANS_{prey}+TRANS_{auto}]$). Hydraulic effects on plankton were neglected because no information was available. Other assumptions were the same as for biomass. Calculations of available and necessary fluxes were also carried out in the same manner as for biomass.

12.2.4
Calculation of Hg Fluxes from Vegetation

The fluxes of Hg from decomposing vegetation to the water column were calculated using an adaptation of the decomposition-water quality model of Thérien and Morrison (1984, 1985). For inputs this model requires inflows and outflows, volume-water level curves, water temperature, and masses of vegetation and soil cover as functions of elevation. The Hg submodel and all coefficient values used in the model were from Morrison and Thérien (1991ab, 1994). In this model, the Robert-Bourassa reservoir is divided into five zones, of which three are hydraulically active while the other two are relatively stagnant and exchange water only with the adjacent active zones. A system of differential equations is used for the water quality and Hg variables in each zone, and there are transfers among the zones as well as into and out of the reservoir via hydraulic flows. The equation for the quantity of mass released as a function of time in zone k is formulated as:

d (Volume x Concentration) / dt =   Mass inflow rates to zone k
                                     + Mass rate released from flooded vegetation
                                       to zone k
                                     − Mass outflow rates from zone k

As the cumulative release of Hg from flooded vegetation showed an asymptotic trend (Chapter 7.0) the quantity of Hg released at time t was expressed as :

$$Q(t) = Qmax [1-exp(-K.t)]$$

where Qmax is the ultimate cumulative quantity released and K, the rate constant for the release under the environmental conditions observed. In differential form, the above relationship is written :

$$d Q(t) / dt = K. [Qmax - Q(t)]$$

and permits the calculation of the mass rate of Hg released in the water column. The model was run from 1978 to 1984. Since there were no data on inflowing Hg concentrations, these were neglected and thus the model results indicate the additional quantities above background levels.

12.2.5
Resolution of the Major Hg Fluxes

To resolve the major Hg fluxes that occurred in the reservoir, several assumptions were made:

1. the estimated fluxes of Hg from vegetation were retained as correct
2. the estimated fluxes of biomass and of Hg to plankton were retained as correct

3. the estimated fluxes of biomass and of Hg to piscivorous fish were retained as correct
4. the fluxes of biomass and of Hg estimated as necessary to explain observed predatory fish Hg were retained as correct.

All other fluxes of biomass and Hg were then calculated to satisfy the conditions of conservation of mass.

12.3
Results and Discussion

12.3.1
Biomass and Hg Fluxes

Average available and required fluxes for biomass are shown in Table 12.1, and those for Hg in Table 12.2. Detailed tables can be found in Morrison and Thérien (1995a).

Table 12.1. Average biomass fluxes available from observed standing stocks to be consumed by the subsequent trophic level compared to fluxes required to support the observed standing stocks of the consuming trophic level (kg ha^{-1} yr^{-1})

Year	Available		Required		
	From zooplankton for prey fish	From prey fish for piscivores	From zoo-plankton for prey fish	From prey fish for piscivores	From zoo-plankton for piscivores
1978	3.55	1.83	7.59	6.66	28.05
1979	4.24	1.45	4.67	2.03	10.33
1980	4.25	2.88	24.77	9.60	53.24
1981	7.39	3.05	15.64	29.10	123.56
1982	7.30	3.63	19.34	34.86	148.70
1983	7.23	3.24	11.28	24.46	100.44
1984	7.24	3.41	15.34	28.57	119.57

Prior to flooding in 1978, zooplankton could only furnish ~50% of the biomass (3.55 kg ha^{-1} yr^{-1}) and Hg (0.18 mg ha^{-1}yr^{-1}) necessary to explain the standing stock (7.59 kg ha^{-1} yr^{-1}) and associated Hg content (0.29 mg ha^{-1} yr^{-1}) in the stock of observed prey fish and, and could only furnish ~10% of both to explain levels observed in piscivores (28.05 kg biomass ha^{-1} yr^{-1} and 1.81 mg Hg ha^{-1} yr^{-1}). Observed prey fish could only furnish 27% (1.83 kg ha^{-1} yr^{-1}) of the required biomass

(6.66 kg ha^{-1} yr^{-1}) and 17% (0.13 mg ha^{-1} yr^{-1}) of the required Hg (0.78 mg ha^{-1} yr^{-1}) to explain levels observed in piscivores.

Table 12.2. Average Hg fluxes available from observed standing stocks to be consumed by the subsequent trophic level compared to fluxes required to support the Hg in observed standing stocks of the consuming trophic level (mg ha^{-1} yr^{-1})

| Year | Available | | Required | | |
	From zooplankton for prey fish	From prey fish for piscivores	From zooplankton for prey fish	From prey fish for piscivores	From zooplankton for piscivores
1978	0.18	0.13	0.29	0.78	1.81
1979	0.21	0.29	1.32	1.59	4.58
1980	0.21	0.87	4.46	6.15	16.78
1981	0.37	1.29	4.69	25.04	59.34
1982	0.37	1.82	6.23	38.41	90.46
1983	0.36	1.67	3.46	23.66	54.42
1984	0.36	1.82	4.79	28.60	66.42

Following flooding, the discrepancies between required and available fluxes remained approximately the same for biomass from zooplankton to observed prey fish. For all other fluxes the discrepancies increased, those for Hg increasing immediately while biomass fluxes increased after a couple of years. During the last four years considered, zooplankton could only furnish 6%-10% (0.36-0.37 mg ha^{-1} yr^{-1}) of Hg required (3.46-6.23 mg ha^{-1} yr^{-1}) to explain observed prey fish Hg levels, and < 1% of Hg required (54.42-90.46 mg ha^{-1} yr^{-1}) to explain piscivore Hg levels. In terms of biomass over the same period, zooplankton could only explain 5%-7% (7.23-7.39 kg ha^{-1} yr^{-1}) of biomass required to support observed piscivore biomass levels (100.44-148.70 kg ha^{-1} yr^{-1}). Observed prey fish during this period could only explain 10%-13% (3.05-3.63 kg ha^{-1} yr^{-1}) of the biomass (24.46-34.86 kg ha^{-1} yr^{-1}) and 5%-7% (1.29-1.82 mg ha^{-1} yr^{-1}) of the Hg in piscivorous fish (23.66-38.41 mg ha^{-1} yr^{-1}).

Given these results, fish samples must not accurately reflect the fish populations. A large part of this problem could be due to nets used for fish sampling. Piscivores were probably well represented in the samples since they tend to be large enough to be captured quantitatively, but small prey species were underrepresented in the samples. The morphoedaphic index was developed for estimation of exploitable populations (i.e., large fish) so the overall estimation of standing stocks of large fish is probably correct within a certain margin of error, and thus the piscivores are probably well estimated. Since there must have been enough biomass flow to support the observed piscivores standing stocks and enough Hg flow to explain the Hg in these standing stocks, the actual standing stocks of prey fish must have been an order of magnitude larger than those observed, at least for the period 1981-1984. In addition, some of the prey species must have had Hg concentrations greater than those observed in whitefish. Some recent data on stomach contents support these conclusions (Lalumière and Dussault 1992; Chapter 10.0).

There must also have been another vector of transmission of biomass and Hg to prey fish species in addition to zooplankton, and this vector must have been more important than zooplankton for both components. The data on zooplankton populations are both extensive and intensive, and the estimation of fluxes from the plankton used a well-documented and verified model (Thérien et al. 1982; Morrison et al. 1987; Morrison and Thérien, 1987). These fluxes were incapable of supporting the observed prey stocks and prey Hg, and as already pointed out, the observed prey stocks were only a fraction of the stocks that had to be present. While some of the Hg discrepancies could be due to an underestimation of plankton Hg concentrations, biomass flux discrepancies cannot be explained. The benthos, therefore, must have played the role of the major food source for prey species, and must have been the major route of Hg contamination. Other studies (Lalumière and Dussault 1992; Chapters 9.0 and 10.0) also support these conclusions.

12.3.2
Fluxes of Hg from Vegetation

The calculated quantities for the major processes are shown in Table 12.3, and the temporal evolution of the cumulative rates are shown in Figure 12.1. Note that in 1983, the water level in the reservoir increased to a maximum, flooding some new territory and resulting in additional Hg release. Total release of non-methyl Hg from flooded vegetation was calculated to be 1.00 g ha^{-1} for the period of 1978-1984, with most of the release occurring in the first year. Of this, 0.62 g ha^{-1} was methylated over this time period while 0.31 g ha^{-1} was demethylated. Losses to biota, seston and sediments were estimated to be 0.03 g ha^{-1}, while losses to outflow were 0.26 g ha^{-1}. There was a net overall increase of dissolved non-methyl Hg of 0.40 g ha^{-1}. From dissolved methylmercury (MeHg), losses to biota were 0.23 g ha^{-1} while those to effluents were 0.06 g ha^{-1}, giving a net overall increase of 0.02 g ha^{-1}. The net accumulation of total Hg in biota was thus 0.26 g ha^{-1} over this period, while 0.32 g ha^{-1} of total Hg were lost to the effluent.

These quantities may seem large, but the actual effects on dissolved concentrations would be small. For instance, given that the average depth of Robert-Bourassa reservoir is 10 m, the net increase from 1978-1984 in inorganic Hg would have been 4.0 ng L^{-1}, while for MeHg it would have been 0.2 ng L^{-1}. Since

Table 12.3. Global quantities of Hg released from flooded vegetation and its fate for the period 1978-1984 (g ha^{-1})

Compartment	Release (non-methyl)	Methylation	Demethylation	Transfer to biota, seston and sediments	Loss in effluent	Net quantities
Non-methyl Hg in water column	+1.00	-0.62	+0.31	-0.03	-0.26	+0.30
MeHg in water column		+0.62	-0.31	-0.23	-0.06	+0.02
Total Hg in biota, seston and sediments				+0.26		+0.26
Total Hg in effluent*					+0.32	+0.32

* Incremental value in addition to total in inflows.

release was greatly diminished at the end of this period, quantities in the water column would have been subsequently reduced by continued losses, with hydraulic losses, alone, being as high as 63% per renewal time (6 months). Because of the natural variability of the concentrations of Hg in reservoirs (Table 8.1; Chapter 8.0), the change in Hg level in the water column may remain undetected. This would even be more so if Hg in solution was adsorbed by suspended particulate matter and if determinations of Hg were made on filtered water samples. The data reported showed high variability with ratios of maximal to minimal concentrations ranging from 2 to 5 in the case of total Hg and up to 28 for MeHg. This variability may have hidden any transient increase in Hg effectively present at that time. Also, the quantities estimated to be transferred to biota were of the same order of magnitude as that calculated to be in fish standing stocks.

12.3.3
Resolution of the Major Mercury Fluxes

Given the assumptions made, average fluxes calculated for the period 1981-1984 are shown in Figure 12.2. The fluxes for 1979 and 1980 are not included because of the extremely high release fluxes from vegetation (see Figure 12.1).

There was a net loss from the system during this period because the major release had already occurred while losses in the effluent were continuing. Also, the system was not at a steady-state, so the fluxes did not all cancel out. In fact the two fish compartments showed net accumulation, especially predators.

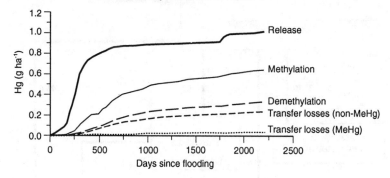

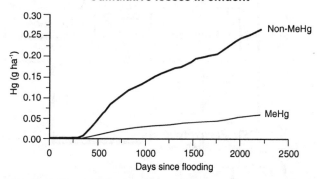

Fig. 12.1. Cumulative rates and losses to effluent of Hg released from flooded vegetation in the Robert Bourassa reservoir over the period 1978-1984.

The greatest single flux was that from the "BENTHOS-PERIPHYTON-SEDIMENTS" compartment to prey fish, while the next largest was a flux into the same compartment from water. Note that this latter flux would not necessarily imply only dissolved Hg, but much, if not most, would be associated with particulates. It is also interesting to note that the flux of Hg recycled from fish to the water column has been shown to be about the same order of magnitude as the flux from the water column to the "BENTHOS-PERIPHYTON-SEDIMENTS" compartment. This may suggest investigating the effects of intensive or selective fishing on Hg cycling in the system and on Hg concentration in fish, as was done elsewhere (Verta 1990b).

The fluxes through the "PLANKTON" compartment were based on a fixed concentration of 50 ng g^{-1} wet weight. This concentration was measured in plankton long after flooding and so the actual fluxes may have been several times higher. The flux from plankton to prey fish, however, was so small that were it to

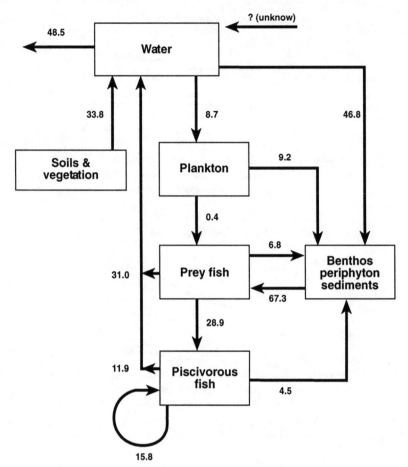

Fig. 12.2. Major Hg fluxes (mg Hg ha^{-1} yr^{-1}) in the Robert-Bourassa reservoir for the period 1981-1984.

be an order of magnitude larger it would still be relatively small as compared to the other major flux entering prey fish from the benthos. The fluxes entering and leaving the "BENTHOS-PERIPHYTON-SEDIMENTS" and "PREY FISH" compartments were inferred from quantities necessary to explain predator Hg. There are thus uncertainties in these fluxes, but these are mostly conservative. The calculated values are the minima and do not take into account possible significant accumulation in these compartments. There is also the possibility of some error in the calculated predator Hg which would then affect these fluxes. However, even an over-estimation of 50% would not greatly alter the qualitative importance of the flux from the benthos into prey fish.

Finally, we considered the Hg flux from water to be the major source to the "BENTHOS-PERIPHYTON-SEDIMENTS" compartment, but in fact part could have been through direct consumption of the flooded vegetation by benthic organisms. In effect, the Hg flux to the "PREY FISH" compartment from the "BENTHOS-PERIPHYTON-SEDIMENTS" compartment is the global flux for the years 1981-1984. The flux indicated (67.3 mg Hg ha^{-1} yr^{-1}) gives no indication as to the origin of the flux from benthos, periphyton or sediments. If part of this flux originated from direct consumption of the flooded vegetation from the "SOIL & VEGETATION" compartment, then the resulting flux effectively originating from the "BENTHOS-PERIPHYTON-SEDIMENTS" compartment would be reduced accordingly. If this was the case, it would not alter the Hg flux indicated from the "SOIL & VEGETATION" to the "WATER" compartments of Figure 12.2 since such consumption would occur on the vegetation matrix holding Hg that would not have been released earlier through the water column. An observation that somewhat limits this latter pathway is that non-piscivorous fish Hg has been declining over the last decade. There has not, however, been a noticeable reduction due to degradation, of the submerged vegetation (e.g., needles are still clearly evident on submerged spruce trees). The actual vector to this compartment would probably not be dissolved Hg as such, but more likely Hg associated with particulates.

12.4
Conclusions

We have shown that an entirely pelagic food chain could not account for observed Hg in fish, but that the greatest vector of Hg transfer to fish must be the benthos. This conclusion is supported independently by analyses of fish stomach contents (Lalumière and Dussault 1992; Chapter 10.0). The fluxes estimated here were inferred from fish data, because no quantitative data exist on benthic populations following flooding, although there are recent data on Hg concentrations in benthos (Chapter 9.0). It is unlikely that estimates of benthic population densities could ever be obtained for reservoirs such as Robert-Bourassa. Trees and vegetation were left in place for flooding and thus the habitat available for colonization is 3-dimensional, making grab sampling impossible and rendering scubadiving dangerous. Nonetheless, we were able to estimate the biomass and Hg fluxes coming from the benthos. In spite of data uncertainties, the order of magnitude of these fluxes is probably correct.

The quantity of Hg estimated to have been released from flooded materials is possibly slightly lower than the actual amount, because the contribution of humus was not taken into account. Data for release from humus has only recently become available (Chapter 7.0). Lichen is the dominant ground cover in the area, and while the vegetative part was included in the calculations, the humus could have contributed up to an additional 50 mg Hg ha^{-1}.

With flooded vegetation and soil being the source of Hg and benthos being the main vector of transmission to fish, the prospects for mitigating the impacts on fish Hg in future reservoirs by controlling the flux to fish are limited. The removal of vegetation and soil cover would remove the Hg source as well as the favorable habitat for benthic and periphyton colonization, but because of the magnitude of the surface areas involved (on the order of 10^5 ha) the costs would be staggering if such a project was even possible. Since the flux of Hg recycled from fish to the water column has been shown to be significant, this may suggest investigating the effects of intensive or selective fishing on Hg cycling in the system and on Hg concentration in fish as a more feasible alternative.

Acknowledgements

Funding for this work was provided by Hydro-Québec. Fruitful discussions with Dominique Roy of Société d'énergie de la Baie James and Roger Schetagne and Richard Verdon of Hydro-Québec helped in the elaboration of the approaches used in this study.

Mercury Toxicity
for Wildlife Resources

13 Assessment of the Ecotoxic Risk of Methylmercury Exposure in Mink (*Mustela vison*) Inhabiting Northern Québec

Marcel Laperle, Julie Sbeghen and Danielle Messier

Abstract

Chronic exposure in wildlife is subjected to several ecological, biological and behavioral factors, whose cumulative effects would reduce rather than increase the mercury (Hg) burden. Since clinical studies using captive animals cannot replicate these natural factors, they tend to overestimate the risk. Therefore, in attempting to evaluate the risk of contaminants to wildlife, it is essential to integrate results of field and laboratory studies. The present paper briefly reviews the exposure results obtained from a clinical *in vitro* study using domesticated mink together with the main conclusions on mortality, birth and fertility in order to determine the lowest observable effect level (LOEL) from the laboratory study. The paper then reports known field Hg concentrations observed in wild mink from various regions of North America compared to their counterparts in northern Québec in order to establish their natural variability regarding their exposure to methylmercury (MeHg) in this northern environment. Following that, it summarizes the numerous bio-ecological characteristics of wild mink that have an influence on the behavior and natural ability of the species as related to its feeding habits. This then permits us to evaluate the risk to wild mink in northern Québec considering their bio-ecological situation and their exposure levels measured in the wild as opposed to those observed in subjects exposed to known diets in the laboratory.

13.1
Introduction

Diagnosing effects of chronic MeHg exposure in the field is difficult, compared to cases of mass mortality due to acute or sub-acute exposure (Borg et al. 1969; Fimreite et al. 1970). Chronic exposure is mitigated by several ecological, biological and behavioral factors (Wren 1986). The amount and regularity of MeHg consumed, the duration of the exposure as well as the variability of the prey in the diet and their size are key elements to consider. The population dynamics, the behavior according to sexual differences, the biological functions as well as the home range of a species are other determinants. Furthermore, in piscivorous mammals, the half-life of Hg in the organism seems shortened by a demethylating capacity that appears more effective than in non-piscivorous species. Such a process has been demonstrated experimentally in mink (Jernolov et al. 1976). Fur also provides an efficient extraction route for MeHg during mink growth following birth, molt or during the development of the winter fur in the fall. Finally, the ecological suitability of a given habitat to support the wildlife species studied, other than occasional visits, needs to be considered.

It is also necessary to acknowledge the fundamental difference in approach between the toxicologist and the ecotoxicologist in appraising the potential threat posed by chemical exposures (WHO 1989). The former is interested in individual risks as it concerns the human health, while the latter is more preoccupied by risks at the population level.

The present paper aims to assess MeHg risk for wild mink populations in northern Québec. To do so, Hg levels measured in various tissues of domesticated mink exposed to controlled conditions in the laboratory are reviewed. Clinical and pathological results derived from this *in vitro* study for given levels of Hg in these tissues are then used to determine the lowest observed effect level (LOEL). Hg concentrations observed in various wild mink tissues from different North American regions are than compared in order to assess the risk to wild mink populations in northern Québec.

13.2
Materials and Methods

First of all, Hg levels measured in vital organs of domesticated mink exposed to known diets of MeHg are reported from an *in vitro* laboratory experiment. In brief, the chronic exposure laboratory study conducted at the Université de Montréal used natural sources of MeHg from fish captured in northern Québec's reservoirs. Diets containing 0.1, 0.5 and 1.0 $\mu g\ g^{-1}$ of total Hg were fed to three groups, each containing 50 female domesticated mink for a period of 704 days (Bélanger and Larivière 1997). Although only total Hg was measured in the diets, it was assumed that most of the Hg assimilated by the mink was in the form of MeHg since it is mostly this form which is present in fish and which is most easily assimilated by wildlife (Bloom 1992; DesGranges et al. 1998). No confounding factors are expected, since fish from that territory do not contain other toxicants whose levels are thought to induce synergistic or antagonistic effects.

Hg levels measured in dead and sacrificed animals were obtained after the first 200 days since the earliest clinical signs occurred during that period. Results of ultrastructural analyses of the renal cortex as well as those associated with the reproductive function of females mated over two reproductive years were kept. Finally, main pathological conditions of dead animals during the study were also used. Hg analyses on the biological samples in the clinical study were performed by the Laboratoire de toxicologie du Québec using standardized methods (Chapter 2.0). All Hg data in biological tissues are reported in wet weight. Using the clinical effects detected at known Hg concentrations over time allowed the setting of the lowest observable effect level (LOEL).

Hg concentrations were also obtained in wild mink from northern Québec and are presented along with similar data from other North American regions to establish natural variability. All exposure data from the various regions were from mink captured in winter, corresponding to the period when the clinical study was initiated in mid December 1993, so that animals for the *in vitro* and the *in vivo* collections would be physiologically comparable. Due to a lack of consistency among field data related to sexual differences, data from all individuals were pooled unless otherwise mentioned. Data from Kucera (1986) were transformed to wet weight using the multiplication factors provided by the author. Mink carcasses from northern Québec were obtained from Cree trappers during the winter of 1993-94 (Bélanger and Larivière 1997). In conclusion, risk to wild mink inhabiting northern Québec is derived by integrating Hg concentrations in vital organs from both *in vitro* and *in vivo* results taking into consideration the environmental factors encountered in their natural habitat.

13.3
Results and Discussion

13.3.1
In vitro Exposure to Methylmercury

Hg levels in tissues of female domesticated mink submitted to *in vitro* chronic MeHg exposure (Bélanger and Larivière 1997) are summarized in Table 13.1. As confirmed in the literature, Hg bioaccumulates at higher concentrations with increasing exposure levels (Bélanger and Larivière 1997), thus explaining the concentrations measured in the groups of mink exposed from 0.1 to 1.0 $\mu g\ g^{-1}$ total Hg. However, mean Hg concentrations for all groups were highest on the 95th day of exposure. Concentrations decreased thereafter (169-200 days) to values comparable to those attained after the first 50 days of exposure. Since these lower values occurred in the spring, part of it could be associated with the molting of the winter fur coat resulting in a transfer of a portion of the Hg burden toward the new growing summer fur. Other extraction routes, such as Hg transfer to the fetuses through blood exchange and physiological demethylation (Jernelov et al. 1976), would also be involved in this decrease.

Table 13.1. Total mercury concentrations ($\mu g \cdot g^{-1}$ wet weight, number of analysis given between parenthesis) reached in female domesticated mink exposed to known diets of methylmercury in fish (from Bélanger and Larivière 1997)

Hg in the Diet	$0.1\,\mu g \cdot g^{-1}$		$0.5\,\mu g \cdot g^{-1}$		$1.0\,\mu g \cdot g^{-1}$		
Tissue	Mean	Range	Mean	Range	Mean	Range	Days of Exposure
Liver	7.4 (4)	6.7-8.06	29.9 (4)	21.6-38.2	48.4 (4)	44.4-52.4	52
	21.8 (2)	15.4-28.2	44.6 (2)	41.8-47.4	73 (8)	65.3-80.7	95
	8.63 (5)	1.63-18.8	no data	n.a.	53.2 (2)	43.9-62.6	169-200
Kidney	6.57 (4)	4.4-8.7	19.7 (4)	17.4-22.0	30.1 (4)	26.4-33.7	52
	8.18 (2)	5.7-10.7	41.3 (1)	n.a.	53.6 (8)	42.4-64.8	95
	4.65 (5)	3.1-7.8	no data	n.a.	45.3 (2)	44.4-46.2	169-200
Brain	1.6 (4)	1.4-1.8	7.87 (4)	5.7-10.1	13.3 (4)	9.5-17.1	52
	2.7 (2)	1.5-3.9	9.9 (2)	7.2-12.5	18.8 (8)	12.50-25.1	95
	1.2 (4)	0.9-1.7	no data	n.a.	14.3 (2)	11.1-17.4	169-200

n.a.: not available.

No external sign of Hg toxicity was observed and no related mortality occurred in the groups of mink exposed to 0.1 and 0.5 $\mu g\ g^{-1}$. However, 66% of the females exposed to a diet of 1.0 $\mu g\ g^{-1}$ died after three months of exposure in the first reproductive year, most of them showing clinical signs of neurointoxication with encephalopatic lesions (Bélanger and Larivière 1997). Mortality occurred after 330 days of exposure to a diet of 1.0 $\mu g\ g^{-1}$ in the second generation of females. The encephalopatic lesions could be Hg induced, but have also been due to other stresses such as extremely cold ambient temperature, a situation that prevailed at the time of death for females belonging to the first generation (Bélanger and Larivière 1997). Females of the most exposed group, surviving the first year of mortality, lived through the second reproductive year and to the end of the study after 704 days of exposure. This suggests that individuals of a population may have the genetic ability to better manage MeHg stress.

The chronic exposure to low doses of MeHg in the two generations of females for the three diets did not affect the mean duration of the pregnancy period (49 days) nor the mean number of births (5.7 young) per female, although fertility of the female in the group exposed to 1.0 $\mu g\ g^{-1}$ was significantly reduced (Bélanger and Larivière 1997). The authors also concluded that young mink exposed *in utero* and through lactation in the three exposed groups did not show significant differences in growth nor survival.

The clinical study also attempted to relate Hg exposure to histopathologic modifications in the kidney. Results indicate that the urinary physiology of domesticated mink can adapt to daily intake of MeHg well above levels normally encountered by wild mustelids. Despite morphological changes, mostly tubular damage without obvious clinical effects, kidneys continued to perform adequately as revealed by results of blood and urinary biochemistry (Bélanger and Larivière 1997). The authors concluded that mink adapt to these aggressions through cell regeneration with evidence observed even in the group exposed to 1.0 µg g^{-1}.

In summary then, the observable effects occurred at the following diet concentrations:

- mortality (2/3 of the animals) with neurotoxic signs at a daily exposure of 1.0 µg g^{-1} after three month or more;
- reduction of the female fertility at a daily exposure of 1.0 µg g^{-1};

The lowest observable effect level (LOEL) would thus be a daily diet of 1.0 µg g^{-1}.

Many of these results have also been recognized in other studies. For instance, Aulerich et al. (1974) mentioned that a diet with 1.0 ppm wet weight can be fatal to mink in approximately two months. Wobeser et al. (1976ab) suggested that 10 ppm Hg in the brain is a critical level for most mammals. This threshold may be to low since after 52 days of exposure, such levels were reached in the brain of a few individuals at 0,5 µg g^{-1} in the diet (without clinical signs of toxicity). Nonetheless, most individuals in the group fed a diet of 1.0 µg g^{-1}, in which mortality occurred, reached such levels (Table 13.1). Wobeser et al. (1976b) also reported various histological changes in the kidneys of mink exposed to daily intakes of 1.1 and 1.8 µg g^{-1}. O'Connor and Nielsen (1981) found that Hg poisoning in otter (*Lutra canadensis*) would occur in areas where MeHg levels in their food items are regularly 2 ppm or greater. They concluded that in mink, clinical signs of mercurialism and death would result in roughly half the time seen in the otter with comparable Hg content, that is within two to three months. Finally, Kucera (1986) concluded that although a small percentage of mink in the Agreement area of northern Manitoba accumulated near toxic levels of Hg in their tissues, there did not appear to be any detrimental effect on the population or the fur harvest of the area.

13.3.2
Mercury Concentrations in Tissues Observed in the Wild

Table 13.2 summarizes mercury levels found in various tissues of wild mink harvested in winter from different regions of North America (O'Connor and Nielsen 1981; Kucera 1986; Poole et al. 1995; Bélanger and Larivière 1997).

Table 13.2. Total mercury concentrations ($\mu g \cdot g^{-1}$ wet weight) in tissues of wild mink from different regions of North America

Location	Tissue	N	Mean	Range	Exposure	Reference
Eastern USA	liver	28	0.96	0.21-4.1	wild	O'Connor and Nielsen 1980
Northwest Territories	liver	109	1.76	1.16-3.30 *	id	Poole et al. 1995
Manitoba south	liver	168	1.93	0.12-7.99	id	Kucera 1986
	kidney	168	1.62	0.17-5.89		
	brain	167	0.41	0.03-1.43		
Manitoba north natural	liver	68	2.98	0.15-25.5	id	Kucera 1986
	kidney	68	2.06	0.15-10.54		
	brain	70	0.61	0.04-2.56		
Manitoba north flooded	liver	77	5.72	0.21-30.6	id	Kucera 1986
	kidney	77	4.5	0.25-20.7		
	brain	77	1.42	0.07-7.4		
Québec north natural	liver	38	4.85	1,17-15.60	id	Bélanger and Larivière 1997
	kidney	38	2.45	0.73-6.32		
	brain	37	0.93	0.27-2.57		

n.a.: not available.
* : community means.

Mean Hg levels in mink tissues show the following geographical gradient: flooded environments in northern Manitoba > natural environments in northwestern Québec > natural environments in northern Manitoba > natural environments in southern Manitoba > natural environments in Northwest Territories > natural environments in mid-eastern USA.

Hg concentrations in mink livers from the mid-eastern Atlantic states in USA and from the Northwest Territories in Canada, at both longitudinal extremities of the continent, are fairly low. Both areas seem to be less affected by long range atmospheric transport of anthropogenic origin, whose contributions may range from 30 to 60% of the total annual fluxes (Fitzgerald and Clarkson 1991). They are followed by Manitoba, which is located west of the Great Lakes, possibly escaping the main trajectory of incoming anthropogenic Hg passing over the mid-continental USA. Fitzgerald and Clarkson (1991) report that "the impact of eolian fluxes to the oceans is manifested clearly by the higher concentrations of Hg in North Atlantic waters relative to the Pacific".

The differences between southern and northern Manitoba may be related to water chemistry, where lakes on Precambrian Shield dominate in the north with higher humic content that is indicative of higher fish Hg concentrations (Verta 1990a). In northern Manitoba lakes, differences are related to the site of the capture: fish from the Churchill river diversion in northern Manitoba feeding the hydroelectric power plants on the Nelson river have experienced a 3- to 4-fold increase in their Hg burden following flooding, as demonstrated in the studies undertaken under the Canada-Manitoba Agreement (1987) on Hg.

Mink from flooded northern Manitoba lakes have average Hg concentrations in their tissues somewhat comparable to those measured in their counterparts living in natural environment of northern Québec, even if mean Hg levels in fish from the former are twice as high as those found in the latter. On the other hand, mink from the natural environment in northern Québec have mean Hg concentrations in the liver higher than those from unflooded lakes within the diversion route of the Churchill-Nelson complex. This seems to follow the ranking of Hg levels found in fish from those two regions. Obviously, some environmental or biological modifying forces are acting within the flooded lakes, since the increased Hg concentrations in mink along the Churchill river diversion did not follow the ranking obtained in fish from the different regions within the overall diversion area (Kucera 1986). It must be noted that severe erosion is taking place along the new shorelines of the flooded lakes resulting in high water turbidity. This might be an ecological constraint for mink.

Previous studies in northern Québec reveal that although the Hg burden in lake sediment steadily increased over the last 60 years, there is no latitudinal gradient in Hg concentrations (Louchouarn et al. 1993), a situation also reported by Poole et al. (1995) in Northwest Territories for heavy metals. Furthermore, Hg levels found in fish species inhabiting the various drainage basins of northern Québec are comparable (Chapter 5). This also explains the results of a study carried out on the fish eating Osprey (*Pandion haliaetus*), where no significant latitudinal difference was found in the Hg levels measured in the blood of young flightless eaglets captured in nests located near natural lakes and rivers of these drainage basins (Des-Granges et al. 1994). The same conclusion was reached for a number of waterfowl species captured in these drainage basins (Chapter 6.0).

As Hg concentrations decreased progressively with distance downstream of the Churchill river diversion route (Kucera 1986), this suggests that a similar situation exists for mink living further inland from the shorelines of a given lake or reservoir. Such a situation was used to discriminate the high and low PCB risk areas for mink living in bordering and adjacent regions of the Great Lakes (Wren 1991). On the other hand, it does not preclude fish with higher Hg concentrations from moving out of flooded areas into accessible streams throughout the diversion route during the spawning period, possibly contributing to the wide range of Hg concentrations observed in mink harvested in that area of northern Manitoba.

Finally, total Hg concentrations in various mink tissues, as confirmed in several other studies, show the following gradient: liver > kidney > brain. Hg concentrations in these tissues also show much natural variation. These variations can be explained by the numerous modifying factors mentioned in the next subchapter as well as those observed in fish Hg levels from individual lakes to which territorial wildlife respond. In northern Québec, mean Hg levels in fish often vary by a factor of 3 to 4 for neighboring lakes (Chapter 5.0; Verdon et al. 1991).

13.3.3
Ecological, Biological and Ethological Determinants

As summarized by Eagle and Whitman (1987), although mink live near aquatic environments, they have few physical adaptations as amphibian mammals. They are primarily nocturnal and their eyes are incompletely adapted to underwater hunting, while the limited oxygen capacity of their lungs restrains their searching and swimming abilities. Therefore, open water is an unsuitable hunting habitat for mink, while shorelines with dense emergent vegetation are their principal hunting grounds. Hence, riverine wetlands along streams and ponds are prime habitat. This explains why their home range size is usually reported in linear rather than areal measurements.

A dependable food source is also an important habitat key for the territorial mink. Males wander more than females, their home ranges being almost twice as large. This trait might explain why the sedentary female normally accumulates more Hg than the male, which encounters a greater animal diversity during its inland wandering (Kucera 1986). In their pursuit of fish, mink prey on the most sluggish ones, that they trap or corner along stream banks. Fish comprises between 7 to 50% of their diet (Eagle and Whitman 1987), the larger proportion being observed during winter, when certain types of food such as migratory birds are no longer available. The habitat preferences of the mink also provide indication as to the species and size of fish usually preyed upon: coarse and slow moving fish of small to medium size.

Population dynamics, more precisely the turnover rate of a population as it affects the age of the animal, determines the exposure period over its life time. In the case of mink, the turnover rate occurs in less than three years. In their study of MeHg levels in wild mink in the north-eastern United States, O'Connor and Neilsen (1981) found that 90% of the mink from their sample were not older than 2 years (n = 128), 64% being less than one year old. Poole et al. (1995) mentioned that mink harvested in their collection were composed of 57% juveniles. Similar findings were reported by Kucera (1986) in Manitoba, where over 85% were less than 2 years (n = 403) and 70% were young of the year. Finally, mean age of 38 minks obtained from Cree trappers in northern Québec during the winter of 1993-94 was less than one year (Bélanger and Larivière 1997).

In the northern part of their range, particularly over the Precambrian Shield, mink encounter severe ecological constraints. There, riverine wetland used preferentially by mink and some of its prey, muskrat (*Ondatra zybethica*) and breeding waterfowl, make up only 4% of the terrestrial environment (Julien and Laperle 1985), while less than 1% of these had good potential as productive wetland (Lehoux 1975). Good mink habitat may provide from 3.75 to over 40 pelts ha^{-1} (Wren 1991). In northern Québec, harvest is usually less than 1 pelt 100 ha^{-1} (Native Harvesting Research Committee 1982).

Hydroelectric reservoirs in the Precambrian Shield are usually characterized by an absence of wetland along their shorelines. Furthermore, their operations impose ecological constraints on the short- and medium-term development of new riparian vegetation. Thus water level fluctuations, that run counter to natural rhythms, delay ice break-up and cause loss of contact between the aquatic environment and the bordering terrestrial habitats that provide shelter.

These constitute negative factors for several wildlife species, including mink (DesGranges et al. 1994). These factors will be less operative in the Boyd-Sakami diversion that transfers water from the Opinaca reservoir to the Robert-Bourassa reservoir as its water regime follows a more natural cycle. Streams accessible to fish from reservoirs during spawning and located upstream from the drawdown zone need to be considered even if the exposure period is of short duration. Finally, the subarctic winter conditions prevailing in northern Québec seal lacustrine water with a solid ice cover during 6 months of the year, rendering them inaccessible to mink.

13.3.4
Risk to Wild Mink

Female domesticated mink exposed to a control diet of 0.1 μg g^{-1} of Hg show mean total Hg concentrations in all tissues higher than those measured in wild mink harvested in natural environments of northern Québec. These results combined with the environmental modifying factors listed above indicate that the risk to mink populations inhabiting natural environments of northern Québec would be very low despite the increase of the Hg burden in the environment over the last 60 years.

Along most reservoir shorelines, even without Hg data in mink from their proximity, the risk appears weak since adult territorial mink have little possibility of finding suitable permanent habitat. Dispersing juveniles could possibly come in contact with fish prey in remote shallow bays away from the main body of a reservoir. Given the density of mink in the Precambrian Shield, these occurrences should not involve a significant part of the local population.

However, shorelines of the Boyd-Sakami diversion may offer mink more suitable habitat. Fish inhabiting the lake shorelines affected by this diversion have experienced an increase in their Hg burden like those of the Churchill river diversion. Fish along the Boyd-Sakami diversion have average Hg levels 5 times higher than those living in natural lakes and rivers (Schetagne et al. 1996). Reservoir fish attempting to spawn in streams accessible from the reservoir can be an additional source of MeHg exposure to mink along their shorelines. In such situations, fish ascending from a nearby reservoir would be diluted with the endemic population of the stream. Furthermore, incoming fish spawning in spring or fall, depending upon the species, would stay only for the duration of reproduction, moving back to the reservoir thereafter. Such movements are usually limited to the first rapids of a stream. Therefore, the added exposure would be less than that encountered by mink inhabiting the shorelines of the Boyd-Sakami diversion. We therefore limit our evaluation to that situation which represents the worst scenario.

Mean Hg levels measured in the feathers of the Osprey eaglets along that diversion route were intermediate between those found in their counterparts living near reservoirs and natural lakes, no statistical differences being found for Osprey raised near two of the stations where fish Hg levels are monitored since 1978 within the Boyd-Sakami diversion as compared to those raised near natural environments (DesGranges et al. 1994).

Since mean Hg levels measured in the tissues of wild mink from natural environments of northern Québec are below the mean levels obtained for the laboratory mink fed a diet of $0.1 \ \mu g \ g^{-1}$ of Hg, the average Hg concentration in the diet of wild mink must be $< 0.1 \ \mu g \ g^{-1}$. Assuming that the Hg levels in mink of the Boyd-Sakami area would increase proportionally to the increase in fish (factor of 5), which was not observed in mink along the Churchill river diversion route of northern Manitoba, then the average Hg concentration in the diet of wild mink of the Boyd-Sakami area would be $< 0.5 \ \mu g \ g^{-1}$. This level remaining below the diet of $0.5 \ \mu g \ g^{-1}$, at which no clinical or other signs of Hg intoxication were observed in domesticated mink, and being well below the diet of $1.0 \ \mu g \ g^{-1}$ which may be considered the LOEL based on the exposure experiment reported above, the mink populations inhabiting the shorelines of Boyd-Sakami would not be at risk.

This evaluation is conservative considering the numerous mitigating factors described earlier, allowing wildlife to accumulate Hg in a more gradual fashion in nature (Wren 1991). Since the cumulative effect of these factors would tend to reduce rather than increase the daily dose of Hg received by wild mink, toxicity values derived from laboratory animals overestimate the risk to wild populations.

13.4
Conclusions

Therefore, it does not appear that mink populations inhabiting the Precambrian Shield of northern Québec to be at risk from MeHg exposure, even those living in the neighborhood of the Boyd-Sakami diversion of the La Grande hydroelectric complex, nor along streams accessible to spawning fish coming from nearby reservoirs for brief periods. As for mink populations inhabiting areas near most reservoir shorelines, the risk also appears weak since adult territorial mink have little possibility of finding suitable permanent habitats.

Acknowledgements

The present work was done using results derived from research funded by the Ministère de l'Environnment et de la Faune du Québec, the James Bay Mercury Committee and Hydro-Québec. The mink carcasses from northern Québec were provided by the Cree Trappers Association.

14 Breeding Success of Osprey under High Seasonal Methylmercury Exposure

Jean-Luc DesGranges, Jean Rodrigue, Bernard Tardif and Marcel Laperle

Abstract

Osprey (*Pandion haliaetus*) commonly breed in northern Québec, often near reservoirs where 4 to 6 fold increases in fish mercury (Hg) levels have been noted following flooding. With the exception of eggs, which are usually laid when reservoirs are still ice covered, total Hg levels were significantly higher in the feathers of adult Osprey and nestlings living close to reservoirs than in those living in a natural environment. Despite a much higher total Hg exposure for Osprey fishing in reservoirs, the number of young fledged was not statistically different between nests located near reservoirs and those located near lakes and rivers. Our results suggest that growing feathers, either in molting adults or in nestlings, provide a good excretion route for total Hg. Nevertheless, the good breeding performance of Osprey nesting near reservoirs cannot be explained solely by the partial molt that adults undergo in summer. Future research on methylmercury (MeHg) exposure through fish consumption and on post-fledging risks related to high Hg exposure during the preceding fall migration must address the physiological mechanisms, such as demethylation by internal organs, that allow piscivorous birds to regulate MeHg in their bodies.

14.1
Introduction

Hg occurs naturally in fish, mainly in the form of MeHg (Bloom 1992). Consequently, mammals and birds whose diet is predominantly fish may be at an increased health risk, although direct evidence is lacking (Monteiro and Furness 1995).

The use of organomercurial fungicides as seed dressings in Europe and western Canada in the late 1960s led to the death of large numbers of granivorous birds as well as of birds of prey feeding on the corpses (WHO 1989). The outcome of these cases is similar to the acute Iraqi human Hg poisoning of 1970, which occurred as a result of the ingestion of bread made with cereal treated with salts of MeHg fungicides (WHO 1990).

In its last appraisal of the effects of Hg on biological organisms, WHO (1989) cautioned that the interpretation of the results of laboratory experiments on birds should take into account the fact that practically all studies have been carried out using gallinaceous birds, which poorly represent birds as a whole. Several of these studies were also carried out with MeHg salts not naturally found in the environment.

Several studies have reported on the demethylation ability of some organs of piscivorous mammals and birds regulating the MeHg concentrations (Norheim and Froslie 1978; Wren 1986; Thompson and Furness 1989; Thompson et al. 1990). Demethylation is a significant detoxification route for MeHg in birds of prey (Norheim and Froslie 1978), so that, in spite of their high exposure due to their position as the final link in the food chain, they tolerate MeHg better than had previously been believed. Moreover, growing tissues made of keratin, such as feathers, beak, claws and the scales covering the tarsus, provide an efficient excretion route for Hg because during the growth of such tissues, MeHg binds strongly with the disulfide bonds of keratin (Furness et al. 1986).

This study examines chronic Hg exposure in Osprey during the breeding season. This species is piscivorous and generally eats larger fish than other piscivorous bird species. Also, the nesting habits of Osprey, such as nesting high up in trees, enables them to escape the constraints imposed by the reservoirs in their habitat, thus making it an ideal choice in this kind of study. Furthermore, Osprey commonly breed in northern Québec, often near reservoirs (DesGranges 1995) where total Hg levels in fish increased 4-6 fold after flooding (Verdon et al. 1991). This paper presents some of the results of an extensive study on Hg exposure in Osprey nesting near hydroelectric reservoirs (DesGranges et al. 1994), and discusses the biological and behavioral characteristics of adult Osprey that may explain our findings.

14.2
Study Area and Methods

The study area is located in northern Québec between latitudes 49°N and 56°N. It incorporates the following drainage basins: 1) the Nottaway, Broadback and Rupert rivers flowing into southern James Bay; 2) the Eastmain, Opinaca and La Grande rivers flowing into northern James Bay, sections of which were altered to create the La Grande hydroelectric complex; and 3) the Grande Baleine and Petite Baleine rivers flowing into southern Hudson Bay (Figure 1.1).

Hg exposure data were collected from 1989 to 1991, in the course of a joint Environment Canada and Hydro-Québec study (DesGranges et al. 1994). Each year, two surveys were carried out during the breeding season: the first during incubation (June) and the second during the pre-fledging period (August), when young Osprey were 35-45 days old. Reproductive parameters were recorded and biological tissues, including samples of eggs as well as feathers of chicks and adults, were collected for measuring total Hg levels.

The data collected were subdivided into two groups to test the effects of Hg on Osprey. The eastern La Grande reservoirs, which at the time of the study were 8 to 12 years old, provided data on high Hg exposure in impounded waters, as revealed by the total Hg levels of fish (Brouard et al. 1990; Chapter 11.0). Lakes and rivers in the La Grande, Nottaway-Broadback-Rupert and Grande Baleine watersheds were combined and represent the natural environment .

Tissue collection and preservation are described in detail in DesGranges et al. (1994). Chemical analyses for total Hg (NAQUADAT 80601) were conducted by the Analex Inc. Laboratory in Laval, Québec, in accordance with the protocols described in Chapter 2.0.

14.3
Results

Most Osprey initiated egg laying toward the end of May in southern James Bay, in the first week of June in northern James Bay and in the second week of June in southern Hudson Bay with the different initiation dates being due to north-south climatic differences (DesGranges et al. 1994). The average number of eggs laid per clutch was 2.8. Slightly more than half (54%) of the eggs laid produced fledged young, resulting in an average of 1.5 fledglings per active nest. There was no statistical difference between reservoirs and natural habitats in either mean clutch size ($Z = 0.04$; $P = 0.97$) or in mean number of chicks fledged per active nest ($Z = 0.66$; $P = 0.51$) (Table 14.1).

Total Hg levels were significantly higher ($Z \geq 3.64$; $P < 0.01$) in the feathers of nestling and adult Osprey nesting close to reservoirs than in those living in a natural environment (Table 14.2). Total Hg in eggs, however, was not different between the two groups (mean = 0.2 mg kg^{-1}; $Z = 1.20$; $P = 0.23$) (Table 14.2). On average, feathers contained 16.5 mg kg^{-1} of total Hg in adult Osprey in natural habitats as compared to 58.1 mg kg^{-1} in adults near reservoirs. A similar pattern was observed in feathers of 35-45 day-old chicks, but concentrations tended to be lower (7.0 mg kg^{-1} in natural habitats compared to 37.4 mg kg^{-1} near reservoirs) than those found in adults. In both cases, individual levels varied widely; the highest concentration detected was 193 $mg\,kg^{-1}$ in the feathers of an adult from the Robert-Bourassa reservoir.

Table 14.1. Osprey nesting data (mean ± SD) in northwestern Québec, 1989-1991

	Hydroelectric reservoirs	Lakes and rivers	Total	Mann-Whitney	
				|Z|	P
Clutch size	2.8 ± 0.7	2.8 ± 0.5	2.8 ± 0.6	0.04	0.97
(no. eggs/active nest)	(n = 53)	(n = 61)	(n = 114)		
% of eggs that fledged	52 %	57 %	54 %		
Fledging success	1.4 ± 1.1	1.6 ± 1.3	1.5 ± 1.1	0.66	0.51
(no. fledglings/active nest)*	(n = 45)	(n = 36)	(n = 81)		

* Nests in which one egg was collected for toxicological analysis were excluded from this calculation.

Table 14.2. Total mercury exposure ($mg\,kg^{-1}$, mean ± SD, n ; min - max) in feathers (dry weight) and in eggs and stomach contents (wet weight) of Osprey in northwestern Québec, 1989-1991

	Hydroelectric reservoirs	Lakes and rivers	Mann-Whitney	
			|Z|	P
Adult feathers	58.1 ± 51.3	16.5 ± 12.8	3.64	< 0.01
	(n = 31; 5.3-193.0)	(n = 29; 1.2-68.0)		
Eggs	0.2 ± 0.1	0.2 ± 0.1	1.20	0.23
	(n = 18; 0.1-0.5)	(n = 33; 0.1-0.5)		
Stomach contents	0.8 ± 0.33	0.3 ± 0.2	2.40	0.02
	(n = 6; 0.4-1.3)	(n = 6; 0.1-0.6)		
Nestling feathers	37.4 ± 20.1	7.0 ± 4.3	9.62	< 0.01
	(n = 78; 5.5-101.0)	(n = 63; 2.1-26.5)		

Although fragmentary (n = 12 stomach contents), total Hg analyses conducted on fish fed to young Osprey were 2.7 times higher in reservoirs than natural habitats (Table 14.2). Stomach content analyses showed that 90% of the total Hg is ingested in the form of MeHg (DesGranges et al. 1998).

14.4
Discussion

At northern latitudes, Osprey are migratory and initiate nesting soon after their arrival in the spring, even if lakes are still ice covered, apparently finding food in unfrozen sections of nearby rivers (Poole 1989). The western reservoirs of the La Grande complex are generally ice-free by June 9 (Hydro-Québec, unpublished data), by which time most Osprey clutches are complete. It is not unexpected, therefore, that Osprey nesting close to reservoirs lay eggs with total Hg levels similar to those found in eggs from nests in natural environments. Furthermore, since Hg exposure is minimal prior to egg laying, hatching is unaffected and thus should proceed normally for both groups of birds. Fimreite (1971, 1974) observed a significant reduction in hatching success of eggs containing between 0.5 and 1.5 mg kg^{-1} of MeHg, with deleterious effects on egg development when the concentration exceeded 1.0 mg kg^{-1}. In our study, total Hg levels in fresh eggs never exceeded 0.55 mg kg^{-1}, with 96% of eggs having concentrations less than 0.5 mg kg^{-1}.

Total Hg levels were significantly higher in feathers of adults and nestlings from reservoirs as compared to those living in a natural environment. Adult and nestling feathers collected near reservoirs contained 3.5 and 5 times more total Hg, respectively, than those from natural habitats. In adults, only feathers found in or underneath nests were analyzed. Since we do not know exactly when during the breeding season those feathers were lost, we can only note that total Hg exposure in adults and nestlings show a similar profile. The higher total Hg levels found in the feathers of adults as compared to nestlings can be explained by the greater quantities of fish consumed by adults, the longer exposure period for adults (3 months vs 1.5 months) (Lindberg and Odsjo 1983; Furness et al. 1990), and the fact that adults undergo only a partial molt during the summer versus a complete body molt and growth of new feathers in nestlings. Since adults molt only a few feathers, the total Hg is concentrated in the few new feathers that grow in, whereas nestlings grow a complete set of new body feathers and therefore the total Hg is distributed more widely and evenly, resulting in lower concentrations per feather (Johnels et al. 1979; Braune and Gaskin 1987).

Despite much higher Hg exposure for Osprey fishing in reservoirs, the number of young that fledged per active nest was not different between nests located near reservoirs and those located near lakes and rivers. Furthermore, fledging success in this study was similar to that obtained in other North American and European studies (Saurola 1983; Seymour and Bancroft 1983; Stocek and Pearce 1983; Poole 1989). Thus, our results suggest that the exposure level to Hg from feeding on reservoir fish is not at critical levels for the adults. These results suggest that growing feathers, either in molting adults or in nestlings, provide a good excretion route for MeHg. For example, feathers from chicks 5 to 7 weeks old contain approximately 86% of the total Hg burden, excluding other tissues containing keratin for which total Hg was not measured (DesGranges et al. 1998).

Two factors lead us to believe that the high concentrations of total Hg measured in the adult feathers collected at nests located near reservoirs were accumulated the preceding year and from the same area. First, Osprey exhibit strong fidelity to their natal areas (Henny 1983), and with 75% of nests (n = 56) in our study being used in two consecutive years it is likely that at least one of the parents was probably the same as in the preceding year (DesGranges et al. 1994). Second, the new feathers that grow in to replace feathers molted during the summer get their Hg signature from the Hg ingested during the breeding season. Moreover, some of the feathers that we collected are molted annually, such as primaries (Prévost 1983), thus these feathers had to have come from birds inhabiting the same area the preceding year.

We observed that Osprey can successfully reproduce under chronic exposure to MeHg. Post-fledging risks related to high Hg exposure were not, however, examined in this study. Young may experience toxicological problems once their feathers have stopped growing at the age of 45 days, about 10 days before leaving the nest. During this period, as well as during the 10 to 20 days they remain in the vicinity of their nest before undertaking the migration south, post-fledged young must rely solely on physiological mechanisms other than excretion via feathers to regulate the MeHg. Depending on how long they remain near the reservoirs prior to migration, substantial amounts of Hg can accumulate in their vital organs. For instance, a 72-day old juvenile caught on Long Island, New York, some 1 500 km from its nesting site (the Robert-Bourassa reservoir), had a higher concentration of total Hg in its liver (23.4 mg kg^{-1}) than the highest level (8.5 mg kg^{-1}) detected in any of the 4 pre-fledged young caught on this reservoir.

14.5
Conclusions

The high Hg exposure of Osprey nesting along reservoirs has not affected breeding performance, nor does it appear to have affected population levels. For both adults and nestlings, growing feathers is a good excretion route for Hg. The partial molt adult Osprey undergo in summer cannot alone explain the good breeding performance we observed in Osprey nesting near reservoirs and under high Hg exposure. How, under these conditions, they are able to sustain a healthy population must be related to the capacity of internal organs to demethylate, a mechanism as yet poorly understood in this species.

Acknowledgements

We are grateful to the several Canadian Wildlife Service employees who took part in field work or helped with tissue preparation prior to Hg analyses. Several of the ideas in this paper grew out of discussions with Birgit Braune of the National Wildlife Research Centre (CWS-Hull) and with Claude Langlois and Martin Pérusse of the Vice-présidence Environnement of Hydro-Québec. Environment Canada and Hydro-Québec jointly funded this study.

15 Synthesis

Roger Schetagne, Marc Lucotte, Normand Thérien, Claude Langlois and Alain Tremblay

The objective of this chapter is to present a comprehensive review of the mercury (Hg) issue in natural aquatic ecosystems and hydroelectric reservoirs in northern Québec. This synthesis is based on the findings of over ten years of studies reported by research teams from three universities, one governmental agency and Hydro-Québec.

15.1
Mercury in Natural Ecosystems of Northern Québec

15.1.1
Sources of Mercury

Although there are no direct industrial or municipal sources, the presence of Hg in natural lakes and terrestrial systems is ubiquitous throughout northern Québec, at sites located hundreds to more than a thousand kilometers away from the closest industrial centers. Airborne Hg of natural origin has progressively been accumulating in the organic layers of terrestrial soils since the beginning of soil formation after the last glaciation, some 5 to 8 thousand years ago. In addition to this natural Hg, anthropogenic Hg has also started to accumulate in remote soils of northern Québec during the last century. In the humic horizon of unaltered soils, Hg, essentially in its inorganic form, the methylated form representing usually less than 1%, appears to remain very stable, being strongly attached to humified organic matter. However, a fraction of the newly deposited Hg is entrained to the lacustrine systems by surficial runoff, with organic material, without fully penetrating the humic horizons.

In lacustrine ecosystems, sediments constitute the main reservoir of Hg. The quantity of Hg brought to a lake appears directly proportional to the amount of carbon leached from the surrounding soils in the drainage basin. Like in humic horizons of soils, Hg is quite stable in lacustrine sediments, being strongly bound to organic matter mostly of terrigenous origin, and little diagenic remobilization is observed once it is sedimented. This latter characteristic makes it possible to give historic interpretations of the deposition of this heavy metal in remote regions. Since the onset of the industrial era at the end of the last century, northern Québec has witnessed 2- to 3-fold increases in atmospheric Hg deposition rates as revealed by lake sediments samples taken between latitudes 45°N and 56°N.

A series of independent indices with fairly constant values measured between latitudes 45°N and 54°N in Québec, such as the carbon normalized anthropogenic enrichment factor in lake sediments, the Hg concentrations in various inferior and superior plants, the carbon normalized Hg content in the humic horizon of soils and the Hg/Al and Hg/Si ratios in lichens, suggest that the entire region is submitted to the deposition of airborne Hg of a uniform intensity and common source, most probably originating from the industrial area of the Great Lakes. North of latitude 54°N, atmospheric Hg appears to originate from an additional source, as indicated by contrasted Hg/Al and Hg/Si ratios. This remote region may be under the influence of Hg either volatilized from the nearby Hudson Bay or transported from Eurasia. Similarly, measurements of Hg concentrations in a number of fish species and a variety of birds have failed to reveal any latitudinal gradient of Hg concentrations between latitudes 45°N and 56°N.

15.1.2
Sediments and the Water Column

In northern Québec, lake sediments usually bear Hg concentrations ranging from 50 to 300 ng g^{-1} dw, which is 4 to 5 orders of magnitude higher than Hg concentrations (dissolved plus particulate) in the water column, which are usually well below 10 ng L^{-1}. Even though the concentrations in lake sediments are of the same magnitude as those in soils, aquatic organisms will bioaccumulate much more Hg than their equivalent counterparts (in terms of trophic position in the food chain) in the terrestrial ecosystems. This is due to the much higher bioavailability of that heavy metal in aquatic environments, principally because the inorganic forms of Hg are methylated via microbial activity at various levels of the aquatic system. In lake sediments, methylmercury (MeHg) usually represents less than 2% of the total Hg content. Although only a small fraction of the total Hg burden in a lake is transformed into MeHg, the form accumulated by organisms, it is sufficient to account for an increase factor in the order of 150 from low trophic level plankton (25 ng g^{-1} dw) to large piscivorous fish (0.6 mg kg^{-1} ww). MeHg is readily accumulated in organisms as a result of both its strong affinity for proteins and low elimination rates, especially in fish.

15.1.3
Invertebrates

Total Hg concentrations in invertebrates of over 20 natural lakes of northern Québec cover a wide range of values, from 25 to 575 ng g^{-1} dw in plankton and from

31 to 790 ng g^{-1} dw in insect larvae. In northern Québec, differences in deposition rates among lakes do not appear to be a factor as they seem to be relatively uniform over the whole region. The fraction of Hg under the methyl form increases along the invertebrate food chain and biomagnification factors of about 3 are observed from one trophic level to another.

Statistical analyses indicate that variations in concentrations of total and MeHg in invertebrates could be explained by feeding behavior (trophic level) and, to a lesser extent, by water quality parameters such as color and dissolved organic carbon, and water temperature.

15.1.4
Fish

In over 180 sampling stations located in natural lakes and rivers of northern Québec, total Hg concentrations in fish were found to be relatively high compared to other regions of North America. Concentrations measured for non-piscivorous species, such as longnose sucker (*Catostomus catostomus*) and lake whitefish (*Coregonus clupeaformis*), are always well below the Canadian marketing standard of 0.5 mg·kg^{-1}, while those obtained for piscivorous species often exceeded this standard. Inter-lake variability within a same region is important for all fish species, as estimated mean concentrations often vary by factors of 3 to 4 for neighboring bodies of water. Mean concentrations for 400-mm non-piscivorous fish ranged from 0.05 to 0.30 mg kg^{-1} ww, while those of piscivorous species (walleye *(Stizostedion vitreum)* or pike *(Esox lucius)* of 400 and 700 mm of total length respectively) ranged from 0.30 to 1.41 mg kg^{-1} ww, from one lake to another. Highest concentrations in fish (of all species surveyed) were usually found in bodies of water with high organic content, as described by color and concentrations of tannins, as well as total and dissolved organic carbon, where bioavailability at the base of the food chain would be greater.

An average biomagnification ratio of about 5 was found between mean concentrations in standardized-length lake whitefish (400 mm), a benthic feeder, and pike (700 mm) or walleye (400 mm), which are piscivorous fishes. Our data show that the biomagnification factor between these fish species is quite variable from one lake to the next. This variability may be explained by differences in fish community structure, resulting in differences in diet for the top predators, and consequently, in differences in Hg concentrations.

Wide distributions of Hg concentrations are observed in individual fish of the same size from a given lake. For example, lake trout (*Salvelinus namaycush*) of about 575 mm captured the same year in Hazeur lake show Hg concentrations ranging from 0.67 to 1.28 mg kg^{-1}. Individual physiology, growth rate or feeding habit, may explain such high individual variability. Stomach content analyses show that pike from the western part of the La Grande region prey on a wide variety of fish of different trophic levels, which further highlights the importance of diet in the process of bioaccumulation of Hg.

15.1.5
Aquatic Birds

The importance of diet, and consequently of trophic level, is also demonstrated by the concentrations obtained for aquatic birds of northern Québec which clearly show a general progression of total Hg, in muscle, liver and feathers, from herbivorous species to piscivorous ones. Concentrations in muscle are typically below or about 0.05 mg kg^{-1} for herbivorous species, while they range from 0.16 to 0.21 mg kg^{-1} (ww) for benthivorous species and from 0.8 to 1.6 mg kg^{-1} for partly or strictly piscivorous species, showing the influence of the aquatic system on wildlife, via the methylation and increased bioavailability of Hg through the aquatic food chain. In fact, for species having similar feeding habits, concentrations in fish or aquatic birds of northern Québec are quite equivalent.

15.1.6
Mammals

Hg concentrations measured in terrestrial mammals of northern Québec further demonstrate the link between the aquatic and terrestrial ecosystems, with fish acting as the transfer vehicle, as levels again vary with diet, with piscivorous species showing maximum concentrations. Indeed, muscle concentrations in strictly herbivorous species, such as hare (*Lepus americanus*) or caribou (*Rangifer caribou*), exhibit levels around or below 0.05 mg kg^{-1}. Corresponding concentrations in carnivores, such as ermine (*Mustela erminea*), marten (*Martes americana*) and red fox (*Vulpes fulva*) range from 0.15 to 0.30 mg kg^{-1}, while concentrations in partly piscivorous species such as mink (*Mustela vison*) vary around 2.5 mg kg^{-1}.

The importance of diet is also demonstrated in marine mammals collected off the coast of Hudson Bay, for which muscle concentrations in benthivorous or partly piscivorous species, such as the ringed seal (*Phoca hispida*) and the bearded seal (*Erignathus barbatus*), showed Hg levels usually well below the Canadian marketing standard of 0.5 mg kg^{-1} (ww), ranging between 0.1 and 0.7 mg kg^{-1}, while those in the mostly piscivorous beluga (*Delphinapterus leucas*) ranged from 0.9 to 6.2 mg kg^{-1}. Mean liver concentrations ranged from 2 to 5 mg kg^{-1} in seals and reached 20 mg kg^{-1} in beluga.

All these measures show that health risks associated with the consumption of Hg-rich resources from northern Québec are essentially limited to piscivorous species, be they fish, aquatic birds, or terrestrial and marine wildlife.

Although the links between inorganic airborne Hg, deposited either directly on lakes or via runoff from the surrounding terrestrial environment, its methylation in aquatic systems and the MeHg biomagnification from invertebrates to fish and fish eating wildlife, are well understood, the effect of recent 2- to 3-fold increases in atmospheric Hg deposition rates on fish Hg levels, and consequently wildlife Hg levels, still remains unknown. Although no temporal trend in fish Hg levels was observed for all species monitored in 5 natural lakes for periods of up to twelve years, this time span is too short, considering the great variability of fish Hg levels, to confirm or refute any direct relationship.

Considering the reduction observed in Hg levels in fish and other aquatic organisms from industrially contaminated sites once the inputs of Hg was reduced (as in the cases of the English Wabigoon and Saguenay rivers), one may suppose that there is a relationship between airborne Hg deposition and concentrations in fish and wildlife. This relationship would imply first, that Hg concentrations in fish and other wildlife of northern Québec have not always been so elevated, and second, that levels would decrease if anthropogenic releases of Hg in the atmosphere would be reduced. As a matter of fact, the recent closure of the industries of former East Germany has led to a dramatic decrease in the Hg atmospheric deposition rates onto the lakes of Sweden, as evidenced by the sharp decrease in the Hg concentrations in the most recent sediments of the lakes of the region. In that particular case, the expected decrease in the Hg concentrations in fishes would clearly demonstrate that the Hg burden in aquatic organisms is dependant upon the local atmospheric deposition rates of the heavy metal.

15.2
The Mercury Issue at the La Grande Hydroelectric Development Complex

Monitoring of fish Hg levels at the La Grande complex was initiated in the late 1970s, before the impoundment of the first reservoir. Like in other reservoirs in Canada and abroad, Hg levels in fish of La Grande complex reservoirs increased strikingly and rapidly in the early 1980s immediately following impoundment. Because this increase could not have resulted from a sudden and drastic increase of Hg from atmospheric fallout, as Hg levels in adjacent unaltered natural lakes remained stable, and because the link between flooded organic materials and fish Hg levels had been evidenced in other reservoirs, research programs were initiated in the early 1990s.

15.2.1
Methylation and Passive Transfer from the Flooded Soils and Vegetation to the Water Column

Both *in vitro* and *in vivo* studies demonstrated the impact of flooding vegetation and organic soils on the methylation and release of Hg to the water column and subsequent transfer through the aquatic food chain.

In vitro experiments revealed significant releases of both total and MeHg to overlying waters and allowed the evaluation of the magnitude, relative importance and duration of Hg releases to the water column from different soil and vegetation types under different conditions of dissolved oxygen, pH and temperature of the overlying water. Diffusion rates obtained at 20°C show that virtually all the releasable Hg would be released within a year, and most within the first few months. Regardless of the environmental conditions applied, total amounts of Hg released after approximately 6 months of flooding, range from 4 to 40 ng g^{-1} for different vegetation types, and from 4 to 45 ng m^{-2} for soil humus samples. Corresponding amounts of MeHg range from 1 to 5 ng g^{-1} for vegetation and from 0.3 to 1 ng m^{-2} for soil humus. For the green part of vegetation, a significant proportion of the initial total Hg content is released to the water phase in about 6 months, from 20% to up to 50-100%, depending on vegetation type. For flooded soils, which represent a far greater biomass of organic matter, only a small fraction of the initial Hg pool is released to the water after one year of flooding. In fact, due to the great variability of the Hg burden in samples, the actual percentage of the initial Hg pool which is released can not be determined.

In situ measurements in flooded soils confirm that only a small fraction of the Hg pool of flooded soils is released to the water column. Indeed, after a decade or more, most flooded soils, except those eroded in the drawdown zone at the rim of the reservoirs, show no significant loss of their Hg burden. This observation seems related to the fact that little structural changes occur within the flooded organic matter aside from the partial biodegradation of the green organic matter flooded. As a matter of fact, even if the bacterial activity in the reservoirs is intense enough to provoke a strong oxygen demand in the flooded soils and in the bottom waters for a few years following impoundment, losses in carbon burdens in these flooded soils is not measurable, being smaller than the inherent variability of these burdens encountered in nearby soils.

The progressive methylation of Hg through time associated with the bacterial activity in the flooded soils is the major observable change in the soils after their flooding. Whereas less than 1% of the total Hg in natural soils was under the organic form, this amount reached maximum values of approximately 10 and 30% after 13 years for peatlands and podzolic soils respectively. For peatlands and podzolic soils, initial methylation rates are similar after the first year but are prolonged in the podzolic soils. An explanation proposed for that observation is that the organic matter of podzolic soils appears more sensitive to degradation after flooding than that of peatlands which is already water saturated under pre-flood conditions with the exception of the living ground cover.

The methylation rates appear similar in shallow areas and in deep areas (10 to 20 meters of depth) lying below the photic zone but still above the seasonal thermocline. Once Hg is methylated in the flooded soils, no net demethylation can clearly be observed during winter months or through time, up to 15 years. Thus, it seems that once formed, most of the MeHg accumulates in the flooded soils, probably due to its strong affinity for terrigenous organic matter.

Passive release of total or MeHg from the flooded soils, confirming *in vitro* observations have only been measured *in situ* over non-eroded soils located in shallow areas at the periphery of the reservoirs, with relatively long water residence times. Elsewhere in reservoirs, the anticipated Hg releases of the order of a few ng m^{-2} (MeHg) to a few tens of ng m^{-2} (total Hg) over a six month period, according to the *in vitro* experiments, are not detectable by current analytical methods as increases in total Hg or MeHg concentrations in the water column. Indeed, the dilution rates are much to great in these very large bodies of water. Moreover, a fraction of the Hg released in the water column may also be adsorbed on suspended particulate matter and as such transferred up the food chain, which further reduces the possibility of detection of the passive release of dissolved Hg in the water column. In shallow areas with reduced dilution rates, such as the LA-40 impoundment, an average increase factor of about 5 has been measured for dissolved MeHg concentrations compared to natural lakes, with mean values reaching 0.3 ng L^{-1}, confirming that a certain fraction of that heavy metal initially bound to the terrestrial system is leached to the water column during the first few months of impoundment. It should be noted that calculations for the Robert-Bourassa reservoir, derived from *in vitro* flooding experiments, generate comparable water column values as the net increase for the first 5 years would have been on the order of 0.2 ng L^{-1} for MeHg. Furthermore, *in situ* measurements show that the proportion of total dissolved Hg which is in the methylated form is, on average, 4 times greater in reservoirs than in natural lakes (12% vs 3%).

Despite the release to the water column of a significant fraction of available Hg originally contained in the green part of the flooded vegetation, the overall burden of Hg in the flooded environment remains almost intact through time. The fact that a total loss in the initial Hg burden of probably less than 5% is sufficient to greatly increase the Hg contamination of the entire food chain of the reservoirs demonstrates how Hg, mostly under its methylated form, is highly efficiently transferred and bioaccumulated through the food chain.

15.2.2
Mercury Increases in Organisms at the Base of the Aquatic Food Chain

Increases in MeHg concentrations, from natural lakes to reservoirs, are relatively constant, usually varying from factors of 3 to 5, for the dissolved fraction (< 0.45 μm), the fine particulate matter (0.45 to 63 μm), the zooplankton (> 150 or > 210 μm), as well as for non-piscivorous and piscivorous fish of standardized length. For example, concentrations measured in invertebrates of 5 different reservoirs of the La Grande complex range from 45 to 680 ng g^{-1} in insect larvae and from 350 to 550 ng g^{-1} in zooplankton (> 150 μm), representing an average increase factor of 3 compared to natural lakes.

Furthermore, comparable biomagnification factors exist from one trophic level to another in invertebrates, zooplankton and insects, as well as in fish, in both La Grande complex reservoirs and neighboring natural lakes. Thus, results show that an increase in the bioavailability of Hg at the base of the food chain in reservoirs is reflected throughout the food chain up to non-piscivorous and piscivorous fish. Moreover, differences in stable isotopes analyses between natural lakes and reservoirs for fine particulate matter and zooplankton suggest that Hg-rich organic detritus of terrigenous origin in suspension in the water column is being ingested by zooplankton, and then transferred to fish which feed on zooplankton.

The importance of the role played by both the dissolved inorganic Hg and the dissolved MeHg, directly released to the water column of the reservoirs, on the contamination of the aquatic food chain remains uncertain. Because of the absence of an adequate series of data concerning Hg concentrations in pure phytoplankton samples, the role played by these organisms (although expected to be the most influenced by the dissolved phase) in the transfer of Hg through the food chain remains unknown. However, the absence of significant increases in MeHg in the 63 to 210 μm fraction (which mostly contains phytoplankton and some micro-zooplanktonic organisms) in the LA-40 impoundment compared to natural lakes, suggests that phytoplankton may not represent a key step in the transfer of Hg from the water to higher organisms in the food chain. This assumption would be corroborated by the calculation of fluxes of biomass and of Hg in the Robert-Bourassa reservoir, using data obtained from the environmental effects monitoring program (RSE). Indeed, the results of this exercise show that the pelagic food chain alone, from dissolved Hg to phytoplankton, zooplankton and fish, was insufficient to explain both the biomass and Hg concentrations found in non-piscivorous fish of this reservoir. Calculations suggested that the greatest vector of Hg transfer to fish must have been through the littoral invertebrates, i.e., both aquatic insects and zooplankton. Fish diet studies, as well as *in situ* biomass and Hg concentration measurements carried out in the Robert-Bourassa and Laforge 1 reservoirs, as well as in the LA-40 impoundment, corroborate this finding.

On the other hand, it is probable that dissolved Hg readily released from the flooded soils to the supernatant water column rapidly re-binds to fine particulate matter in suspension, thus rendering this Hg more bioavailable to filter feeding organisms at the base of the food chain. Although probably not dominant, the role of pelagic zooplankton, enhanced by the filtering of Hg-rich detritus of terrigenous origin, may not be negligible. Indeed, complementary studies to the RSE monitoring program using pelagic nets, as opposed to the standard littoral nets used for fish monitoring, have revealed an abundance of small cisco *(Coregonus artedii)* in the Robert-Bourassa reservoir, which are strictly plankton feeders.

15.2.3
Additional Active Transfer from the Flooded Soils to the Aquatic Food Chain

In addition to the passive release from the flooded vegetation and soils to the water column and through the planktonic food chain, Hg may also be transferred to the aquatic food web by active mechanisms implicating biotic and abiotic processes. First, as MeHg is accumulated in the flooded soils, insect larvae burrowing in the first centimeters, feeding on partially degraded Hg-rich organic matter, rapidly bioaccumulate significant concentrations of the heavy metal and transfer it to higher aquatic organisms as they emerge into the water column. Estimates derived from *in situ* measurements in La Grande complex reservoirs suggest that, for shallow areas, the Hg burden may be up to 6 times greater in aquatic insects than in zooplankton. For pelagic areas however, the Hg burden in zooplankton is probably greater, as the density of benthic organisms is reduced, the periphyton layer is absent and as the lower temperatures are less favorable to methylation.

Second, during the first few years of impoundment of reservoirs in northern Québec, most of the shallow and exposed flooded soils of the drawdown zone are progressively eroded. In particular, the organic horizon of podzolic soils is susceptible to erosion by wave and ice action. A rapid sorting of the eroded soil particles then occurs, maintaining for some time the Hg-rich fine organic particles in suspension in the water column. Filter feeding organisms ingesting these particles may constitute a prime way for Hg to enter the aquatic food chain of reservoirs. These particles may eventually settle back on the surface of other flooded soils, where they may become in turn MeHg-rich food available for benthic feeders.

Third, the release of nutrients resulting from the bacterial degradation of flooded terrigenous organic matter stimulates autochthonous production. Degradation of the resulting fairly labile organic matter, as opposed to the ligno-cellulosic compounds of flooded soils, promotes additional methylation. This process may be of particular importance in shallow areas, with relatively long water residence times, where the combined effect of light penetration and mineralization of nutrients leads to the development a layer of periphyton, benthic algae growing in symbiosis with the bacterial colonies, which promotes the methylation of Hg and constitutes a prime source of MeHg-rich food for zooplankton and insect larvae. Zooplankton collected in these areas continue to show high concentrations after 13 or 15 years of flooding, contrary to pelagic zooplankton which showed pre-impoundment Hg concentrations after about 8 years of flooding.

15.2.4
Mercury Increases in Fish

15.2.4.1
Reservoirs

In La Grande reservoirs, concentrations in all fish species increased rapidly after impoundment to levels 3 to 7 times those measured in surrounding natural lakes. Maximum concentrations in lake whitefish and longnose sucker, non-piscivorous species, were reached 5 to 9 years after impoundment. Maximum concentrations in 400-mm lake whitefish ranged from 0.4 to 0.5 mg kg^{-1} (total Hg, ww), from one reservoir to another, slightly surpassing the Canadian marketing standard of 0.5 mg kg^{-1} only in the Robert-Bourassa reservoir.

For piscivorous species such as northern pike and walleye, concentrations for standardized-length fish (400 mm for walleye and 700 mm for pike) increased for a longer period of time, peaking 10 to 13 years after impoundment. Maximum concentrations obtained for northern pike, varying from 1.7 to 4.2 mg kg^{-1} depending on the reservoir, are 3 to 8 times greater than the Canadian marketing standard of 0.5 mg kg^{-1}.

Fish stomach content studies showed that non-piscivorous fish from both natural lakes and reservoirs have a similar diet and that the total Hg increases observed in reservoir fish are more related to changes in the concentrations of MeHg in the organisms of the food web than in changes in feeding habits, again stressing the importance of increased methylation at the base of the food chain. Lake whitefish of both environments shift from a zooplankton dominated diet to one dominated by benthos with increasing fish size. Cisco and longnose sucker feed mainly on zooplankton and benthos respectively, regardless of size. In the western part of the La Grande complex, small cisco is the most common prey found in piscivorous fish stomachs. Pike of this region also feed on a variety of species, from strictly non-piscivorous species to strictly piscivorous species. As they grow in size, pike feed less on coregonids and more on piscivorous fishes, such as walleye, pike and burbot. These latter species may represent up to 60% of the ingested biomass for large pike (> 400 mm). Such a feeding behavior exposes them to large concentrations of MeHg. In the eastern part of the territory, the piscivorous species diet is highly dominated by lake whitefish, a behavior contributing to lower Hg concentrations.

15.2.4.2
Downstream from Reservoirs

Monitoring of fish Hg levels at the La Grande complex has also shown that Hg is exported downstream from reservoirs. Studies at the La Grande complex indicate that Hg is mostly exported by Hg-rich organic debris as well as by plankton, aquatic insects or small fish and suggest that the magnitude of increase in downstream fish Hg levels, as well as the distance over which it occurs, depends on: (1) the importance of downstream tributaries diluting the Hg-rich debris or organisms (food source for fish) originating from the reservoir and, (2) the presence or absence of large bodies of water downstream (lakes or other reservoirs) permitting the sedimentation of organic material or the biological uptake of the Hg-rich organisms. The importance of large bodies of water downstream is further corroborated by the absence of a cumulative effect in fish Hg levels passing from the Caniapiscau reservoir to the La Grande 4 reservoir, via the Fontanges and Vincelotte reservoirs. The results obtained from this series of reservoirs strongly suggest that in the case of a chain of large deep reservoirs, the downstream effect of one reservoir, on fish Hg levels, is limited to the first reservoir located immediately below. Furthermore, spatial distributions of fish Hg levels within a given receiving reservoir show that this effect, in the case of large reservoirs, is limited to the area located near the input from the upstream reservoir.

15.2.5
Duration of Increased Mercury Levels in Fish

After peaking 5 to 9 years after impoundment, concentrations in standardized-length lake whitefish then significantly and gradually declined in all reservoirs, actually reaching background levels measured in neighboring lakes after 10, 11 and 17 years in the case of the La Grande 4, Caniapiscau and Robert-Bourassa reservoirs respectively. These data, in addition to the similar evolution measured in longnose sucker, strongly suggest that concentrations return to levels normally found in natural lakes of the region after 10 to 25 years for all non-piscivorous species.

Concentrations in piscivorous species, such as walleye and pike, started to decline significantly in all La Grande complex reservoirs after 15 years. In other reservoirs of northern Québec and Labrador, concentrations in pike have returned within the range of concentrations measured in natural lakes in all reservoirs older than 20 to 30 years. A similar duration has also been observed in Finland and Northern Manitoba (Verta et al. 1986b; Strange and Bodaly 1997). Although the age of the La Grande complex reservoirs varied only from 10 to 17 years at the time of the latest monitoring campaign, the general evolution of Hg levels in piscivorous species appears to follow the same time trend.

A number of key processes in the increase of Hg levels in fish in reservoirs either have a temporary effect or are greatly reduced after a few years of impoundment. These processes include (1) the passive release of Hg to the water column from the flooded vegetation and soils via degradation of terrigenous organic matter; (2) the release of nutrients stimulating autochthonous production with its labile organic matter promoting methylation; (3) the erosion of organic material in the drawdown zone releasing Hg-rich organic particles for filter feeding organisms; (4) the active transfer from burrowing aquatic insects; (5) the development of a periphytonic layer at the water-flooded soil interface promoting methylation and active transfer of Hg by aquatic insects and grazing zooplankton.

15.2.5.1
Degradation of Terrigenous Organic Matter and Release of Nutrients

Water quality monitoring at the La Grande complex has shown that intensive decomposition of flooded organic matter is short lived due to the rapid depletion of readily decomposable organic matter. Water quality modifications due to this decomposition, such as dissolved oxygen depletion and increases of CO_2 and phosphorus concentrations (release of nutrients) are virtually over after 8 to 14 years after impoundment depending on the reservoir's hydraulic and morphological characteristics. *In vitro* flooding experiments confirmed that, at temperatures characteristic of the waters of the La Grande complex, virtually all the releasable Hg would be liberated within a few years, a good portion of which would be released within the first year. These experiments further suggest that only a fraction of the green part of the flooded vegetation biomass (including the ground cover) decomposes, releasing significant amounts of their initial Hg burden, the remainder being composed of substances resistant to biochemical degradation. Ligneous components of flooded vegetation remained virtually unaffected after many decades, as spruce tree trunks had lost less than 1% of their biomass after 55 years of flooding in the Gouin reservoir located in Québec.

15.2.5.2
Erosion

Field observations at the La Grande complex reservoirs have shown that the erosion of organic material in the drawdown zone was highest during the first few years and was virtually over within a 5 to 10 year period. Thus, the increased availability of Hg-rich fine organic particles in suspension for filter feeding organisms is also short lived.

15.2.5.3
Active Transfer from Burrowing Insects and Periphyton

Furthermore, the area of active biological transfer of MeHg, from flooded organic soils to water column, by benthic organisms is also greatly reduced after a few years, as organic material is removed in areas of the drawdown zone exposed to ice and wave action. At the La Grande complex, where surficial materials are mostly derived from glacial till, large proportions of the rim of the reservoirs are thus rapidly brought to sand, gravel and rocks, greatly reducing both the passive and active transfer of Hg from the flooded soils to the water column. As a result, zooplankton and benthic organisms collected from stomach contents of white suckers of the Desaulniers reservoir, located close to the Robert-Bourassa reservoir, had MeHg concentrations equivalent to those of natural lakes, 17 years after impoundment. Hg concentrations in zooplankton collected in the pelagic zone of a number of La Grande complex reservoir also had MeHg levels equivalent to those of zooplankton of surrounding natural lakes 8 to 10 years after impoundment.

Thus, after a rapid upsurge of Hg through the reservoir food chain, through both diffusion and active biological transfer from flooded soils (by zooplankton filtering fine particulate organic debris or grazing on periphyton growing on the flooded soils, as well as by invertebrates burrowing in or feeding on these soils), the intensity of these processes diminishes in time. After a certain period of time the rate of transfer of Hg from the flooded vegetation and soils to the aquatic food chain of reservoirs becomes equivalent to the rate of transfer from sediments to the aquatic food chain of lakes, with the result that Hg concentrations in fish gradually return to levels typical of surrounding natural lakes.

15.2.6
Morphological and Hydrologic Factors Influencing the Evolution of Mercury Levels in Fish in Reservoirs

Monitoring at the La Grande complex indicates that all reservoirs show a similar trend in fish Hg levels, although they exhibit differences in the magnitude of the post-impoundment increases, as well as in the rates of increase and subsequent decrease. Our results suggest that these differences among reservoirs may be explained, at least in part, by a limited number of physical and hydrologic characteristics, related to the principal factors identified as responsible for the increase of fish Hg levels, such as: the extent of land area flooded, the annual volume of water flowing through, the filling time and the proportion of flooded area in the drawdown zone. These characteristics may be used to obtain a preliminary evaluation of the order of magnitude of the increase of Hg levels in fish of proposed reservoirs.

15.2.6.1
Land Area Flooded to Annual Volume of Water Ratio

The ratio between the land area flooded (in km^2) and the annual volume of water (in km^3) flowing through the reservoir (LAF/AVW ratio) should be a good indicator of the magnitude of the post-impoundment increase of Hg levels in fish. The land area flooded indicates the amount of organic matter stimulating bacterial methylation and also the passive and active release of Hg. The annual volume of water flowing through a reservoir is also considered a key factor because: (1) it is an indicator of the diluting capacity of the Hg released into the water column, (2) it plays a role in the extent of oxygen depletion (as anoxic conditions are believed to promote methylation), (3) it determines the extent of export of Hg out of a reservoir, and (4) it plays a role in the extent of export of nutrients out of a reservoir, which reduces the autochthonous bioproduction, thus reducing the additional methylation and release of Hg brought about by the degradation of the particularly labile phytoplankton and periphyton.

15.2.6.2
Flooded Land Area in Drawdown Zone

The proportion of the total land area flooded located in the drawdown zone would be a good indicator of the magnitude and duration of the active biological transfer of MeHg from the flooded soils to fish. On the one hand, it would be a good indicator of the erosion of organic material in the drawdown zone increasing availability of Hg-rich fine organic particles in suspension for filter feeding organisms. On the other hand, it would be an indicator of the duration of the transfer of Hg from burrowing aquatic insects, as well as from zooplankton grazing on the periphyton layer. This biological transfer could play a significant role during a prolonged period (at least 15 years based on *in situ* measurements) in shallow areas, protected from wave action, where organic matter has not been eroded. At the La Grande complex reservoirs, the flooded soils are generally very thin and are rapidly eroded and subsequently deposited in deeper colder areas, less favorable to bacterial methylation. This erosion reduces the surface of flooded soils where biological transfer of Hg by invertebrates still takes place. Thus, for reservoirs such as those found at the La Grande complex, where the organic layer of flooded soils is quickly removed by wave action, the greater the proportion of the flooded land area located in the drawdown zone, the quicker the return to background fish Hg levels because of the reduction of the Hg transfer from the flooded soils to fish through the benthic organisms.

15.2.6.3
Filling Time

The time it takes to fill a reservoir is also considered an important factor in determining post-impoundment peak Hg levels in fish, as *in vitro* studies have shown that the release of Hg from flooded organic material to the water column is very rapid, with most of the Hg being released in the first few months. In the case of a reservoir filled over a number of years, Hg is continuously liberated over a longer period of time, but at a slower rate. Water quality modifications, indicative of bacterial decomposition, peaked after 2 to 3 years in reservoirs filled within a year (Robert-Bourassa and Opinaca), but peaked after 6 to 10 years in the Caniapiscau reservoir, filled in 3 years. Thus, considering all other factors equal, the longer the filling time, the lower would be the peak Hg levels in fish, but the longer would be the period necessary to return to concentrations typical of neighboring lakes. The filling time, as it relates to the rate of flooding or increasing water levels, is also a good indicator of the rate of erosion of the flooded organic soils during filling. In a large reservoir filled slowly, such as the Caniapiscau, the water level in the drawdown zone rising only a few centimeters per day may permit the complete erosion, during the filling period, of thin podzolic soils in areas subjected to important wave action. For the Caniapiscau reservoir, this is thought to have contributed to shortening the period of elevated concentrations in lake whitefish.

15.2.7
Risk to Wildlife

Two terrestrial animals, mink (*Mustela vison*) and Osprey (*Pandion haliaetus*), feeding occasionally or exclusively on fish, have served as models for the study of potential effects of increasing Hg levels in reservoir fish on terrestrial animals.

15.2.7.1
Mink Experiment

Mink, a partly piscivorous mammal, is a species widely distributed in northern Québec. Wild specimens obtained from Cree trappers and domesticated specimens were used to carry out *in vivo* and *in vitro* exposure studies respectively.

An *in vitro* exposure experiment was conducted, in which female domesticated mink were exposed to daily diets containing 0.1, 0.5 and 1.0 μg g^{-1} of total Hg. These diets were prepared with reservoir fish, so that the Hg was essentially assimilated under the MeHg form. Mink exposed to daily diets of 0.1 and 0.5 μg g^{-1} of total Hg showed no effects, while those exposed to a daily diet of 1.0 μg g^{-1} showed reductions in fertility, as well as mortality (2/3 of the animals) with neurotoxic signs after 3 months or more.

Mean concentrations in all tissues of wild mink caught in natural habitats of northern Québec are all below those measured in the group exposed to daily diets of 0.1 µg g^{-1}. The ecotoxicologic risk from MeHg to mink populations inhabiting natural habitats of northern Québec is thus very low, despite the increase of the Hg burden in the environment over the last 60 years, since no clinical or other signs of Hg intoxication were observed at daily diets of 0.1 and 0.5 µg g^{-1}.

Along most reservoir shorelines, the risks also appear weak, despite important increases in fish Hg levels, since adult territorial mink have little possibility of finding suitable permanent habitat. However, shorelines of the Boyd-Sakami diversion, which shows little water level fluctuations, may offer mink more suitable habitat. Fish along this diversion have shown average Hg levels 5 times higher than those living in natural lakes and rivers. Since mean Hg levels in the tissues of wild mink from northern Québec are well below the mean levels obtained for the laboratory mink fed a diet of 0.1 µg g^{-1} of Hg, the average Hg concentration in the diet of wild mink must be well below 0.1 µg g^{-1}. Assuming that Hg levels in the diet of mink of the Boyd-Sakami area would increase proportionally to the increase in fish (factor of 5), which was not observed for mink along the comparable Churchill-Nelson diversion route in northern Manitoba, then the average Hg concentration in the diet of wild mink inhabiting the Boyd-Sakami area should be below 0.5 µg g^{-1}. This level remaining below the diet of 0.5 µg g^{-1}, at which no clinical or other signs of Hg intoxication were observed in domesticated mink, the risks for the mink population inhabiting the Boyd-Sakami shorelines would also be low. Furthermore, this assumed level in the diet of wild mink of this area is well below the dose of 1.0 µg g^{-1} in the diet at which the first effects were observed. This evaluation appears conservative since environmental factors, such as the 6 month long ice covered period of northern Québec lakes and ponds rendering fish inaccessible to mink, whose cumulative effect would tend to reduce rather than increase Hg exposure to mink.

Reservoir fish attempting to spawn in streams accessible from reservoir can be an additional source of MeHg exposure to mink along their shorelines. In such situations, fish ascending from a nearby reservoir would be diluted with endemic populations of the stream. Furthermore, incoming fish spawning in spring or fall depending of the species would stay only for the duration of reproduction, moving back to reservoir thereafter. Such movements are usually limited to the first rapids of a stream. Therefore, the added exposure would be less than that encountered by mink inhabiting the shorelines of the Boyd-Sakami diversion.

15.2.7.2
Breeding Success of Osprey

The breeding success of Osprey nesting near La Grande complex reservoirs was also studied, as they are top predators relying on highly efficient sight and neuro-motor co-ordination to feed their young, functions particularly sensitive to Hg intoxication. Results for Osprey nesting in the region of the La Grande complex show that, with the exception of eggs, which are usually laid when reservoirs are still ice covered, total Hg levels were significantly higher in tissues of adult Osprey and nestling living close to reservoirs than in those inhabiting natural environments. Adult and nestling feathers collected near reservoirs contained 3.5 and 5 times more total Hg, respectively, than those from natural habitats. On average, feathers contained 16.5 mg kg^{-1} dw of total Hg in adult Osprey in natural habitats as compared to 58.1 mg kg^{-1} in adults near reservoirs. A similar pattern was observed in feathers of 35-45 day-old chicks, but concentrations tended to be lower than those found in adults (7.0 mg kg^{-1} dw in natural habitats compared to 37.4 mg kg^{-1} near reservoirs). Despite a much higher total Hg exposure for Osprey fishing in reservoirs, the number of young fledged was not statistically different between nests located near reservoirs and those located near natural lakes and rivers.

Our results suggest that growing feathers, either in molting adults or in nestlings, provides a good excretion route for total Hg. For example, feathers from chicks 5 to 7 weeks old contain approximately 86% of the total Hg burden, excluding other tissues containing keratin for which total Hg was not measured. Nevertheless, the good breeding performance of Osprey nesting near reservoirs cannot be explained solely by the partial molt that adults undergo in summer. Several studies have reported that demethylation is a significant detoxification route for MeHg in birds of prey, so that, in spite of their high exposure due to their position as the final link in the food chain, they tolerate MeHg better than had previously been believed.

16 Conclusion and Prospects

Roger Schetagne

Although all the questions regarding the source and fate of mercury (Hg) in natural and modified aquatic ecosystems of northern Québec have yet to be answered, the findings of over 10 years of studies, as well as of over 20 years of monitoring, in the area have enabled a comprehensive assessment of the biogeochemical processes involved. They have also permitted the definition of the importance and duration of the phenomenon of increased fish Hg levels in reservoirs. Except for a limited number of species, the risk it represents to fish consuming wildlife remains to be evaluated.

Although our studies have demonstrated that the post-impoundment increases of Hg levels in reservoir fish are temporary, lasting in the order of 10 to 30 years depending on the fish species and the reservoirs morphological and hydrologic characteristics, the increases are such that, during a number of years following impoundment, the regular consumption of reservoir piscivorous fish may lead to levels of Hg exposure that exceed public health advisories. As mentioned in the introduction, the public health issue was not addressed in this monograph on Hg in northern Québec. Within the *"James Bay Mercury Agreement"* binding the Government of Québec, Hydro-Québec and the native Cree, the health aspects were the responsibility of the Cree Board of Health and Social Services of James Bay. The exposure to Hg of the Cree was monitored and remedial measures were implemented to allow them to continue their hunting and fishing activities (Chevalier et al. 1997). Measures included subsidizing fisheries in areas where fish were less contaminated, or enhancement of wildlife habitats to improve hunting success of species less contaminated, such as non-piscivorous waterfowl.

Future hydroelectric development must favor schemes that limit Hg increases in fish and must provide for risk management programs for fish consumers. Above and beyond such considerations, the solution to reducing health risks related to the consumption of fish, caught either in natural lakes or in young reservoirs, ultimately lies in the reduction of anthropogenic mercury emissions, responsible for increased atmospheric transport and deposition of Hg on the terrestrial and aquatic ecosystems of remote and otherwise pristine regions.

The following future research prospective is aimed at facilitating the assessment and the management of the environmental and human health risk related to the Hg issue for future projects.

16.1
Development of Models Predicting Fish Mercury Levels in Reservoirs

In developing mechanistic models to predict the evolution of fish Hg levels in future reservoirs, the major morphological and hydrologic characteristics of the proposed reservoirs should be taken into account, such as: the extent of land area flooded, the annual volume of water flowing through, the filling time and the proportion of flooded land area in the drawdown zone. These characteristics are related to the principal processes identified as responsible for the increase of fish Hg levels: (1) passive release of Hg to the water column from the flooded vegetation and soils via decomposition of terrigenous organic matter; (2) release of nutrients stimulating autochthonous production of labile organic matter promoting additional methylation and transfer of Hg to aquatic organisms; (3) active transfer by burrowing aquatic insects; (4) erosion of organic material in the drawdown zone increasing availability of Hg-rich fine organic particles in suspension for filter feeding organisms; (5) development of periphytonic layer at the water-flooded soil interface promoting active transfer by aquatic insects and grazing zooplankton. The challenge lies in the quantification and the relative importance of each of these processes. Such models should also incorporate water quality, nature and density of flooded vegetation, density, thickness and organic content of flooded soils, as well as elements of the aquatic food chain, such as plankton and benthos biomass and fish community structure.

The measurement of the actual rates of transfer of Hg from the flooded environment to the aquatic food chain, by each of the principal processes, appears to be a virtually impossible task for large reservoirs such as those of the La Grande complex. Nevertheless, the integration of the extensive data accumulated by over a decade of studies carried out at this complex, by a simplified model containing major fluxes of biomass and of Hg along the food chain may prove to be a useful tool. As is often the case with these types of models, the simulation of different plausible scenarios may lead to the determination of the relative importance of the processes, without the necessity of actual quantification. Preliminary work carried out with the La Grande data seem promising.

16.2
Potential Mitigation Measures to Reduce the Temporary Increase in Fish Mercury Levels

The results of our studies first suggest that mitigation measures to reduce the temporary increase in fish Hg levels in reservoirs should be aimed at the 5 principal processes mentioned above by which Hg is transferred from the flooded environment to the aquatic food chain. Since these processes are all related to the flooded vegetation and soils, measures aimed at the removal or isolation of the flooded organic matter by either mechanical removal, burning or covering with clean clay have been identified in the literature. Other measures suggested in the literature are either aimed at reducing the bioavailability or methylation of Hg, such as liming, keeping fine sediments in suspension in the water column to adsorb the Hg and genetic manipulation of bacterial populations, or at reducing the bioaccumulation of Hg, such as the addition of selenium or of nutritive salts. A review of these measures by Sbeghen (1995) has failed to reveal any realistic solution, for various reasons including potential harmful side effects and economic and technical impracticability of all measures when it comes to their scale of application.

Measures related to the operation of reservoirs, aimed at increasing erosion of flooded organic material to reduce active biological transfer of Hg, or at exposing flooded matter to air in summer by lowering water levels to reduce bacterial populations responsible for the methylation of Hg may merit further analysis, although such operations usually entail very high costs in hydroelectric reservoirs. For the northern Québec context, the temporary nature of the phenomenon must also be considered, as well as the low productivity of the aquatic environments, where the long term annual sustainable fish yield in lakes, or for that matter in reservoirs more than 10 years old, is in the order of 2 kg of fish per hectare for all fish species present. For non-piscivorous fish, which have lower Hg concentrations, this yield drops to about 1 kg per hectare.

Studies at the La Grande complex nevertheless suggest that, in order to reduce health risks related to the temporary increase in Hg levels in fish resulting from the creation of hydroelectric reservoirs, future development plans should favor schemes that result in lower ratios of land area flooded to annual volume of water flowing through a reservoir, as well as development schemes in northern regions where temperature, density of vegetation and organic content of soils are usually lower. Although reservoirs with extensive drawdown may lead to lower Hg levels in fish, they should not necessarily be favored because of their potential negative effects on fish production, particularly considering the temporary nature of the increase in fish Hg levels.

16.3
Environmental Risk

It does not appear that mink populations inhabiting the Precambrian Shield of northern Québec are at risk from MeHg exposure, even those living in the neighborhood of the reservoirs or diversion routes of La Grande hydroelectric complex.

Despite a much higher total Hg exposure for Osprey fishing in reservoirs, the number of young fledged was not statistically different between nests located near reservoirs and those located near lakes and rivers. The high Hg exposure of Osprey nesting near reservoirs has not affected breeding performance, nor does it appear to have affected population levels. Under these conditions, the fact that they are able to sustain a healthy population must be related to the capacity of their internal organs to demethylate Hg, a mechanism yet poorly understood in this species.

As the risk for terrestrial animals associated with the temporary increase in fish Hg levels in reservoirs is limited to piscivorous species, very few other species of northern Québec, except maybe the otter, Mergansers, or Loons, would be at risk. Mergansers, although they are strictly piscivorous, prey on much smaller fish than Osprey, so that they would be exposed to much lower MeHg concentrations. The same holds for Loons, with the added restriction that reservoirs with high water level fluctuations, usually running counter to natural conditions, are not used by the species. Nevertheless, the risk for these species remains to be addressed. The same applies to otters, although open bodies of water such as reservoirs or diversion routes are not considered prime habitats.

References

Most of the unpublished manuscripts can be obtained from *Centre de documentation, Hydro-Québec, 75 René-Lévesque Ouest, 18ᵉ étage, Montréal (Québec) Canada H2Z 1A4.* fax : 514-289-4932

Aastrup M, Johnson J, Bringmark E, Bringmark H, Iverfeldt Å (1991) Occurence and transport of mercury within a small catchment area. Water Air Soil Pollut 56: 155-167

Abernathy AR, Cumbie PM (1977) Mercury accumulation by largemouth bass (*Micropterus salmoides*) in recently impounded reservoirs. Bull Environ Contam Toxicol 17: 595-602

Allard B, Arsenie I (1991) Abiotic reduction of mercury by humic substances in aquatic system; an important process for the mercury cycle. Water Air Soil Pollut 56: 457-464

Allard M, Stokes PM (1989) Mercury in crayfish species from thirteen Ontario lakes in relation to water chemistry and smallmouth bass (*Micropterus dolomieui*) mercury. Can J Fish Aquat Sci 46: 1040-1046

AMAP (1998) AMAP assessment report : Arctic pollution issues. Arctic Monitoring and Assessment Program, Oslo, Norway

APHA, AWWA, WPCF (1989) Standard methods for the examination of water and wastewater,17th edn. American Public Health Association, American Water Works Association, Water Pollution Control Federation, Washington, DC

Arakel AV, Hongjun T (1992) Heavy metal geochemistry and dispersion pattern in coastal sediments, soil, and water of Kedron Brook floodplain area, Brisbane, Australia. Environ Geol Water Sci 20: 219-231

Aulerich RJ, Ringer RK, Iwamoto S (1974) Effects of dietary mercury on mink. Arch Environ Contam Toxicol 2: 43-51

Back RC, Visman V, Watras CJ (1995) Micro-homogenization of individual zooplankton species improves mercury and methylmercury determinations. Can J Fish Aquat Sci 52: 2470-2475

Barrow NJ, Cox VC (1992) The effect of pH and chloride concentration on mercury sorption. I. By goethite. J Soil Sci 43: 295-304

Beak Consultants (1978) Heavy metals project. Mackenzie delta and estuary. For Imperial Oil Ltd, Calgary, Alberta

Becker PH (1992) Egg mercury decline with the laying sequence in charadriiformes. Bull Environ Contam Toxicol 48: 762-767

Bégin M (1997) Méthylation du mercure dans les sols inondés des réservoirs hydroélectriques de la Baie-James. Dissertation, Université du Québec à Montréal

Bélanger D, Larivière N (1997) Développement et validation de biomarqueurs d'effets physiopathologiques précoces chez certains piscivores relativement à leur exposition au méthylmercure. Faculté de médecine vétérinaire, Université de Montréal

Bellrose FD (1980) Ducks, geese and swans of North America, 3ʳᵈedn. Stacpole Books, Harrisburg, Pennsylvania

Benoît G (1995) Evidence of the particle concentration effect for lead and other metals in fresh waters based on ultraclean technique analyses. Geochim Cosmochim Acta 59: 2677-2687

Berthelsen BO, Steinnes E, Solberg W, Jingsen L (1995) Heavy metal concentrations in plants in relation to atmospheric heavy metal deposition. J Environ Qual 24: 1018-1026

Bissonnette P (1975) Extent of mercury and lead uptake from lake sediments by chironomids. In: Drucker H, Widung RE (eds) Biological implications of metals in the environment. Proceedings of the Hanford Life Science Symposium, Richland, Washington

Bjorklund I, Borg H, Johansson K (1984) Mercury in Swedish lakes: its regional distribution and causes. Ambio 13: 118-121

Björnberg A, Håkanson L, Lundbergh K (1988) A theory on the mechanisms regulating the bioavailability of mercury in natural waters. Environ Pollut 49: 53-61

Blake S (1985) Method for the determination of low concentrations of mercury in fresh and saline waters. Environment TR 229. Water Research Centre, Marlow, UK

Bligh EG, Armstrong FAJ (1971) Marine mercury pollution in Canada. CM 1971/E34. International Council for the Exploration of the Sea (ICES), Charlottenlund, Denmark

Bloom NS (1989) Determination of picogram levels of methylmercury by aqueous phase ethylation, followed by cryogenic gas chromatography with cold vapour atomic fluorescence detection. Can J Fish Aquat Sci 46: 1131-1140

Bloom NS (1992) On the chemical form of mercury in edible fish and marine invertebrate tissue. Can J Fish Aquat Sci 49: 1010-1017

Bloom NS, Crecelius EA (1983) Determination of mercury in seawater at subnanogram per liter levels. Mar Chem 14: 49-59

Bloom NS, Fitzgerald WF (1988) Determination of volatile mercury species at the picogram level by low-temperature gas chromatography with cold-vapour atomic fluorescence detection. Anal Chim Acta 208: 151-161

Bloom NS, Horvat M, Watras CJ (1995) Results of the international aqueous mercury speciation intercomparison exercise. Water Air Soil Pollut 80: 1257-1268

Bodaly RA, Hecky RE, Fudge RJP (1984) Increases in fish mercury levels in lakes flooded by the Churchill River diversion, northern Manitoba. Can J Fish Aquat Sci 41: 682-691

Bodaly RA, Rudd JWM, Fudge RJP, Kelly CA (1993) Mercury concentrations in fish related to size of remote Canadian Shield lakes. Can J Fish Aquat Sci 50: 980-987

Bodaly RA, St Louis VL, Paterson MJ, Fudge RJP, Hall BD, Rosenberg DM, Rudd JWM (1997) Bioaccumulation of mercury in the aquatic food chain in newly flooded areas. In: Sigel A, Sigel H (eds) Mercury and its effects on environment and biology. Marcel Dekker, New York, pp 259-287

Booserman RW (1985) Distribution of heavy metals in aquatic macrophytes from Okefenokee swamp. Symposia Biologica Hungarica 29: 31-40

Bordage D (1988) Suivi des couples nicheurs de canard noir en forêt boréale, 1987. Service canadien de la faune, région du Québec, Série de rapports techniques no 35. Sainte-Foy, Québec

Borg K, Wanntorp H, Erne K, Hanko E (1969) Mercury poisoning in Swedish wildlife. Viltrevy 6: 301-379

Boudreault J, Roy D (1985) Réseau de surveillance écologique du complexe La Grande 1978-1984; macroinvertébrés benthiques. Société d'énergie de la Baie James, Montréal, Québec

Bowlby JN, Gunn JM, Liimatainen VA (1988) Metals in stocked lake trout (Salvelinus namaycush) in lakes near Sudbury, Canada. Water Air Soil Pollut 39: 217-230

Braune BM, Gaskin DE (1987) Mercury levels in Bonaparte's gulls (Larus philadelphia) during autumn molt in the Quoddy region, New Brunswick, Canada. Arch Environ Contam Toxicol 16: 539-549

Brooks-Rand (1990) Operational manual for the CVAFS-2 mercury analyzer. Brooks-Rand Ltd, Seattle, Washington

Brouard D, Demers C, Lalumière R, Schetagne R, Verdon R (1990) Évolution des teneurs en mercure des poissons du complexe hydro-électrique La Grande, Québec (1978-1989); rapport synthèse. Hydro-Québec et Groupe Environnement Shooner inc, Montréal, Québec

Brouard D, Doyon JF, Schetagne R (1994) Amplification of mercury concentration in lake whitefish (*Coregonus clupeaformis*) downstream from the La Grande 2 reservoir, James Bay, Québec. In: Watras CJ, Huckabee JW (eds) Mercury pollution; integration and synthesis. Lewis Publishers, CRC Press, Boca Raton, Florida, pp 369-380

Bruce WJ, Spencer KD, Arsenault E (1979) Mercury content data for Labrador fishes, 1977-1978. Fisheries and Marine Service Data Report no 142. Research and Resource Services Directorate, St John's, Newfoundland

Cabana G, Tremblay A, Kalff J, Rasmussen JB (1994) Pelagic food chain structure in Ontario lakes: a determinant of mercury levels in lake trout (*Salvelinus namaycush*). Can J Fish Aquat Sci 51: 381-389

Campbell PGC, Tessier A, Bisson M, Bougie R (1985) Accumulation of copper and zinc in the yellow water lily *Nuphar variegatum*: relationships to metal partitioning in the adjacent lake sediments. Can J Fish Aquat Sci 42: 23-32

Canada-Manitoba Agreement (1987) Canada-Manitoba Agreement on the study and monitoring of mercury in the Churchill River diversion; summary report. Winnipeg, Manitoba

Caron B (1997) Biogéochimie du mercure dans les écosystèmes naturels terrestres du Québec septentrional. Dissertation, Université du Québec à Montréal

Chartrand N, Schetagne R, Verdon R (1994) Enseignements tirés du suivi environnemental au complexe La Grande. Comptes rendus du 18e Congrès de la Commission internationale des Grands Barrages (CIGB), Paris, pp 165-190

Chevalier G, Dumont C, Langlois C, Penn A (1997) Mercury in northern Québec: role of the Mercury Agreement and status of research and monitoring. Water Air Soil Pollut 97: 53-61

Cloutier L, Dufort F (1979) Effet de la mise en eau du réservoir Desaulniers sur les communautés benthiques et les insectes adultes. Programme SEBJ no 75 et no 85. Laboratoire de la Société d'énergie de la Baie James, Montréal, Québec

Cocking D, Rohrer M, Thomas R, Walker J, Ward D (1995) Effects of root morphology and Hg concentration in the soil on uptake by terrestrial vascular plants. Water Air Soil Pollut 80: 1113-1116

Cope WG, Wiener JG, Rada RG (1990) Mercury accumulation in yellow perch in Wisconsin seepage lakes: relation to lake characteristics. Environ Toxicol Chem 9: 931-940

Cox JA, Carnahan J, DiNunzio J, McCoy J, Meister J (1979) Source of mercury in fish in new impoundments. Bull Environ Contam Toxicol 23: 779-783

Delisle CE (1978) Vue d'ensemble sur le mercure dans l'environnement québécois. Service de la protection de l'environnement, Ministère des pêches et de l'environnement, Ottawa, Ontario

DesGranges JL (1995) Osprey. In: Gauthier J, Aubry Y (eds) Les oiseaux nicheurs du Québec: atlas des oiseaux nicheurs du Québec méridional. Association québécoise des groupes d'ornithologues, Société québécoise de protection des oiseaux, Service canadien de la faune de la région du Québec, Montréal, Québec, pp 360-353

DesGranges JL, Rodrigue J, Tardif B, Laperle M (1994) Exposition au mercure de balbuzards nichant sur les territoires de la Baie James et de la Baie d'Hudson. Service canadien de la faune, région du Québec, Série de rapports techniques no 220. Sainte-Foy, Québec

DesGranges JL, Rodrigue J, Tardif B, Laperle M (1998) Mercury accumulation and biomagnification in ospreys (*Pandion haliaetus*) in James Bay and Hudson Bay regions of Québec. Arch Environ Contam Toxicol 35: 330-341

Deslandes JC, Belzile L, Doyon JF (1993) Réseau de suivi environnemental du complexe La Grande, phase 1(1991-1992); étude des rendements de pêche. For Hydro-Québec, Montréal, Québec

Deslandes JC, Guénette S, Fortin R (1994) Evolution of fish communities in environments affected by the development of the La Grande complex, phase I (1977-1992); summary report. For Hydro-Québec, Montréal, Québec

Deslandes JC, Guénette S, Prairie Y, Roy D, Verdon R, Fortin R (1995) Changes in fish populations affected by the construction of the La Grande complex (phase 1), James Bay region, Quebec. Can J Zool 73: 1860-1877

Dmytriw R, Mucci A, Lucotte M, Pichet P (1995) The partitioning of mercury in the solid components of dry and flooded forest soils and sediments from a hydroelectric reservoir, Québec (Canada). Water Air Soil Pollut 80: 1099-1103

Doyon JF (1995a) Réseau de suivi environnemental du complexe La Grande, phase 1; recherches exploratoires sur le mercure (1992). For Hydro-Québec, Montréal, Québec

Doyon JF (1995b) Réseau de suivi environnemental du complexe La Grande, phase 1 (1993); évolution des teneurs en mercure des poissons et études complémentaires. For Hydro-Québec, Montréal, Québec

Doyon JF, Belzile L (1998) Réseau de suivi environnemental du complexe La Grande, phase 1 (1977-1996); suivi des communautés de poissons et étude spéciale sur le doré (secteur ouest du territoire). For Hydro-Québec, Montréal, Québec

Doyon JF, Tremblay A (1997) Réseau de suivi environnemental du complexe La Grande, phase I (1996); évolution des teneurs en mercure et études complémentaires (secteur ouest). For Hydro-Québec, Montréal, Québec

Doyon JF, Tremblay A, Proulx M (1996) Régime alimentaire des poissons du complexe La Grande et teneurs en mercure dans leur proies (1993-1994). For Hydro-Québec, Montréal, Québec

Doyon JF, Schetagne R, Verdon R (1998a) Different mercury bioaccumulation rates between sympatric populations of dwarf and normal lake whitefish (Coregonus clupeaformis) in the La Grande complex watershed, James Bay, Québec. Biogeochemistry 40: 203-216

Doyon JF, Bernatchez L, Gendron M, Verdon R, Fortin R (1998b) Comparison of normal and dwarf populations of lake whitefish (Coregonus clupeaformis) with reference to hydroelectric reservoir in northern Quebec. Arch Hydrobiol Spec Issues Advanc Limnol 50: 97-108

Driscoll CT, Yan C, Schofield CL, Munson R, Holsapple J (1994) The mercury cycle in fish in the Adirondack lakes. Environ Sci Technol 28: 137A-143A

Driscoll CT, Blette V, Yan C, Schofield CL, Munson R, Holsapple J (1995) The role of dissolved organic carbon in the chemistry and bioavailability of mercury in remote Adirondack lakes. Water Air Soil Pollut 80: 499-508

Duchemin É, Lucotte M, Canuel R, Chamberland A (1995) Production of the greenhouse gases CH_4 and CO_2 by hydroelectric reservoirs of the boreal region. Global Biogeochem Cycles 9: 529-549

Duchemin É, Lucotte M, Canuel R (1996) Source of organic matter responsible for greenhouse gas emissions from hydroelectric complexes of the boreal region. Fourth International Symposium on the Geochemistry of the Earth's Surface, Ilkley, UK, pp 393-396

Dumont P, Fortin R (1978) Quelques aspects de la biologie du grand corégone Coregonus clupeaformis des lacs Hélène et Nathalie, territoire de la Baie James. Can J Zool 56: 1402-1411

Eagle TC, Whitman JS (1987) Mink. In: Novak M, Baker JA, Obbard MG, Mullork B (eds) Wild furbearers: management and conservation in North America. Ontario Trappers Association, Toronto, Ontario, pp 615-624

Environment Canada (1979) Analytical methods manual. Inland Waters Directorate, Ottawa, Ontario

EPRI (1996) Protocol for estimating historic atmospheric mercury deposition. TR-106768-3297. Electrical Power Research Institute, Palo Alto, California

Evans CA, Hutchinson TC (1996) Mercury accumulation in transplanted moss and lichens at elevation sites in Québec. Water Air Soil Pollut 90: 475-488

Farella N (1998) Impacts du déboisement sur les sols et les sédiments de la région Rio Tapajós (Amazonie brésilienne) illustrés par des biomarqueurs. Dissertation, Université du Québec à Montréal

Fimreite N (1971) Effects of dietary methylmercury on ring-necked pheasants. Canadian Wildlife Service, Occasional Paper no 9

Fimreite N (1974) Mercury contamination of aquatic birds in northwestern Ontario. J Wildl Manage 38: 120-131

Fimreite N, Fyfe RW, Keith JA (1970) Mercury contamination of Canadian Prairie seed-eaters and their avian predators. Can Field-Nat 83: 269-276

Fitzgerald WF, Clarkson TW (1991) Mercury and monomethylmercury: present and future concerns. Environ Health Perspect 96: 159-166

Fitzgerald WF, Engstrom DR, Mason RP, Nater EA (1998) The case for atmospheric mercury contamination in remote areas. Environ Sci Technol 32: 1-7

Fleurbec (1987) Plantes sauvages des lacs, rivières et tourbières. Fleurbec, Saint-Augustin, Québec

Forrester JW (1968) Principles of systems. MIT Press, Cambridge, Massachusetts

Furness RW, Muirhead SJ, Woodburn M (1986) Using bird feathers to measure mercury in the environment: relationship between mercury content and moult. Mar Pollut Bull 17: 27-30

Furness RW, Lewis SA, Mills JA (1990) Mercury levels in the plumage of red-billed gulls (*Larus novaehollandiae scopulinus*) of known sex and age. Environ Pollut 63: 33-39

Furutani A, Rudd JWM (1980) Measurement of mercury methylation in lake water and sediment samples. Appl Environ Microbiol 40: 770-776

Garzon CE (1984) Water quality in hydroelectric projects; consideration for planning in tropical forest regions. World Bank Technical Paper no 20. Washington, DC

Gaskin DE, Frank R, Holdrinet M, Ishida K, Walton CJ, Smith M (1973) Mercury, DDT and PCB in harbour seals (*Phoca vitulina*) from the Bay of Fundy and Gulf of Maine. J Fish Res Board Can 30: 471-475

Gauthier J, Aubry Y (eds) (1995) Les oiseaux nicheurs du Québec: atlas des oiseaux nicheurs du Québec méridional. Association québécoise des groupes d'ornithologues, Société québécoise de protection des oiseaux, Service canadien de la faune de la région du Québec, Montréal, Québec

Gauthier L, Nault R, Crête M (1989) Variations saisonnières du régime alimentaire du caribou. Naturaliste can 116: 101-112

Gauthier-Guillemette, GREBE (1990) Complexe Grande-Baleine; avant-projet phase II; étude de l'avifaune et du castor; écologie de la sauvagine (été 1989). For Hydro-Québec, Montréal, Québec

Gauthier-Guillemette, GREBE (1992a). Complexe Nottaway-Broadback-Rupert; la sauvagine; vol 1: densité, abondance et habitat de la sauvagine. For Hydro-Québec, Montréal, Québec

Gauthier-Guillemette, GREBE (1992b) Complexe Nottaway-Broadback-Rupert; les oiseaux aquatiques; vol 5: habitats, abondance et répartition des huarts, des râles, de la grue du Canada et des autres espèces d'oiseaux aquatiques. For Hydro-Québec, Montréal, Québec

Gauthier-Guillemette, GREBE (1992c) Complexe Nottaway-Broadback-Rupert; les oiseaux aquatiques; vol 4: habitats, abondance et répartition de la sterne pierregarin (*Sterna hirundo*). For Hydro-Québec, Montréal, Québec

Gilmour CG, Henry EA (1991) Mercury methylation in aquatic systems affected by acid deposition. Environ Pollut 71: 131-169

Glass GE, Sorensen JA, Schmidt KW, Rap GR Jr (1990) New sources identification of mercury contamination in the Great Lakes. Environ Sci Technol 24: 1059-1069

Godbold DL, Hüttermann A (1988) Inhibition of photosynthesis and transpiration in relation to mercury-induced root damage in spruce seedlings. Physiol Plant 74: 270-275

Goodwin BW (1989) An introduction to systems variability analysis. In: Saltelli A et al (eds) Risk analysis in nuclear waste management. ECSC, EEC, EAEC, Brussels and Luxembourg, pp 57-68

Grimard E (1996) Traçage comparé des réseaux trophiques du réservoir LG-2 et de lacs naturels à l'aide d'isotopes stables du carbone et de l'azote; implications pour la contamination par le mercure. Dissertation, Université du Québec à Montréal

Grondin A (1994) Géochimie du mercure dans les sols et les macrophytes du moyen nord québécois. Dissertation, Université du Québec à Montréal

Grondin A, Lucotte M, Mucci A, Fortin B (1995) Mercury and lead profiles and burdens in soils of Quebec (Canada) before and after flooding. Can J Fish Aquat Sci 52: 2493-2506

Haines TA, Brumbaugh WG (1994) Metal concentration in the gill, gastrointestinal tract, and carcass of white suckers (*Catostomus commersoni*) in relation to lake acidity. Water Air Soil Pollut 73: 265-274

Haines TA, Komov VT, Jagoe CH (1994) Mercury concentration in perch (*Perca fluviatilis*) as influenced by lacustrine physical and chemical factors in two regions of Russia. In: Watras CJ, Huckabee JW (eds) Mercury pollution; integration and synthesis. Lewis Publishers, CRC Press, Boca Raton, Florida, pp 397-408

Håkanson L (1980) The quantitative impact of pH, bioproduction and Hg-contamination on the Hg-content of fish (pike). Environ Pollut 1: 285-304

Håkanson L, Andersson T, Nilsson A (1990a) Mercury in fish in Swedish lakes; linkages to domestic and European sources of emission. Water Air Soil Pollut 50: 171-191

Håkanson L, Nilsson A, Andersson T (1990b) Mercury in the Swedish mor layer; linkages to mercury deposition and sources of emission. Water Air Soil Pollut 50: 311-329

Hall BD, Bodaly RA, Fudge RJP, Rudd JWM, Rosenberg DM (1997) Food as the dominant pathway of methylmercury uptake by fish. Water Air Soil Pollut 100: 13-24

Hall BD, Rosenberg DM, Wiens AP (1998) Methylmercury in aquatic insects from an experimental reservoir. Can J Fish Aquat Sci 55: 2036-2047

Hammer UT, Huang PM, Liaw W (1982) Bioaccumulation of mercury in aquatic ecosystems. Canadian Technical Report of Fisheries and Aquatic Sciences 1163: 69-82

Harris RC, Spencer PD (1998) Bioenergetic simulations of mercury uptake and retention in walleye (*Stizostedion vitreum*) and yellow perch (*Perca flavescens*). Water Qual Res J Can 28: 217-236

Hayne DW, Ball RC (1956) Benthic productivity as influenced by fish predation. Limnol Oceanogr 1: 162-175

Health and Welfare Canada (1985) Chemical contaminant guidelines for fish and fish products. Law and regulations on food and drugs. Health and Welfare Canada, Ottawa, Ontario

Hecky RE, Bodaly RA, Ramsey DJ, Strange NE (1986) Enhancement of mercury bioaccumulation in fish by flooded terrestrial material ecosystems. Appendix no 6. Canada-Manitoba Agreement on the Study and monitoring of mercury in the Churchill River diversion, Winnipeg, Manitoba

Hecky RE, Bodaly RA, Strange NE, Ramsey DJ, Anema C, Fudge RJP (1987) Mercury bioaccumulation in yellow perch in limnocorrals simulating the effects of reservoir creation. Canadian Data Report of Fisheries and Aquatic Sciences no 628

Hecky RE, Ramsey DJ, Bodaly RA, Strange NE (1991) Increased methylmercury contamination in fish in newly formed freshwater reservoirs. In: Suzuki T (ed) Advances in mercury toxicology. Plenum Press, New York, pp 33-52

Henderson A (1992) Literature on air pollution and lichens. Lichenologist 24: 399-406

Henny CJ (1983) Distribution and abundance of nesting ospreys in the United States. In: Bird DM (ed) Biology and management of bald eagles and ospreys. Harpell Press, Sainte-Anne-de-Bellevue, Québec, pp 175-186

Hessen DO (1992) Dissolved organic carbon in a humic lake: effects on bacterial production and respiration. Hydrobiologia 229: 115-123

Hewett SW, Johnson BL (1992) Fish bioenergetics model 2; an upgrade of a generalized bioenergetics model of fish growth for microcomputers. WIS-SG-92-250. Madison Sea Grant Institute, University of Wisconsin, Madison, Wisconsin

Horvat M, Bloom NS, Liang L (1993a) Comparison of distillation with other current isolation methods for the determination of methyl mercury compounds in low-level environmental samples: Part I - Sediments. Anal Chim Acta 281: 135-152

Horvat M, Liang L, Bloom NS (1993b) Comparison of distillation with other current isolation methods for the determination of methyl mercury compounds in low-level environmental samples: Part II - Water. Anal Chim Acta 282: 153-168

Hoyer M, Burke J, Keeler G (1995) Atmospheric sources, transport and deposition of mercury in Michigan: two years of event precipitation. Water Air Soil Pollut 80: 199-208

Huckabee JW, Elwood JW, Hildebrand SG (1979) Accumulation of mercury in freshwater biota. In: Nriagu JO (ed) The biogeochemistry of mercury in the environment. Elsevier/North-Holland Biomedical Press, Amsterdam, The Netherlands, pp 277-302

Hudson RJM, Gherini SA, Watras CJ, Porcella D (1994) Modeling the biogeochemical cycle of mercury in lakes: the mercury cycling model (MCM) and its application to the MTL study lakes. In: Watras CJ, Huckabee JW (eds) Mercury pollution; integration and synthesis. Lewis Publishers, CRC Press, Boca Raton, Florida, pp 473-523

Hurley JP, Watras CJ, Bloom NS (1994) Distribution and flux of particulate mercury in four stratified seepage lakes. In: Watras CJ, Huckabee JW (eds) Mercury pollution; integration and synthesis. Lewis Publishers, CRC Press, Boca Raton, Florida, pp 69-82

Iverfeldt Å (1991) Occurrence and turnover of atmospheric mercury over nordic countries. Water Air Soil Pollut 56: 251-265

Jackson TA (1988a) The mercury problem in recently formed reservoirs of northern Manitoba (Canada): effects of impoundment and other factors on the production of methyl mercury by microorganisms in sediments. Can J Fish Aquat Sci 45: 97-121

Jackson TA (1988b) Accumulation of mercury by plankton and benthic invertebrates in riverine lakes of northern Manitoba (Canada): importance of regionally and seasonally varying environmental factors. Can J Fish Aquat Sci 45: 1744-1757

Jackson TA (1991) Biological and environmental control of mercury accumulation by fish in lakes and reservoirs of northern Manitoba, Canada. Can J Fish Aquat Sci 48: 2449-2470

Jackson TA (1997) Long-range atmospheric transport of mercury to ecosystems, and importance of anthropogenic emissions: a critical review and evaluation of the published evidence. Environ Rev 5: 99-120

Jensen J, Adare K, Shearer R (1997) Rapport de l'évaluation des contaminants dans l'Arctique canadien. Ministère des Affaires indiennes et du Nord canadien, Ottawa, Ontario

Jernolov A, Johansson A, Sorenson L, Svenson A (1976) Methylmercury degradation in mink. Toxicology 6: 315-321

Johansson K (1985) Mercury in sediment in Swedish forest lakes. Vehr Internat Verein Limnol 22: 2259-2363

Johansson K, Aastrup M, Andersson A, Bringmark L, Iverfeldt Å (1991) Mercury in Swedish forest soils and waters: assessment of critical load. Water Air Soil Pollut 56: 267-281

Johansson K, Andersson A, Andersson T (1995) Regional accumulation pattern of heavy metals in lake sediments and forest soils in Sweden. Sci Tot Environ 160/161: 373-380

Johnels AG, Tyler G, Westermark T (1979) A history of mercury levels in Swedish fauna.
    Ambio 8: 160-168
Johnston TA, Bodaly RA, Mathias JA (1991) Predicting fish mercury levels from physical
    characteristics of boreal reservoirs. Can J Fish Aquat Sci 48: 1468-1475
Jonasson PM (1965) Factors determining population size of *Chironomus anthracinus* in lake
    Esrom. Internat Ver Theor Angew Limnol 13: 139-162
Jones ML, Cunningham GL, Marmorek DR, Stokes PM, Wren C, DeGrass P (1986) Mercury
    release in hydroelectric reservoirs. Canadian Electrical Association, Montréal, Québec
Jones RI (1992) The influence of humic substances on lacustrine planktonic food chains.
    Hydrobiologia 229: 73-91
Jorgensen SE (1980) Lake management. Water development, supply and management Series; vol
    14, Pergamon Press, New York
Julien M, Laperle M (1985) Surveillance écologique du complexe La Grande; synthèse des
    études sur la sauvagine. Société d'énergie de la Baie James, Montréal, Québec
Kari T, Kauranen P (1978) Mercury and selenium contents of seals from fresh and brackish
    waters in Finland. Bull Environ Contam Toxicol 19: 273-280
Kelly CA, Rudd JWM, Bodaly RA, Roulet NP, St Louis VL, Heyes A, Moore TR, Schiff S,
    Aravena R, Scott KJ, Dyck B, Harris R, Warner B, Edwards G (1997) Increases in fluxes of
    greenhouse gases and methyl mercury following flooding of an experimental reservoir.
    Environ Sci Technol 31: 1334-1344
Kidd KA, Hesslein RH, Fudge RJP, Hallard KA (1995) The influence of trophic level as
    measured by $\delta^{15}N$ on mercury concentrations in freshwater organisms. Water Air Soil Pollut
    80: 1011-1015
Kitchell JF, Stewart DJ, Weininger D (1977) Applications of a bioenergetics model to yellow
    perch (*Perca flavescens*) and walleye (*Stizotedion vitreum vitreum*). J Fish Res Board Can 34:
    1922-1935
Korthals ET, Winfrey MR (1987) Seasonal and spatial variations in mercury methylation and
    demethylation in an oligotrophic lake. Appl Environ Microbiol 53: 2397-2404
Kucera E (1986) Mercury in mink, otter and small mammals from the Churchill River diversion,
    Manitoba. Technical appendices no 15. Canada-Manitoba Agreement on the study and
    monitoring of mercury in the Churchill River diversion, Winnipeg, Manitoba
Lalumière R, Dussault D (1992) Résultats des pêches exploratoires effectuées en 1991 dans le
    réservoir de La Grande 2. For Hydro-Québec, Montréal, Québec
Landers DH, Ford J, Gubala C, Monetti M, Lasorsa BK, Martinson J (1995) Mercury in
    vegetation and lake sediments from the US Arctic. Water Air Soil Pollut 80: 591-601
Landers DH, Gubala C, Verta M, Lucotte M, Johansson K, Lockhart WL (1997) Using lake
    sediment mercury flux ratios to evaluate the regional and continental dimensions of mercury
    deposition in Arctic and boreal ecosystems. Atmos Environ 32: 918-928
Lange TR, Royals HE, Connor LL (1993) Influence of water chemistry on mercury
    concentration in largemouth bass from Florida lakes. Trans Am Fish Soc 122: 74-84
Langlois C, Langis R, Pérusse M (1995) Mercury contamination in northwest Québec
    environment and wildlife. Water Air Soil Pollut 80: 1021-1024
Lathrop RC, Rasmussen PW, Knauer DR (1991) Mercury concentrations in walleyes from
    Wisconsin (USA) lakes. Water Air Soil Pollut 56: 295-307
Lebeau D (1996) Accumulation du mercure par les communautés périphytiques: contribution
    dans le transfert du mercure vers la chaîne alimentaire dans les lacs et réservoirs
    hydroélectriques du nord québécois. Dissertation, Université du Québec à Montréal
Lee YH (1987) Determination of methyl- and ethylmercury in natural waters at sub-nanogram
    per liter using SCF-adsorbent preconcentration procedure. Int J Environ Anal Chem 29: 263-
    276

Lee YH, Hultberg H (1990) Methylmercury in some Swedish surface waters. Environ Toxicol Chem 9: 833-841

Lee YH, Iverfeldt Å (1991) Measurement of methylmercury and mercury in run-off, lake and rain waters. Water Air Soil Pollut 56: 309-321

Lee YH, Mowrer J (1989) Determination of methylmercury in natural waters at the sub-nanogramms per litre level by capillary gas chromatography after adsorbent preconcentration. Anal Chim Acta 221: 259-268

Lee YH, Hultberg H, Andersson I (1985) Catalytic effect of various metal ions on the methylation of mercury in the presence of humic substances. Water Air Soil Pollut 25: 391-400

Legendre L, Legendre P (1984) Écologie numérique, 2e edn. Masson, Paris et Presses de l'Université du Québec, Québec

Lehoux D (1975) Mise en application du système de classification des terres pour la sauvagine dans la région de la Baie James. Environnement Canada, Service canadien de la faune, Ottawa, Ontario

Lewis S, Becker PH, Furness RW (1993) Mercury levels in eggs, tissues, and feathers of herring gulls *Larus argentatus* from the German Wadden Sea coast. Environ Pollut 80: 293-299

LGL ltd (1993) Mercury concentrations in fishes within the Smallwood Reservoir-Churchill River complex, Labrador 1977-1992. For Department of Fisheries and Oceans, St John's, Newfoundland

Lindberg P, Odsjo T (1983) Mercury levels in feathers of peregrine falcon (*Falco peregrinus*) compared with total mercury content in some of its prey species in Sweden. Environ Pollut (Ser B) 5: 297-318

Lindberg SE (1996) Forests and the global biogeochemical cycle of mercury: the importance of understanding air/vegetation exchange processes. In: Baeyens W, Ebinghaus R, Vasiliev O (eds) Global and regional mercury cycles: sources, fluxes and mass balances. Kluwer Academic, Dordrecht, The Netherlands

Lindqvist O (ed) (1991) Mercury in the Swedish environment; recent research on causes, consequences and corrective methods. Water Air Soil Pollut 55: 1-262

Lockhart WL, Wilkinson P, Billeck BN, Hunt RV, Wagemann R, Brunskill GJ (1995) Current and historical inputs of mercury to high-latitude lakes in Canada and to Hudson Bay. Water Air Soil Pollut 80: 603-610

Lodenius M, Seppänen A, Herranen M (1983) Accumulation of mercury in fish and man from reservoirs in northern Finland. Water Air Soil Pollut 19: 237-246

Loeb SL, Reuter JE, Goldman CR (1981) Littoral zone production of oligotrophic lakes; the contribution of phytoplankton and periphyton; periphyton in freshwater ecosystems. In: Developments in hydrobiology. Dr W Junk Publishers, The Hague, The Netherlands, pp 161-166

Louchouarn P, Lucotte M, Mucci A, Pichet P (1993) Biogeochemistry of mercury in hydroelectric reservoirs of northern Québec, Canada. Can J Fish Aquat Sci 50: 269-281

Lucotte M, d'Anglejan B (1985) A comparison of several methods for the determination of iron hydroxides and associated orthophosphates in estuarine particulate matter. Chem Geol 48: 257-264

Lucotte M, Mucci A, Hillaire-Marcel C, Pichet P, Grondin A (1995a) Anthropogenic mercury enrichment in remote lakes of northern Québec (Canada). Water Air Soil Pollut 80: 467-476

Lucotte M, Tremblay A, Grimard E, Hillaire-Marcel C, Meili M (1995b) Traçage isotopique des structures des réseaux trophiques de lacs naturels et de réservoirs hydroélectriques. Chaire de recherche en environnement, Université du Québec à Montréal. For Hydro-Québec, Montréal, Québec

Magnin E (1977) Ecologie des eaux douces du territoire de la Baie James. Société d'énergie de la Baie James, Montréal, Québec

Malley DF, Lawrence SG, MacIver MA, Findlay WJ (1989) Range of variation in estimates of dry weight for planktonic crustacea and rotifera from temperate north american lakes. Technical Report no 21. Department of Fisheries and Oceans, Central and Arctic Region, Winnipeg, Manitoba

Mathers RA, Johansen PH (1985) The effects of feeding ecology on mercury accumulation in walleye (*Stizostedion vitreum*) and pike (*Esox lucius*) in lake Simcoe. Can J Zool 63: 2006-2012

Mattingly RL, Cummins KW, King RH (1981) The influence of substrate organic content on the growth of a stream chironomid. Hydrobiologia 77: 161-165

Maystrenko YG, Denisova AI (1972) Method of forecasting the content of organic and biogenic substances in the water of existing and planned reservoirs. Soviet Hydrology, Selected Papers 6: 515-540

McMurtry MJ, Wales DL, Scheider WA, Beggs GL, Dimond PE (1989) Relationship of mercury concentrations in lake trout (*Salvelinus namaycush*) and smallmouth bass (*Micropterus dolomieui*) to the physical and chemical characteristics of Ontario lakes. Can J Fish Aquat Sci 46: 426-434

MEF, MSSS (1995) Guide de consommation du poisson de pêche sportive en eau douce. Ministère de l'environnement et de la faune, Ministère de la santé et des services sociaux, Québec, Québec

Meili M (1991a) Mercury in boreal forest lake ecosystems. Acta Universitatis Upsaliensis, Comprehensive Summaries of Uppsala Dissertations from the Faculty of Science no 336, Uppsala, Sweden

Meili M (1991b) In situ assessment of trophic levels and transfer rates in aquatic food webs, using chronic (Hg) and pulsed (Chernobyl ^{137}Cs) contaminants. Verh Internat Verein Limnol 24: 2970-2975

Meili M, Fry B, Kling GW (1993) Fractionation of stable isotopes (^{13}C, ^{15}N) in the food web of a humic lake. Verh Internat Verein Limnol 25: 501-505

Meister JF, DiNunzio J, Cox JA (1979) Source and level of mercury in a new impoundment. J Am Water Works Ass 71: 574-576

Merritt RW, Cummins KW (1985) An introduction to the aquatic insects of North America. Hunt Publisher, New York

Messier D, Roy D (1987) Concentration en mercure chez les poissons au complexe hydroélectrique de La Grande Rivière (Québec). Naturaliste can 114: 357-368

Mierle G (1990) Aqueous inputs of mercury to precambrian shield lakes in Ontario. Environ Toxicol Chem 9: 843-851

Mierle G, Ingram R (1991) The role of humic substances in the mobilization of mercury from watersheds. Water Air Soil Pollut 56: 349-358

Miller RB (1941) A contribution to the ecology of the Chironomidea of Costello Lake, Algonquin Park, Ontario. Publications of the Ontario Fisheries Research Laboratory no 60, University of Toronto Press, Toronto, Ontario

Miskimmin BM, Rudd JWM, Kelly CA (1992) Influence of dissolved organic carbon, pH, and microbial respiration rates on mercury methylation and demethylation in lake water. Can J Fish Aquat Sci 49: 17-22

Monteiro LR, Furness RW (1995) Seabirds as monitors of mercury in the marine environment. Water Air Soil Pollut 80: 851-870

Montgomery S, Mucci A, Lucotte M, Pichet P (1995) Total dissolved mercury in the water column of several natural and artificial aquatic systems of northern Quebec (Canada). Can J Fish Aquat Sci 52: 2483-2492

Montgomery S, Mucci A, Lucotte M (1996) The application of in situ dialysis samplers for close interval investigations of total dissolved mercury in interstitial waters. Water Air Soil Pollut 87: 219-229

Moore TR, Bubier J, Heyes A, Dyck B, Shay J (1994) Mass of methyl and total mercury in vegetation, peat and pore-water in ombrotrophic peatlands. Third Annual Report, Progress Report no 11. Experimental lakes area project (ELARP), Winnipeg, Manitoba

Moore TR, Bubier JL, Heyes A, Flett RJ (1995) Methyl and total mercury in boreal wetland plants, experimental lakes area, northwestern Ontario. J Environ Qual 24: 845-850

Morrison KA, Thérien N (1987) Importance de la considération des effets hydrauliques par les modèles prévisionnels de la dynamique du plancton. Naturaliste can 114: 381-388

Morrison KA, Thérien N (1991a) Experimental evaluation of mercury release from flooded vegetation and soils. Water Air Soil Pollut 56: 607-619

Morrison KA, Thérien N (1991b) Influence of environmental factors on mercury release in hydroelectric reservoirs. Report no 708 G 608. Canadian Electrical Association, Montréal, Québec

Morrison KA, Thérien N (1994) Mercury release and transformation from flooded vegetation and soils: experimental evaluation and simulation modelling. In: Watras CJ, Huckabee JW (eds) Mercury pollution; integration and synthesis. Lewis Publishers, CRC Press, Boca Raton, Florida, pp 355-365

Morrison KA, Thérien N (1995a) Fluxes of mercury through biota in the LG-2 reservoir after flooding. Water Air Soil Pollut 80: 573-576

Morrison KA, Thérien N (1995b) Changes in mercury levels in lake whitefish (*Coregonus clupeaformis*) and northern pike (*Esox lucius*) in the LG-2 reservoir since flooding. Water Air Soil Pollut 80: 819-828

Morrison KA, Thérien N (1996) Release of organic carbon, Kjeldahl nitrogen and total phosphorus from flooded vegetation. Water Qual Res J Can 31: 305-318

Morrison KA, Thérien N, Marcos B (1987) A comparison of six models for nutrient limitations on phytoplankton growth. Can J Fish Aquat Sci 44: 1278-1288

Mucci A, Lucotte M, Montgomery S, Plourde Y, Pichet P, Van Tra H (1995) Mercury remobilization from flooded soils in a hydroelectric reservoir of northern Québec, Robert-Bourassa: results of a soil resuspension experiment. Can J Fish Aquat Sci 52: 2507-2517

Mundie JH (1957) The ecology of Chironomidae in storage reservoirs. Trans R Entomol Soc Lond 109: 149-232

Munthe J, Hultberg H, Lee YH, Parkman H, Iverfeldt Å, Renberg I (1995) Trends of mercury and methymercury in deposition, run-off water and sediments in relation to experimental manipulations and acidification. Water Air Soil Pollut 85: 743-748

Nater EA, Grigal DF (1992) Regional trends in mercury distribution across the Great Lakes States, north central USA. Nature 358: 139-141

Native Harvesting Research Committee (1982) The wealth of the land: wildlife harvests by the James Bay Cree, 1972-73 to 1978-79. For the Coordinating committee on hunting, fishing and trapping, Montréal, Québec

Nobert M, Beaudet S, Vandal D, Roy L (1992) La fréquentation des routes de la Baie James à des fins de chasse et pêche récréatives. In: Chartrand N, Thérien N (eds) Les enseignements de la phase I du complexe La Grande, Hydro-Québec, Montréal, Québec, pp 190-200

Norheim G, Froslie A (1978) The degree of methylation and organ distribution of mercury in some birds of prey in Norway. Acta Pharmacol Toxicol 43: 196-294

Norstrom RJ, McKinnon AE, de Freitas ASW (1976) A bioenergetics-based model for pollutant accumulation by fish; simulation of PCB and methylmercury residue levels in Ottawa River yellow perch (*Perca flavescens*). J Fish Res Board Can 33: 248-267

Nuorteva P, Nuorteva SL, Suckcharoen S (1980) Bioaccumulation of mercury in blowflies collected near the mercury mine of Idrija, Yugoslavia. Bull Environ Contam Toxicol 24: 515-521

O'Connor DJ, Nielsen SW (1981) Environmental survey of methylmercury levels in wild mink (*Mustela vison*) and otter (*Lutra canadensis*) from the northeastern United States and experimental pathology of methylmercurialism in the otter. In: Chapman JA, Pursley D (eds) Worldwide Furbearers Conference Proceedings, Frosburg, Maryland, pp 1728-1745

Ouellet M, Jones HG (1982) Evidence paléolimnologique de transport atmosphérique longue portée de polluants acidifiants et de métaux traces. Eau du Québec 15: 356-368

Parkman H (1993) Mercury accumulation in zoobenthos: an important mechanism for the transport of mercury from sediment to fish. Acta Universitatis Upsaliensis, Comprehensive Summaries of Uppsala Dissertations from the Faculty of Science no 462, Uppsala, Sweden

Parkman H, Meili M (1993) Mercury in macroinvertebrates from Swedish forest lakes: influence of lake type, habitat, life cycle and food quality. Can J Fish Aquat Sci 50: 521-534

Parks JW, Lutz A, Sutton JA (1989) Water column methylmercury in the Wabigoon/English River-lake system: factors controlling concentrations, speciation, and net production. Can J Fish Aquat Sci 46: 2184-2202

Paterson MS, Fondlay D, Beaty K, Findlay W, Schindler EU, Stainton M, McCullough G (1997) Changes in the planktonic food web of a new experimental reservoir. Can J Fish Aquat Sci 54: 1088-1102

Pinel-Alloul B (1991) Annual variations of the phytoplankton community during impoundment of Canadian subarctic reservoirs. Verh Internat Verein Limnol 24: 1282-1287

Pinel-Alloul B, Méthot G (1984) Étude préliminaire des effets de la mise en eau du réservoir LG-2 (territoire de la Baie James, Québec) sur le seston grossier et le zooplancton des rivières et des lacs inondés. Int Revue Ges Hydrobiol 69: 57-78

Pirrone N, Keeler GJ, Nriagu JO (1996) Regional differences in worldwide emissions of mercury to the atmosphere. Atmos Environ 30: 2981-2987

Plourde Y, Lucotte M, Pichet P (1997) Contribution of suspended particulate matter and zooplankton to MeHg contamination of the food chain in mid-northern Québec (Canada) reservoirs. Can J Fish Aquat Sci 54: 821-831

Poissant L, Casimir A (1997) Water-air and soil-air exchange rate of total gaseous mercury measured at background sites. Atmos Environ 32: 883-891

Ponce RA, Bloom NS (1991) Effects of pH on the bioaccumulation of low level, dissolved methylmercury by rainbow trout (*Oncorhynchus mykiss*). Water Air Soil Pollut 56: 631-640

Poole AF (1989) Ospreys: a natural and unnatural history. Cambridge University Press, Cambridge, Massachusetts

Poole KG, Elkin BT, Bethke RW (1995) Environmental contaminants in wild mink in the Northwest Territories, Canada. Sci Tot Environ 160/161: 473-486

Porvari P, Verta M (1995) Methylmercury production in flooded soils: a laboratory study. Water Air Soil Pollut 80: 765-773

Potter DWB, Cardiff LMA (1974) A study of the benthic macro-invertebrates of a shallow eutrophic reservoir in Southern Wales with emphasis on the Chironomidae (Diptera); their life-histories and production. Arch Hydrobiol 74: 186-226

Potter L, Kidd D, Standiford D (1975) Mercury levels in lake Powell; bioamplification of mercury in man-made desert reservoir. Environ Sci Technol 9: 41-46

Poulin-Thériault (1983) Évolution des berges et de la zone de marnage du réservoir de La Grande 2 et du réservoir Opinaca. For Société d'énergie de la Baie James, Montréal, Québec

Poulin-Thériault (1987) Évolution des caractéristiques physiques de la zone de marnage du détournement Caniapiscau-Laforge (période de 1984-1987). For Société d'énergie de la Baie James, Montréal, Québec

Poulin-Thériault (1994) Réservoirs La Grande 2, La Grande 3, La Grande 4 et Laforge 1; efficacité du déboisement par les agents naturels; compte rendu de visite. For Société d'énergie de la Baie James, Montréal, Québec

Poulin-Thériault, Gauthier-Guillemette (1993) Méthode de caractérisation de la phytomasse appliquée aux complexes Grande-Baleine et La Grande. For Hydro-Québec, Montréal, Québec

Power G, Le Jeune R (1976) Le potentiel de pêche du Nouveau-Québec. Cah Géogr Québec 20: 409-428

Prévost YA (1983) The molt of osprey. Ardea 71: 199-209

Rada RG, Winfrey MR, Wiener JG, Powell DE (1987) A comparison of mercury distribution in sediment cores and mercury volatilization from surface waters of selected northern Wisconsin lakes. Wisconsin Department of Natural Resources, Madison, Wisconsin

Rada RG, Wiener JG, Winfrey MR, Powell DE (1989) Recent increases in atmospheric deposition of mercury to north-central Wisconsin lakes inferred from sediment analyses. Arch Environ Contam Toxicol 18: 175-181

Ramlal PS, Rudd JWM, Hecky RE (1986) Methods for measuring specific rates of mercury methylation and degradation and their use in determining factors controlling net rates of mercury methylation. Appl Environ Microbiol 51: 110-114

Ramlal PS, Anema C, Furutani A, Hecky RE, Rudd JWM (1987) Mercury methylation studies at Southern Indian Lake, Manitoba: 1981-1982. Canadian Technical Report of Fisheries and Aquatic Sciences no 1490

Ramlal PS, Kelly CA, Rudd JWM, Furutani A (1993) Sites of methyl mercury production in remote Canadian Shield lakes. Can J Fish Aquat Sci 50: 972-979

Rask M, Metsälä TR, Salonen K (1994) Mercury in the food chains of a small polyhumic forest lake in southern Finland. In: Watras CJ, Huckabee JW (eds) Mercury pollution; integration and synthesis. Lewis Publishers, CRC Press, Boca Raton, Florida, pp 409-419

Rasmussen PE (1994) Current methods of estimating mercury fluxes in remote areas. Environ Sci Technol 28: 2233-2241

Rasmussen PE, Mierle G, Nriagu JO (1991) The analysis of vegetation for total mercury. Water Air Soil Pollut 56: 379-390

Robertson DE, Sklarew DS, Olsen KB, Bloom NS, Crecelius EA, Apts CW (1987) Measurement of bioavailable mercury species in fresh water and sediments. EPRI/EA-5197. Electric Power Research Institute, Palo Alto, California

Rodgers DW (1994) You are what you eat and a little bit more: bioenergetics-based models of methylmercury accumulation in fish revisited. In: Watras CJ, Huckabee JW (eds) Mercury pollution; integration and synthesis. Lewis Publishers, CRC Press, Boca Raton, Florida, pp 427-439

Rodgers DW, Beamish FWH (1983) Uptake of waterborne methylmercury by rainbow trout (*Salmo gairdneri*) in relation to oxygen consumption and methylmercury concentration. Can J Fish Aquat Sci 38: 1309-1315

Rosenberg DM, Wiens AP (1994) Response of benthic invertebrates to the flooding of Lake 979. Report to Session 5. Experimental lakes area project (ELARP), Winnipeg, Manitoba

Rosenberg DM, Bilyj B, Wiens AP (1984) Chironomidae (Diptera) emerging from the littoral zone of reservoirs with special reference to Southern Indian Lake, Manitoba. Can J Fish Aquat Sci 41: 672-681

Roulet M, Lucotte M (1995) Geochemistry of mercury in pristine and flooded ferralitic soils of a tropical rain forest in French Guiana, South America. Water Air Soil Pollut 80: 1079-1088

Rudd JWM (1995) Sources of methyl mercury to freshwater ecosystems: a review. Water Air Soil Pollut 80: 697-713

Rudd JWM, Turner MA (1983) The English-Wabigoon River system. II. Supression of mercury and selenium bioaccumulation by suspended and bottom sediments. Can J Fish Aquat Sci 40: 2218-2227

Rudd JWM, Bodaly D, Kelly CA, Roulet N, Hecky RE (1992) Experimental lakes area research project (ELARP). First Annual Report. Experimental lakes area project (ELARP), Winnipeg, Manitoba

Rudd JWM, St Louis V, Kelly CA, Heyes A, Sellers T, Fowle B, Beaty K, Moore T, Roulet N, Flett B (1994) Effects of flooding on the MeHg biogeochemistry of wetland 979. Third Annual Report, Progress report no 9. Experimental lakes area project (ELARP), Winnipeg, Manitoba

Ruiz N, Thérien N (1997) Modelo para el cálculo del mercurio desprendido en represas hidroeléctricas considerando la variabilidad inherente a la vegetación y a lòs suelos inundados (*Modelling mercury release in hydroelectric reservoirs considering the inherent variability of the flooded vegetation and soil*). Cuarto Congreso interamericano sobre el medio ambiente (CIMA 97), Publicado por Universidad Simón Bolívar, Caracas, Venezuela

Ryder RA (1965) A method for estimating the potential fish production of north-temperate lakes. Trans Am Fish Soc 94: 214-218

Ryder RA, Kerr SR, Loftus KH, Regier HA (1974) The morphoedaphic index, a fish yield estimator - review and evaluation. J Fish Res Board Can 31: 663-688

SAGE (1983) Analyse de contenus stomacaux de poissons des régions de LG2, Opinaca et Caniapiscau; rapport sur les travaux de 1980, 1981 et 1982. For Société d'énergie de la Baie James, Montréal, Québec

Saurola P (1983) Population dynamics of the osprey in Finland during 1971-80. In: Bird DM (ed) Biology and management of bald eagles and ospreys. Harpell Press, Sainte-Anne-de-Bellevue, Québec, pp 201-206

Savard JP, Lamothe P (1991) Distribution, abundance and aspects of breeding ecology of black scoters, *Melanitta nigra*, and surf scoters, *M. perspicillata*, in northern Québec. Can Field-Nat 105: 488-496

Sbeghen J (1995) Review of mitigative and compensatory measures. For James Bay Mercury Committee, Montréal, Québec

Schetagne R (1992) Suivi de la qualité de l'eau, du phytoplancton, du zooplancton et du benthos au complexe La Grande, territoire de la Baie James. In : Chartrand N, Thérien N (eds) Les enseignements de la phase I du complexe La Grande, Hydro-Québec, Montréal, Québec, pp 13-25

Schetagne R (1994) Water quality modifications after impoundment of some large northern reservoirs. Arch Hydrobiol Beih 40: 223-229

Schetagne R, Roy D (1985) Réseau de surveillance écologique du Complexe La Grande 1977-1984; physico-chimie et pigments chlorophylliens. Société d'énergie de la Baie James, Montréal, Québec

Schetagne R, Doyon JF, Verdon R (1996) Evolution des teneurs en mercure des poissons du complexe hydroélectrique La Grande (1978-1994); rapport synthèse. Hydro-Québec et Groupe-conseil Génivar, Montréal, Québec

Schuster E (1991) The behavior of mercury in the soil with special emphasis on complexation and adsorption processes; a review of the literature. Water Air Soil Pollut 56: 667-680

Scott WB, Crossman EJ (1974) Poissons d'eau douce du Canada. Office des recherches sur les pêcheries du Canada, Bulletin no 184, Ottawa, Ontario

Scotter GW (1967) The winter diet of barren-ground caribou in northern Canada. Can Field-Nat 81: 33-39

SEBJ (1987) The La Grande Rivière hydroelectric complex; the environmental challenge. Société d'énergie de la Baie James, Montréal, Québec

SEEEQ, Environnement Illimité (1993) Complexe Nottaway-Broadback-Rupert; faune ichtyenne; mercure. For Hydro-Québec, Montréal, Québec

Seymour NR, Bancroft RP (1983) The status and use of two habitats by ospreys in northeastern Nova Scotia. In: Bird DM (ed) Biology and management of bald eagles and ospreys. Harpell Press, Sainte-Anne-de-Bellevue, Québec, pp 275-280

Shotyk W (1996) Peat bog archives of atmospheric metal deposition: geochemical evaluation of peat profiles, natural variations in metal concentrations, and metal enrichment factors. Environ Rev 4: 149-183

Smith RG, Armstrong FAJ (1978) Mercury and selenium in ringed and bearded seal tissues from Arctic Canada. Arctic 31: 75-84

SOMER inc (1992) Échantillonnage des sédiments et de la faune pour le dosage des contaminants; guide méthodologique. For Hydro-Québec, Montréal, Québec

SOMER inc (1993) Complexe Grande-Baleine; la contamination du milieu et des ressources fauniques de la zone d'étude du complexe Grande-Baleine. For Hydro-Québec, Montréal, Québec

SOMER inc (1994) Complexe Nottaway-Broadback-Rupert; qualité de l'eau. For Hydro-Québec, Montréal, Québec

Sorensen JA, Glass GE, Schmidt KW, Huber JK, Rapp GR (1990) Airborne mercury deposition and watershed characteristics in relation to mercury concentations in water, sediments, plankton, and fish of eighty northern Minnesota lakes. Environ Sci Technol 24: 1716-1727

Spry D, Wiener JG (1991) Metal bioavailability and toxicity to fish in low-alkalinity lakes: a critical review. Environ Pollut 71: 243-304

St Louis VL, Rudd JWM, Kelly CA, Beaty KG, Bloom NS, Flett RJ (1994) Importance of wetlands as sources of methyl mercury to boreal forest ecosystem. Can J Fish Aquat Sci 51: 1065-1076

St Louis VL, Rudd JWM, Kelly CA, Beaty KG, Flett RJ, Roulet NT (1996) Production and loss of methylmercury and loss of total mercury from boreal forest catchments containing different types of wetlands. Environ Sci Technol 30: 2719-2729

Steinnes E (1995) Mercury. In: Alloway BJ (ed) Heavy metals in soils, 2nd edn. Blackie Academic & Professional, London, UK, pp 245-259

Stocek RF, Pearce PA (1983) Distribution and reproductive success of ospreys in New Brunswick, 1974-1980. In: Bird DM (ed) Biology and management of bald eagles and ospreys. Harpell Press, Sainte-Anne-de-Bellevue, Québec, pp 215-221

Strange NE (1993) Mercury in fish in northern Manitoba reservoirs and associated waterbodies: results from 1992 sampling. Program for monitoring mercury concentrations in fish in northern Manitoba reservoirs. Canada Department of Fisheries and Oceans, Manitoba Hydro, Manitoba Department of Natural Resources and Hydro-Québec, Winnipeg, Manitoba

Strange NE (1995) Mercury in fish in northern Manitoba reservoirs and associated waterbodies: results from 1994 sampling. Program for monitoring mercury concentrations in fish in northern Manitoba reservoirs. Canada Department of Fisheries and Oceans, Manitoba Hydro, Manitoba Department of Natural Resources and Hydro-Québec, Winnipeg, Manitoba

Strange NE, Bodaly RA (1997) Mercury in fish in northern Manitoba reservoirs and associated water bodies; summary report for 1992, 1994 and 1996 sampling. Program for monitoring mercury concentrations in fish in northern Manitoba reservoirs. Canada Department of Fisheries and Oceans, Manitoba Hydro, Manitoba Department of Natural Resources and Hydro-Québec, Winnipeg, Manitoba

Strange NE, Bodaly RA, Fudge RJP (1991) Mercury concentrations of fish in Southern Indian Lake and Issett Lake, Manitoba 1975-1988: the effect of lake impoundment and Churchill River diversion. Canadian Technical Report of Fisheries and Aquatic Sciences no 1824

Sturges WT, Barrie LA (1989) Stable lead isotope ratios in arctic aerosols: evidence for the origin of Arctic air pollution. Atmos Environ 23: 2513-2519

Suns K, Hitchin G, Loescher B, Pastorek E, Pearce R (1987) Metal accumulations in fishes from Muskoka-Haliburton lakes in Ontario (1978-1984). Ontario Ministry of the Environment, Rexdale, Ontario

Surma-Aho K, Paasivirta J, Rekolainen S, Verta M (1986) Organic and inorganic mercury in the food chain of some lakes and reservoirs in Finland. Chemosphere 15: 353-372

Suter GW, Barnthouse LW, Bartell SM, Mill T, Mackay D, Paterson S (1993) Ecological risk assessment. Lewis Publishers, Ann Arbor, Michigan

Swain E, Helwig DD (1989) Mercury in fish from northeastern Minnesota lakes: historical trends, environmental correlates, and potential sources. J Minn Acad Sci 55: 103-109

Swain EB, Engstrom DR, Brigham ME, Henning TA, Brezonik PL (1992) Increasing rates of atmospheric mercury deposition in midcontinental North America. Science 257: 784-787

Thérien N (1990) Mercure dans l'eau et le seston. For Hydro-Québec, Montréal, Québec

Thérien N (1994) Recherche exploratoire sur le transfert du mercure; analyse du mercure dans des échantillons d'eau, contenus stomacaux et petits poissons; échantillonnages de 1992, 1993 et 1994. For Hydro-Québec, Montréal, Québec

Thérien N, Morrison KA (1984) The evolution of water quality in large hydro-electric reservoirs: a model of active and stagnant zones. In: Veziroglu TN (ed) The biosphere: problems and solutions. Studies in Environmental Sciences; vol 25, Elsevier Science Publishers BV, Amsterdam, The Netherlands, pp 287-296

Thérien N, Morrison KA (1985) Modèle prévisionnel de la qualité des eaux du réservoir hydro-électrique LG-2: considération des zones hydrauliques actives et stagnantes. Rev Inter Sci Eau 1: 11-20

Thérien N, Morrison KA (1994) Détermination des flux temporels de mercure dans le réservoir La Grande-2 considérant l'accumulation dans les poissons ainsi que les données de relargage provenant de la décomposition de la végétation et des sols inondés. For Hydro-Québec, Montréal, Québec

Thérien N, Morrison KA (1995) Détermination des flux de matières en fonction du temps pour des échantillons types de sols inondés sous diverses conditions environnementales contrôlées; carbone, azote et phosphore; anhydride carbonique et méthane; mercure volatil, inorganique et méthylique; épaisseur active des sols; biodégradabilité de la tourbe et des sols. For Hydro-Québec, Montréal, Québec

Thérien N, Morrison KA, de Broissia M, Marcos B (1982) A simulation model of plankton dynamics in reservoirs of the La Grande complex. Naturaliste can 109: 869-881

Thompson DR (1996) Mercury in birds and terrestrial mammals. In: Beyer WN, Heinz GH, Redmon-Norwood AW (eds) Environmental contaminants in wildlife; interpreting tissue concentrations. Lewis Publishers, CRC Press, Boca Raton, Florida, pp 341-356

Thompson DR, Furness RW (1989) Comparison of the levels of total and organic mercury in seabird feathers. Mar Pollut Bull 20: 577-579

Thompson DR, Stewart FM, Furness RW (1990) Using seabirds to monitor mercury in marine environments. Mar Pollut Bull 21: 339-342

Toms SR, Matisoff G, McCall PL, Wang X (1995) Models for alteration of sediment by benthic organisms. Project 92-NPS-2. Water Environment Research Foundation, Alexandria, Virginia

Tranvik LJ (1992) Allochthonous dissolved organic matter as an energy source for pelagic bacteria and the concept of the microbial loop. Hydrobiologia 229: 107-114

Tremblay A (1996) Transfert du mercure et du méthylmercure sédimentaire vers la chaîne trophique par les invertébrés d'écosystèmes boréaux. Dissertation, Université du Québec à Montréal

Tremblay A, Lucotte M (1997) Accumulation of total mercury and methylmercury in insect larvae of hydroelectric reservoirs. Can J Fish Aquat Sci 54: 832-841

Tremblay A, Lucotte M, Rowan D (1995) Different factors related to mercury concentration in sediments and zooplankton of 73 Canadian lakes. Water Air Soil Pollut 80: 961-970

Tremblay A, Lucotte M, Rheault I (1996a) Methylmercury in a benthic food web of two hydroelectric reservoirs and a natural lake. Water Air Soil Pollut 91: 255-269

Tremblay A, Lucotte M, Meili M, Cloutier L, Pichet P (1996b) Total mercury and methylmercury contents of insects from boreal lakes: ecological, spatial and temporal patterns. Water Qual Res J Can 31: 851-873

Tremblay A, Lucotte M, Schetagne R (1998a) Total mercury and methylmercury accumulation in zooplankton of hydroelectric reservoirs in northern Québec (Canada). Sci Tot Environ 213: 307-315

Tremblay A, Cloutier L, Lucotte M (1998b) Total mercury and methylmercury fluxes via emerging insects in recently flooded hydroelectric reservoirs and a natural lake. Sci Tot Environ 219: 209-221

Tremblay G, Legendre P, Doyon JF, Verdon R, Schetagne R (1998c) The use of polynomial regression analysis with indicator variables for interpretation of mercury in fish data. Biogeochemistry 40: 189-201

Tsalkitzis E (1995) Methylmercury in golden shiners (*Notemigonus crysoleucas*) and zooplankton from Mouse Lake (Ontario). Dissertation, York University, Toronto

Vallières L, Gilbert L (1992) Haut-Saint-Maurice; aménagement des centrales des Rapides-des-Coeurs et Rapides-de-la-Chaudière; avant-projet, phase 1; étude sur la faune aquatique; effets du marnage sur la faune. For Hydro-Québec, Montréal, Québec

Van Collie R, Visser SA, Campbell PGC, Jones HG (1983) Évaluation de la dégradation de bois de conifères immergés durant plus d'un demi-siècle dans un réservoir. Annales de Limnologie 19: 129-134

Verdon R, Brouard D, Demers C, Lalumière R, Laperle M, Schetagne R (1991) Mercury evolution (1978-1988) in fishes of the La Grande hydroelectric complex, Québec Canada. Water Air Soil Pollut 56: 405-417

Verta M (1984) The mercury cycle in lakes; some new hypotheses. Aqua Fennica 14: 215-221

Verta M (1990a) Mercury in Finnish forest lakes and reservoirs: anthropogenic contribution to the load and accumulation in fish. Publications of the Water and Environment Research Institute no 6, National Board of Waters and the Environment, Helsinski, Finland

Verta M (1990b) Changes in fish mercury concentration in an intensively fished lake. Can J Fish Aquat Sci 47: 1888-1897

Verta M, Rekolainen S, Mannio J, Surma-Aho K (1986a) The origin and level of mercury in Finnish forest lakes. Publications of the Water Research Institute no 65, National Board of Waters, Helsinki, Finland, pp 21-31

Verta M, Rekolainen S, Kinnunen K (1986b) Causes of increased fish mercury levels in Finnish reservoirs. Publications of the Water Research Institute no 65, National Board of Waters, Helsinki, Finland, pp 44-71

Verta M, Mannio J, Iivonen P, Hirvi JP, Järvinen O, Piepponen S (1990) Trace metals in Finnish headwater lakes; effects of acidification and airborne load. In: Kaupp P, Kenthamies K, Anttila P (eds) Acidification in Finland. Springer-Verlag, Heidelberg, Germany, pp 883-908

Wagemann R (1989) Comparison of heavy metals in two groups of ringed seals (*Phoca hispida*) from the Canadian Arctic. Can J Fish Aquat Sci 46: 1558-1563

Wagemann R (1994) Mercury, methylmercury, and other heavy metals in muktuk, muscle, and some organs of belugas (*Delphinapterus leucas*) from the western Canadian Arctic. For the Fisheries joint management committee of the Inuvialuit settlement region. Department of Fisheries and Oceans, Winnipeg, Manitoba

Wagemann R, Stewart REA, Béland P, Desjardins C (1990) Heavy metals in tissues of beluga whales from various locations in the Canadian Arctic and the St Lawrence River. In: Smith TG, St Aubin DJ, Geraci JR (eds) Advances in research on the beluga whale *Delphinapterus leucas*. Can Bull Fish Aquat Sci 224, pp 191-206

Watras CJ, Bloom NS (1992) Mercury and methylmercury in individual zooplankton: implications for bioaccumulation. Limnol Oceanogr 37: 1313-1318

Westcott K, Kalff J (1996) Environmental factors affecting methyl mercury accumulation in zooplankton. Can J Fish Aquat Sci 53: 2221-2228

Westoo G (1967) Determination of methylmercury compounds in foodstuffs. Acta Chem Scand 21: 1790-1800

Wetzel RG (1983) Limnology, 2nd edn. Saunders College Publishing, Montréal, Québec

Wiener JG, Martini RE, Sheffy TB, Glass GE (1990) Factors influencing mercury concentrations in walleyes in northern Wisconsin lakes. Trans Am Fish Soc 119: 862-870

Wobeser G, Neilsen NO, Schiefer B (1976a) Mercury and mink. I. The use of mercury contaminated fish as a food for ranch mink. Can J Comp Med 40: 30-33

Wobeser G, Neilsen NO, Schiefer B (1976b). Mercury and mink. II. Experimental methyl mercury intoxication. Can J Comp Med 40: 34-45

WHO (1989) Mercury: environmental aspects. Environmental health criteria no 86. World Health Organization, Geneva, Switzerland

WHO (1990) Methylmercury. Environmental health criteria no 101. World Health Organization, Geneva, Switzerland

Wren CD (1986) A review of metal accumulation and toxicity in wild mammals: mercury. Environ Res 40: 210-244

Wren CD (1991) Cause-effect linkages between chemicals and populations of mink (*Mustela vison*) and otter (*Lutra canadensis*) in the Great Lakes basin. J Toxicol Environ Health 33: 549-585

Wren CD, MacCrimmon HR, Loescher BR (1983) Examination of bioaccumulation and biomagnification of metals in a precambrian shield lake. Water Air Soil Pollut 19: 277-291

Wren CD, Scheider WA, Wales DL, Muncaster BW, Gray IM (1991) Relation between mercury concentration in walleye (*Stizostedion vitreum vitreum*) and northern pike (*Esox lucius*) in Ontario lakes and influence of environmental factors. Can J Fish Aquat Sci 48: 132-139

Xun L, Campbell NER, Rudd JWM (1987) Measurements of specific rates of net methyl mercury production in the water column and surface sediments of acidified and circumneutral lakes. Can J Fish Aquat Sci 44: 750-757

Zhang L, Planas D, Qian JL (1995) Mercury concentrations in black spruce (*Picea mariana* (Mill) BSP) and lichens in boreal Quebec, Canada. Water Air Soil Pollut 81: 153-161

Printing: Mercedesdruck, Berlin
Binding: Buchbinderei Lüderitz & Bauer, Berlin